Traditional Chinese Architecture Surveying and
Mapping Series:
Shrines and Temples Architecture & Garden Architecture

# THE CITY TEMPLES AND ITS GARDENS
# IN SHANGHAI

Compiled by College of Architecture and Urban Planning, Tongji University
Edited by CHANG Qing , ZHU Yuhui

China Architecture & Building Press

国家出版基金项目
NATIONAL PUBLICATION FOUNDATION

『十二五』国家重点图书出版规划项目

中国古建筑测绘大系·祠庙建筑与园林建筑

# 上海庙园

同济大学建筑与城规学院 编写

常 青 朱宇晖 主编

中国建筑工业出版社

# Contents

Shanghai County Chenghuang Temple and Temple Garden (Yu Garden) 001

Jiading County Chenghuang Temple and Temple Garden
(Qiuxiapu Garden and Shenshi Garden) 131

Qingpu County Chenghuang Temple and Temple Garden
(Qushui Garden) 183

Jiading County Xuegong and Xuegong Garden
(Yingkui Mountain and Huilong Pond) 219

Songjiang County Shantang and Gardens (Zuibaichi Garden) 293

Shanghai County Guozhai and East Garden (Shuyin Building) 353

Songjiang County Yi Garden and Nearby Houses 407

References 471

List of Participants Involved in Surveying and Related Works 473

目　录

上海县城隍庙及庙园（豫园）——〇〇一

嘉定县城隍庙及庙园（秋霞圃与沈氏园）——一三一

青浦县城隍庙及庙园（曲水园）——一八三

嘉定县学宫及学宫苑区（应奎山与汇龙潭）——二一九

松江府城善堂及附园（醉白池）——二九三

上海县郭宅及东园（书隐楼）——三五三

松江府城颐园及邻近宅第——四〇七

主要参考文献——四七〇

参与上海地区庙园建筑测绘及图稿整理的人员名单——四七二

**Item of survey:** Shanghai County Chenghuang Temple and Temple Garden (Yu Garden)
**Location:** No132, Anren Street, Huangpu District, Shanghai
**Age of construction:** Chenghuang Temple: 1397 (year 30 of Hongwu Emperor of Ming Dynasty)
　　　　　　　　　　　Temple Garden (original Yu Garden in Ming Dynasty): 1559 (year 38 of Jiajing Emperor of Ming Dynasty)
**Site area:** More than 30 mu
**Competent organization:** Management Office of Shanghai Yu Garden; Shanghai Taoist Association
**Survey organization:** College of Architecture and Urban Planning, Tongji University; Shanghai Cultural Heritage Management Committee; Management Office of Shanghai Yu Garden
**Time of survey:** 2004; 2005; 2013

测绘项目：上海县城隍庙及庙园（豫园）

地　　址：上海市黄浦区安仁街一三二号

始建年代：城隍庙始建于明洪武三十年（一三九七年）
　　　　　庙园（原为明代豫园）始建于明嘉靖三十八年（一五五九年）

占地面积：三十余亩

主管单位：豫园管理处

测绘单位：同济大学建筑与城规学院
　　　　　上海市文物管理委员会
　　　　　上海豫园管理处

测绘时间：二〇〇四年／二〇〇五年／二〇一三年

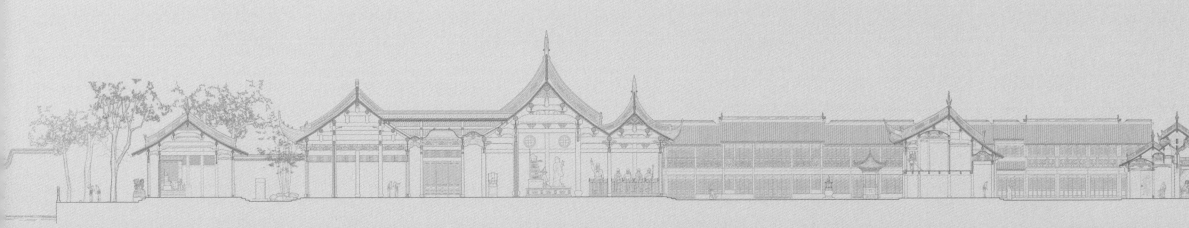

上海县城隍庙及庙园（豫园）

# Shanghai County Chenghuang Temple and Temple Garden (Yu Garden)

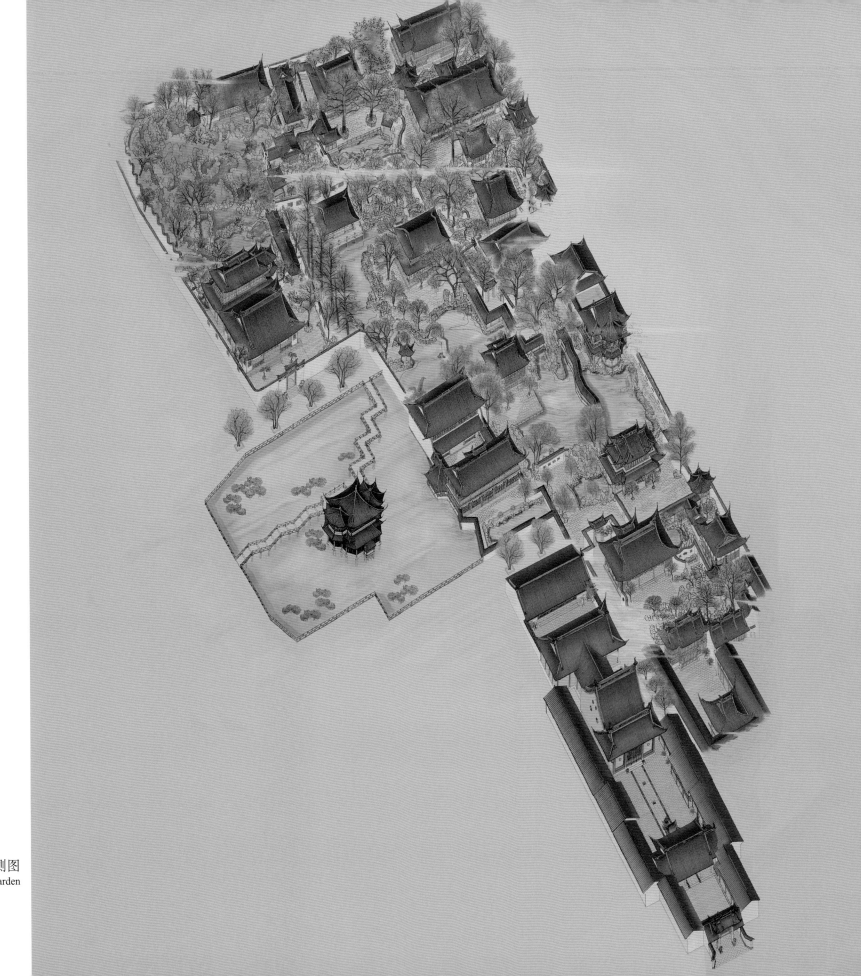

上海县城隍庙（部分复原）及豫园轴测图
Axonometric drawing of Shanghai County Chenghuang Temple and Yu Garden

豫园鱼乐榭院落轴测解析图
Axonometric and exploded drawing of Yu Garden's Yulexie Courtyard

豫园听鹂亭院落轴测解析图
Axonometric and exploded drawing of Yu Garden's Tingliting Courtyard

城隍庙内园轴测解析图

Axonometric and exploded drawing of Chenghuang Temple's Nei Garden

# Introduction

Shanghai County Chenghuang Temple Garden includes West Garden and East Garden. West Garden, namely Yu Garden owned by PAN Yunduan in Ming Dynasty, was built by the local gentry as West Garden of Shanghai Chenghuang Temple during the middle of the Qing Dynasty. East Garden, namely the garden of Shanghai Chenghuang Temple, was originally built in early Qing Dynasty. After the inclusion of Yu Garden owned by the PAN family, two gardens were called the east and the west or the inner and the outer, and finally were collectively called Yu Garden in 1949.

Yu Garden is located in the north-east of Laochengxiang district, Shanghai, which was a continuous neighborhood where the PAN family lived together in middle and late Ming Dynasty. North of Yu Garden is Fuyou Road (formerly Heiqiaobang), Anren Street in the east, Jiujiaochang Road in the west, Middle Fangbang Road (formerly Fangbang) in the south, and Hehua Lake in the middle. South-east of Yu Garden was Shanghai Chenghuang Temple, which was initially built during early Ming Dynasty. Yu Garden now covers an area of nearly 30 mu, including the east expansion in recent years and Nei Garden. In addition to the eastern part of the restoration under the guidance of Mr. CHEN Congzhou, Mr. CAI Dafeng, etc., today's Yu Garden can be divided into five distinct areas, namely Sansuitang (including Dashanshan and Cuixiutang), Wanhualou (including Yulexie and its courtyard), Dianchuntang, Deyuelou, and Nei Garden.

Among the five areas, Sansuitang area has preserved more of the original state of the Ming Dynasty, and its cultural relics are particularly valuable. The lower part of the base of Sansuitang is still the relic of Leshoutang, the main hall of the Ming Dynasty. The Wukang stone rockery built in Ming Dynasty in the northern part, was the third core scene in the garden (the south side of Sansuitang, namely Huxin Pavilion and its water area and the southern rockery, which is now separated from the garden, was the first core scene in the garden. Yuhuatang and its courtyard are the second core scene). After the independence of the area, Sansuitang is opposite to Yangshantang, which was reconstructed in the late Qing Dynasty, and the scene pattern of the hall "facing the mountain across the water"

# 导　言

上海县城隍庙园，包括西园、东园。西园即明代潘氏豫园，清中叶被当地士绅醵资构建为上海县城隍庙西园，东园即原建于清初的上海县城隍庙灵苑，潘氏豫园纳入后，两园以东西或内外相称，一九四九年后合称『豫园』。

豫园位于上海市老城厢东北部，明中后期的簪缨世族——潘氏家族聚族而居的连片街坊内，北倚福佑路（原黑桥浜），东临安仁街，旧日曾西近旧校场路，南近方浜中路（原方浜），环拥荷花一池，东南与始建于明初的上海县城隍庙毗邻。全园现占地近三十亩（含近年东扩部分及内园），除近年由陈从周、蔡达峰等先生指导复原的东部外，今日之豫园可分为畛域分明的五区，即三穗堂（含大假山、萃秀堂）、万花楼（含鱼乐榭院落）、点春堂、得月楼、内园。

其中，三穗堂一区保存明代原状偏多，文物价值尤高，三穗堂台基下部仍为明代全园主要厅堂——乐寿堂遗物，北侧明代武康石大假山曾是全园第三景象核

still exists. Wanhualou area consists of two courtyards, the east and the west, which is an example of a small and compact courtyard with a large pavilion as a center in the middle and late Qing Dynasty, and its scale has a significant jump compared with the Sansuitang area on the west side. Dianchuntang and Deyuelou are both arranged along the axis, most of which were reconstructed in the Tongzhi and Guangxu period of the late Qing Dynasty, which is the product of a large-scale functional building space combined with a small-scale garden space. Nei Garden is based on the layout of 1662 to 1722 in the early Qing Dynasty, and was reconstructed in the late Qing Dynasty and the early Republic of China. The four areas of Wanhualou, Dianchuntang, Deyuelou, and Nei Garden all present a typical feature of early Shanghai style gardens.

The five areas of Yu Garden, which was built in different years, were once divided and owned by several commercial guilds. The five areas are adjacent to each other but have their own systems, and therefore form the complex and diverse 'collection style' of today's Yu Garden.

心（三穗堂南侧、今被隔于园外的湖心亭水域及其南部山林曾是全园第一景象核心，玉华堂院落为第二景象核心），独立成区后与晚清翻建之仰山堂相对，仍存留厅堂『隔水面山』之景象格局；万花楼一区由东、西两个庭院组成，是以清中后期大型楼阁为中心，布置小型紧凑庭院之实例，其尺度较西侧的三穗堂景域有明显跳跃；点春堂与得月楼两区均沿轴线布置，大部为晚清同光年间重建，是大尺度功能化的建筑空间与小尺度园林景象空间结合的产物；内园则在清康熙年间的布局基础上，历经了清末民初的重建——万花楼、点春堂、得月楼、内园四区均集中呈现出典型的早期海派园林风格。

豫园之五区曾分属多个公所，建造年代各异，相互毗邻却又自成系统，构成了今日豫园繁复多元的『集锦式』园林面貌。

## A Heterogeneous of Literati Gardens in Middle and Late Ming Dynasty

When Yu Garden was first built in middle and late Ming Dynasty, it was just a garden next to a house in a riverside wetland. In 1559, PAN Yunduan, who sadly returned to Shanghai because of the failure in imperial examination, began to build a small garden on the vegetable field on the west side of his residence. Three years later, PAN went to Beijing again and finally passed the exam with high score. Since then, PAN had been busy with his political work and had no time to return to his own garden. Therefore, the operation of the small garden had to be intermittent.

Until 1577, as Sichuanyoubuzhengshi, PAN Yunduan resigned and returned to his hometown. For the sake of 'Yu Yue Lao Qin' (honor and respect parents), he spent more than ten years devoting all his wealth and effort to expand the old garden, and named it as 'Yu Garden'. Unfortunately, the garden had not yet been completed, and his old father, PAN En, who retired from his position as an official in imperial palace, died in 1582. However, the four hundred years of history of Yu Garden has just begun.

In the middle and late Ming Dynasty, the clear temperament of the literati gardens in Jiangnan area painted in Yu Garden cannot be separated from the gentry background of the PAN family, whose male members had continuously passed the imperial examination and earned a high position. However, with the investment of huge wealth and the late Ming customs that advocate density and luxury, the atmosphere of secularization in its space was quietly nurtured. Therefore, at the beginning of its establishment, Yu Garden had already slightly escaped the ranks of mainstream literati gardens and became a literati garden with a secular and urban temperament. It seemed that there was a frustration in PAN's career path, and instead he subconsciously wanted to use his exaggerated and luxurious garden to compete with Luxiang Garden in the north-west of the city.

# 明中晚期的文人园林异类

豫园于明中晚期发端之初，仅是建于滨江湿地中的傍宅一园。明嘉靖三十八年（一五五九年），春试落第、黯然南归的潘允端开始在住宅西面的菜畦上营造小园。三年后，潘氏再度进京，终于高中。此后沉浮宦海，无暇回顾，小园的经营只是断续进行。

直至明万历五年（一五七七年），官至四川右布政使的潘允端氏黯然辞归故里，始祭起『豫（愉）悦老亲』的大旗，以十余年之力，倾尽家财、大兴土木，充拓旧园，名之为『豫园』。遗憾的是，园林未成，以左都御史高位致仕的老父潘恩即于万历十年（一五八二年）撒手尘寰，而豫园的四百年沧桑风雨则刚刚揭开帷幕。

明中晚期的豫园保有江南文人园的鲜明底色，离不开园主潘氏一族，父子兄弟连科及第、纵横仕途的士族底蕴；而巨量财富的投入与崇尚繁密奢华的晚明风习，亦使其景象空间世俗化的气息悄然孕育，令豫园稍稍逸出了主流文人园林之行列，于立身之初，就成为颇具世俗与市井气质的文人园林异类，似乎暗藏着潘氏仕途失意，而欲以家园夸奇斗巧，甚至与城西北顾氏露香名园争胜的潜意识。

Fig.1  The Wukang Stone Rockery in the Yu Garden of the Ming Dynasty and the later added Taihu Stone part

This temperament is first manifested by its unprecedented number and density of buildings. After the completion of the Yu Garden, it covered an area of about 70 mu, of which there were more than 20 buildings including five halls and six pavilions, which was completely different from the sparse and low-density layout pattern that Jiangnan literati gardens frequently used before. The second is the combination of its gorgeous garden life and secular functions. The functional orientation of "Yu Yue Lao Qin", that is, spontaneously came with wealth and secular temperament. Moreover, old documents and books such as PAN Yunduan's Yuyuanji, Taicang Wang Shizhen Yuyuanji, and Pan Yunduan's Yuhuatangriji also record the rich and diverse garden life and religious worship in Yu Garden.

As a heterogeneous of literati gardens, the secular temperament of Yu Garden is the symbolic embodiment of the secularization of Chinese traditional gardens after the changes in cultural and artistic atmosphere during 1567-1620, and also the reaction of the eastward shift of the economic, cultural, and artistic focus of the Yangtze River Delta from Suzhou to Songjiang.

这一气质，首先表现为其空前的建筑数量与密度，完工后的豫园占地七十余亩，其中建筑竟有五堂六阁等二十余处之多，全然不同于此前江南文人园多采用的疏朗散淡的低密度布局模式。其次是其绚烂的园居生活与世俗功能的组合——「豫悦老亲」的功能定位，即自带富贵与世俗气质；而如园主潘允端《豫园记》、太仓王世贞《豫园记》、潘氏《玉华堂日记》等文献旧籍，亦均记载了园内丰富多元的园居生活与宗教崇祀。

作为文人园林异类，豫园的世俗气质，正是「隆万风变」（指明代隆庆与万历两朝，一五六七年至一六一九年——本书责编注）之后，中国传统园林世俗化进程的标志性体现，亦是整片长三角经济与文化、艺术重心由苏州府向松江府东移的体现。

图三　今三穗堂台基上层花岗石台口下，仍存留明代乐寿堂青石台基，南面束腰部位尚饰以缠枝纹

图四　万花楼前原明代豫园青石勾栏，疑原即明「乐寿堂（今三穗堂）」南临水处之「石砌栏围，栏外碧水一池」

图二　上海豫园门额，明万历五年（一五七七）丁丑秋八月望太原王稚登（一五三五年至一六一二年）书，此额原应嵌于全园东南角，「环龙桥」东南塊面东的长楼正中

Fig.2  The plaque of Shanghai Yu Garden, Taiyuan WANG Zhideng (1535–1612) written on the August 15th in 1577 , and this plaque was inlaid in the long building in the east of the southeast side of Huanlong Bridge in the southeast corner of the whole garden.

Fig.3  Under the upper Huagang Stone layer platform of today's Sansuitang, there is the Qing Stone platform of Leshoutang in the Ming Dynasty, the south side of which is still decorated with an entwined vines decoration.

Fig.4  The original Qing Stone Goulan of Yu Garden of the Ming Dynasty in front of Wanhualou, which is suspected that the original 'stone fence surrounded by the water' outside Leshoutang (today's Sansuitang) of the Ming Dynasty on the south side of the water.

## The Garden Belongs to Shanghai Chenghuang Temple and Commercial Guilds in Middle Qing Dynasty

Yu Garden in middle and late Ming Dynasty had quickly thrived and suddenly perished. By the end of the Ming Dynasty, Yu Garden had fallen into disrepair. Until the beginning of the Qing Dynasty, only a part of the landscape pattern remained, and the building was wiped out, and the large Taihu Stone rockery in the south of the Hehua Lake may also disappear at this time. Such a prosperous and famous garden, only Wukang Stone rockery, Yulinglong and Qing Stone base of Leshoutang remained. It seemed difficult to find other explanations except for being actively sold by owner descendants.

This kind of decadence continued till the middle of the Qing Dynasty.

Around the middle of the Qing Dynasty, Chenghuang Temples at various levels in counties and cities started to build or receive temple gardens, and the gardens were often expanded. In the name of entertaining the god, they provide ordinary citizens and businessmen with open, complex, and charming urban spaces, showing artistic tastes and aesthetic trends different from traditional literati gardens, particularly the Songjiang county, which is located on the eastern edge of the Jiangnan area (today's Shanghai). Same as Gongshi Garden (Qiuxiapu Garden) as Jiading Chenghuang Temple Garden, Guyi Garden as the garden of the Chenghuang ancestral temple in Nanxiang Town, Jiading County and Qushui Garden as Qingpu Chenghuang Temple, Yu Garden which had lost in the prosperous city center also started her new life.

In 1760, the Shanghai gentry with increasing money and wealth finally bought Yu Garden at a cheap price from the descendants of the Pan family. Yu Garden then subsumed into Shanghai Chenghuang Temple as its West Garden, which was also called the outer garden. And the renovation and expansion had been greatly propelled, so that the area of the whole garden once again restored to more than 70 mus, and the architectures inside were extremely flourishing. Due to its huge scale and large cost of money, the renovation project continued till 1784. Large-scale buildings such as Sansuitang, and Huxin Pavilion and the column bases of Cuixiutang were all relics from this period. In Nei Garden, Jingguan Dating and the square hall Keyiguan are partially and even completely preserved with beams and column bases from the time of their construction during the Kangxi period in the Qing Dynasty, making them the oldest wooden structures in the garden.

# 清中叶的邑庙园林与『商会园林』集群

明中晚期的豫园『其兴也勃焉，其亡也忽焉』，至明末就已寥落失修，清初仅存部分山水格局，建筑则泯灭殆尽，荷花池南的大型太湖石主山亦可能即于此时消失。偌大名园，仅武康石次山、玉玲珑一峰，与主要厅堂乐寿堂之青石台基黯然残存——除了被园主后裔主动拆卖，似很难找到别的解释。

这一颓势一直延续至清中叶。

清中叶的江南各地，各级府县城隍庙纷纷增筑或收纳庙园，并一再增拓，以娱神的名义，为普通商贾市民提供开放、复合、魅力十足的城市园林空间，呈现出与传统文人园不同的艺术趣味、审美趋向——而位居江南东缘的松江府域（约今上海市区）一带，此风尤甚，如归为嘉定县城隍庙园的龚氏园（秋霞圃），归为嘉定县南翔镇城隍行祠之园的古漪园，以及青浦县城隍庙新建的曲水园——沉沦于繁华城市间的潘氏荒园也由此迎来新的生机。

乾隆二十五年（一七六〇年），夹袋日益丰盈的上海士绅终于酿资自潘氏后裔手中廉价购得豫园，归入上海县城隍庙，作为其西园，也称『外园』，并大加整治扩建，使全园面积恢复至七十余亩，堂楼舫榭盛极一时。因其规模过于浩大，资金亦需逐步筹措，故工程一直延续至乾隆四十九年（一七八四年），今存三穗堂、湖心亭等大型建筑及萃秀堂之覆盆柱础均是这一时期的遗物。而内园静观大厅，『可以观』方厅更局部甚至完整地存留着清康熙年间始建之际的梁架、柱础，成为全园最古老的木构。

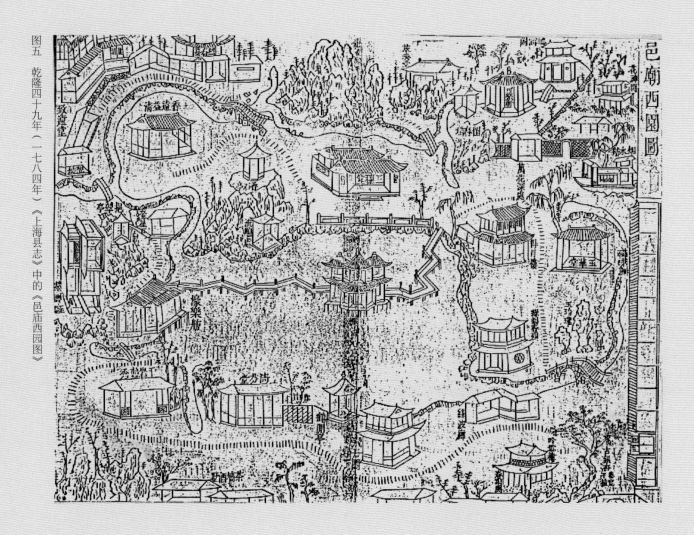

Fig.5 *Yimiaoxiyuantu* (the picture of the West Garden of Shanghai Chenghuang Temple in) *Shanghaixianzhi* in 1784

The newly-built West Garden of Shanghai Chenghuang Temple was even larger than the previous Yu Garden. The huge amount of money spent was actually raised by the gentry and the commercial office that had earned more money and reached their life peak, since In 1684 marine trade was permitted and the nearby custom outside Xiaodongmen was established, which led to the inevitable outcome of commercial office and guilds entering the garden in the future. Sansuitang, the big rockery, Cuixiutang, Yangshantang (the second floor was the Juanyulou) and Shenchitang (Wanhualou) were the offices of pie and bean trade. Dianchuntang (later built) was the office of cotton and sugar trade. Deyuelou was the office of cloth trade. And Qingxuetang (now Jingguandating) in the original East Garden of Chenghuang Temple (Nei Garden) was the office of money trade……

The investment and settlement of the commercial guilds also greatly influenced the overall space and architectural form of the West Garden. The number and density of buildings further increased. The composition of the scene became more and more diversified and centrifugal. In terms of boat sterns, there were Yanshuifang in the east, Qifang in the north-western part and Haolefang in the west part, in order to create scenic spots throughout the garden. The decorative details of the building also started to appear, which later popularized the Dragon Wall in Yu garden. At the same time, with the number of buildings greatly increasing, there was a more significant vertical trend in the garden. For example, Huxin pavilion developed from the simple and misty Fuyiting on the small island of in the Ming Dynasty, into a giant pavilion with complex shapes, engulfing the entire island, and pressing the water. In the north-west corner of the garden, a guild of buildings, including Xichuntai, Qinglou and Hanbilou were connected into a form of three-dimensional traffic, and beyond the 'house-circled landscape', they were gradually showing the tendency of 'overlook the landscape from house', which appeared more often in gardens in the late Qing Dynasty.

By the middle and late Qing Dynasty, the West Garden was gradually divided by the commercial guilds, and became a cluster of several independent and self-contained 'small garden with large building'. Today, Yu Garden's multi-regional and mutual-borrowing 'collection' model was established based on that.

This kind of spatial change, actually corresponds to the rise of the citizen class and the surge of business culture in the coastal town during the middle Qing Dynasty. And the Pan family's private garden in the Ming Dynasty, had become a material carrier of the urban faith that Shanghai residents are increasingly enthusiastic about, and the poetic projection of worldly prosperity.

新建的城隍庙西园，规模甚至更甚于此前的潘氏豫园，而所费财资之巨，皆为康熙二十三年（一六八四年）开海、邻近的小东门外江海关设立以来，家财日渐丰厚的上海士绅与事业臻于巅峰的商业公所集资筹措，这也导致了日后各公所纷纷入驻园内的必然结局——三穗堂、大假山、萃秀堂、仰山堂（二层为卷雨楼）、神尺堂（万花楼）一带为饼豆业公所，点春堂（后建）为花糖洋货公所，得月楼为布业公所，而创建于康熙四十八年（一七〇九年）的原城隍庙东园（内园）晴雪堂（今静观大厅）则为钱业公所……

商业行会的出资与入驻，亦对西园整体空间与建筑形态产生了极大影响。如建筑数量与密度的进一步增加；如景象构成的日渐多元化和离心化，即以船舫而论，就有东部烟水舫、西北部憩舫、西部濠乐舫三处，俨然云水缥缈，处处留情；再如建筑细部的装饰化，后来脍炙人口的豫园龙墙就肇因于此——同时，园内空间有纵向化趋势，楼屋大大增多，如湖心亭之由明代『凫佚亭发展为体形繁复、吞没全岛、压迫水面的庞然巨阁为最典型案例，园西北角熙春台、磐楼、涵碧楼一组楼屋更连缀成片，于『环抱山林』之外，更渐呈晚清园林的『俯水瞰山』之势。

至清中后期，西园更逐渐被各个公所分割，渐成各自独立而自成系统的『大厅小园』。今日豫园多区并立、相互因借的『集锦』模式由此奠定。

这样的景象空间变化，实与清中叶海滨小城中市民阶层的崛起、商业文化的涌动相生相应——明代的潘氏私园，倏然变身为上海县民日益热衷的城隍信仰的物质载体，与世俗荣华的诗意投影。

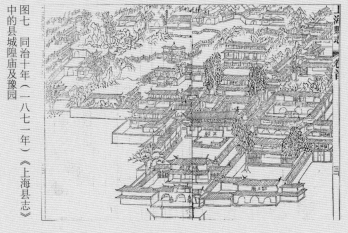

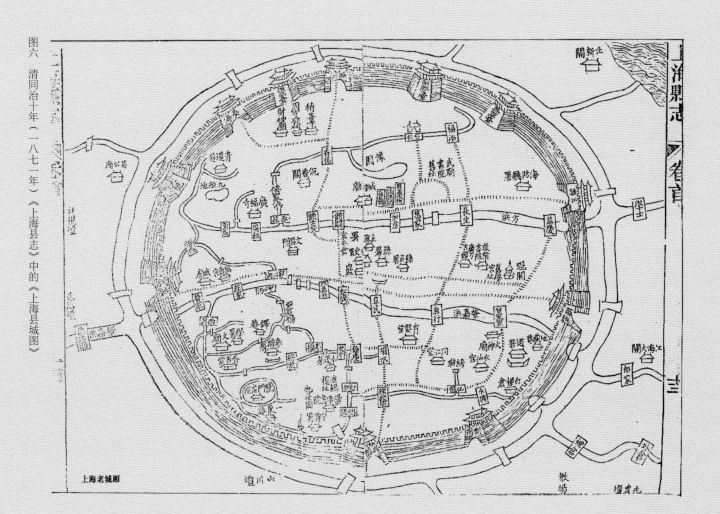

图七 同治十年（一八七一年）《上海县志》中的县城隍庙及豫园

图八 《申江胜景图》中光绪年间（一八七五—一九〇八）的豫园湖心亭

图六 清同治十年（一八七一年）《上海县志》中的《上海县城图》

Fig.6　*The map of Shanghai County* in *Shanghaixianzhi* in 1871

Fig.7　The picture of Shanghai County Chenghuang Temple and Yu Garden in *Shanghaixianzhi* in 1871

Fig.8　Huxin Pavilion of Yu Garden in *Shenjiangshengjingtu* in 1875−1908

## The "Civilian Square" and "Urban Park" in Late Qing Dynasty and Early Republic of China

In May 1842, the smoke of First Opium War relentlessly burned to Wusong Estuary. The West Garden of Chenghuang Temple in the Shanghai County was also influenced. The scenery in the garden was mostly burned and destroyed. After that, commercial guilds settled in the garden started to maintain and regenerate the garden by their owm. Specifically, the famous Wanhualou was rebuilt by pie and bean trade guild in 1843.

Afterwards, Small Sword Society Uprising during the Autumn of 1853 to the Spring of 1855 and the war of Taiping Army during 1853-1864, also influenced the West Garden and the East Garden of Shanghai Chenghuang Temple directly or indirectly. In 1866, Yangshantang (Juanyulou) was rebuilt by pie and bean trade guild. In 1868, Dianchuntang area was rebuilt by cotton and sugar trade guild. In 1892, Deyuelou area was rebuilt by cloth trade guild. Surviving traditional commercial guilds were struggling to maintain the glory of the past, but in the face of established dilapidated facts, gradual weakening of official control and increasingly rich economic temptations, some guilds had gradually changed the garden and landscapes into markets.

In this way, the West Garden of Shanghai Chenghuang Temple began her marketization process which was also closely related to the fate of the city.

# 清末民初的『市民广场』与『城市公园』

清道光二十二年（一八四二年）五月，鸦片战争的硝烟无情地燃至吴淞口，上海县城中的邑庙西园亦难以幸免，园中胜景多遭焚掠。事后驻园各公所自行维修，如万花楼即由饼豆业重建于道光二十三年（一八四三年）。

此后，咸丰三年（一八五三年）秋至五年（一八五五年）春的『小刀会』起事，再至咸丰三年（一八五三年）至同治三年（一八六四年）的『太平军』江南战事，亦直接或间接波及西、东两园。事后的同治五年（一八六六年），仰山堂（卷雨楼）由饼豆业重建；同治七年（一八六八年），点春堂景域由花糖业重建；光绪十八年（一八九二年），得月楼景域由布业重建——实力尚存的传统商业行会勉力维持着昔日的点点荣光，但面对既成的破败事实、逐渐疲弱的官方控制与日益丰厚的经济诱惑，部分公所渐渐变园景为市肆。

上海县城隍庙西园就此启动了它与这座城市的命运休戚相关的圩市化进程。

图十　清末民初的豫园湖心亭景域，人头攒动的「市民广场」

图九　清末豫园湖心亭景域旧照，周边建筑界面低矮

Fig.9  The Huxin Pavilion area of Yu Garden in the late Qing Dynasty, surrounded by low buildings

Fig.10  The Huxin Pavilion area of Yu Garden in the late Qing Dynasty and early Republic of China, just like a crowded 'civilian square'

Till the period of the Republic of China, the northwest, west, and south sides of Hehua Lake, as the main water of the West Garden, finally disintegrated. The whole garden became a large city market, as many historic photos clearly record this process that roughly began during 1831-1861. Originally maintained by cyan cloth trade guild and then transferred into teahouse in 1855, Huxin Pavilion and the water area together constituted a transition space between the noisy city market and the remaining garden space in the east and north sides.

Declining due to the fall of the scholar-class, reviving due to the rise of the citizens and gentry and businessmen, and gradually collapsing due to the collision of civilizations, the evolution of the times, the disintegration of the original gentry-business co-governance system, and the baptism of the wars, Yu Garden in the Ming Dynasty and the West Garden of Shanghai Chenghuang Temple in the middle Qing Dynasty, finally inevitably embarked on the road of marketization and localization. She derived from temples, gardens, and cities as 'trinity', colorful and complex functional form, and eventually completed the change from the context of literati society to the context of civil society, and became a 'civilian square' and 'urban park' for the general public.

到了民国时期，西园主体水面荷花池之西北、西、南三面景域终于彻底解体，沦为市肆，诸多历史照片清晰记录了这一始自咸丰年间的进程。原由青蓝布业维护、至咸丰五年（一八五五年）转让为茶楼的湖心亭则与一汪池水共同构成嘈杂市肆与东、北两面留存几片景域间的过渡空间。

因士大夫阶层的沉沦而衰败，复因城市市民与绅商阶层的勃兴而重振，终因文明的碰撞、时代的演进、原有绅商共治体系的解体、与战火的洗礼而渐渐瓦解的明代潘氏豫园、清中期城隍庙西园，终于不可避免地走上了圩市化却又场所化的道路，衍生出庙、园、市『三位一体』、缤纷错杂的功能形态，进而彻底完成了由文人社会向市民社会语境的转变，成为面向广大市民公众的『市民广场』与『城市公园』。

图十一　清同治十年（一八七一年）《上海县志》中的邑庙内园图

Fig.11  Nei Garden in Yi Temple of *Shanghai xianzhi* in 1871.

# From Defending the City to Defending the Sea, the Diversified and Compatible Characteristic of Shanghai Chenghuang Temple

When Shanghai County was first established in 1291, the Chenghuang Temple of the county had not yet been built. Perhaps for a city facing the ocean and not setting walls, the gods symbolizing the city wall and the moat were not as gracious and necessary as the gods defending the sea, like the famous god of navigation Huo Guang in the Han Dynasty, and Mazu in the Southern Song Dynasty.

In January 1369, the newly enthroned emperor Zhu Yuanzhang couldn't wait to crown the Chenghuang gods (gods defending the city wall and the moat) in Beijing and throughout the country. In June of the next year, in order to strengthen the operability, a decree was issued to simplify the Chenghuang Temples to three levels, namely the capital, the prefecture, the state and county, and finally, the grid management system in the living world was projected into the underworld entirely. In 1373, Qin Yubo (1296-1373), a Jinshi in Yuan Dynasty who just die, was re-crowned as the Chhenghuang god of Shanghai County. In 1872, he was once again re-crowned as Huhaigong, symbolizing that the empire's authority was pushed to seaside.

At first, the newly-appointed Chenghuang god was worshipped at the Songjiang Chenghuangxingci outside the county, which had a huge space gap and lack identification. In 1397, the local official of Shanghai County changed the Jinshanshenxingci worshiping Huo Guang into Shanghai Chenghuang Temple, in which the front hall consecrated Huo Guang, and the back hall consecrated Qin Yubo, thus creating an open and compatible space for worshiping. The temple was located on the north side of Shanghai County. The main river, Fangbang, in front of the gate of the temple, flowed into the East China Sea. Eventually, in Shanghai County, the faith of Chenghuang and the faith of sea were then unified, and the urban spiritual landmark had also been established, which continued up to now.

# 从守城到捍海——多元兼容的上海县城隍庙

元至元二十八年（一二九一年）上海建县之初，并未建立城隍之庙，或许对于一座面向海洋、不设城墙的城市来说，城（城垣）与隍（护城河）这样的神灵，还不及捍海之神、西汉名臣霍光和航海之神、南宋妈祖来得亲切和必要。

明洪武二年（一三六九年）正月，新近登基的太祖朱元璋急不可待地诏令『封京都及天下城隍神』；第二年六月，为强化可操作性，再下诏令，把全国城隍庙宇简化为京都、府与州县三级——终于把阳世的网格化管理体系投射到冥冥阴间，再无遗漏。洪武六年（一三七三年），复追封刚刚去世的元代进士秦裕伯（一二九六年至一三七三年）为上海县城隍，后来的清同治十一年（一八七二年）又追赠其为护海公，帝国的威权被恭推到了海疆。

新任城隍起初仅被奉祀在县城外的松江府城隍行祠，空间悬隔，认同不足。直至洪武三十年（一三九七年），上海知县才脑洞大开地把原奉祀霍光的金山神行祠改为上海县城隍庙，前殿祀霍，后殿祀秦，神怀坦荡，兼容不悖。神庙位于上海县衙北侧，门前的县城干河方浜，宛转流向东海，上海一县的城隍信仰与海神信仰就此合一，城市精神地标也卓然确立，延续至今。

图十二　旧时的上海县城隍庙面对大殿的晚清戏楼（南半为仪门）

Fig.12  Xilou facing the main hall of Chenghuang Temple in the old city of Shanghai (Part of south is Yi Gate.)

Since then, the temple gradually expanded. Till 1535, the torii gate was re built, and the giant granite pillars that have survived up to now may be relics at this time. In 1553, the Shanghai residents who were unbearable of the torture by pirate from the sea, joined forces to build the nine-mile city wall, and until that time, the Chenghuang god was worthy of his name.

In 1573 or later, on the west and north sides of the Chenghuang Temple, Yu Garden owned by the Pan family was built in a high profile and became a wonderful scenery. However, the garden had been scattered in the early Qing Dynasty. Because of that, in 1709, local gentry started to build the East Garden (also called Nei Garden or Lingyuan) in the east of the temple, which provided a recreational place for the Chenghuang god who were 'busy' ever day. To 1760, the large-scale Yu Garden was purchased and built by the enthusiastic gentry as the Chenghuang Temple West Garden (also known as the outer garden), which highlighted the joy of visiting Shanghai Chenghuang Temple Garden. Moreover, it is obviously the result of the city becoming rich due to maritime trade, the release of religious beliefs, the strong demand for public space, and the improvement in quality after the opening of the sea during 1654-1722. And this kind of interaction between temple, garden, and city remains up to now.

The pattern of the East Garden and the West Garden still partly exists today. The Chenghuang Temple also remains four houses along the central axis, including the first gate (the middle room as torii), the second gate (connect with Xilou in the back side), the main hall (with gable and hip roof baosha in the front), and the back palace. The first gate was rebuilt during 1862-1875 while the second gate was rebuilt during 1871-1908. The main hall, using steel-framed cement to imitate wood structure, was reconstructed in 1927. And the back palace with a strong life and surviving the disaster, is the relic rebuilt in the 1748, which has unique value throughout the whole temple.

此后神庙应渐有扩展，至倭寇将兴的嘉靖十四年（一五三五年），重建头门牌坊，留存至今的庞然花岗石柱可能是此时遗物。嘉靖三十二年（一五五三年），不堪倭患折磨的上海县民合力修筑九里城墙，城隍之神这才实至名归，披上袍甲。

明万历前期，城隍庙西、北两侧有潘氏豫园高调兴建，成为其绝妙借景，至清初而颓败已甚。或许有伤于此，清康熙四十八年（一七〇九年），地方人士在庙东兴建东园（又称『内园』『灵苑』），为日日应酬不暇的城隍之神提供了游憩之所；到了乾隆二十五年（一七六〇年），规模宏大的潘氏豫园又被热情如火的士绅们购建为城隍庙西园（又称『外园』），上海县城隍的游园之乐大概甲于天下了——这显然是康熙开海后，城市因海洋贸易而致富，宗教信仰尽情宣泄，公共空间亦需求旺盛、品质提高的结果。这一庙、一园、一市互动的局面遗惠至今。

今东西两园格局尚存，城隍庙亦存中轴一路四进，即头门（明间起牌楼）、二门（背后连接戏楼）、大殿（前接歇山顶抱厦）、寝宫——其中头门、二门为晚清同治、光绪年间重建，钢骨水泥之仿木大殿系民国十六年（一九二七年）重建而成，而命格强悍、劫后余生的寝宫则至晚是乾隆十三年（一七四八年）重建的遗物，于全庙独具价值。

豫园轴测图
Axonometric drawing of Yu Garden

0 5 10 20m

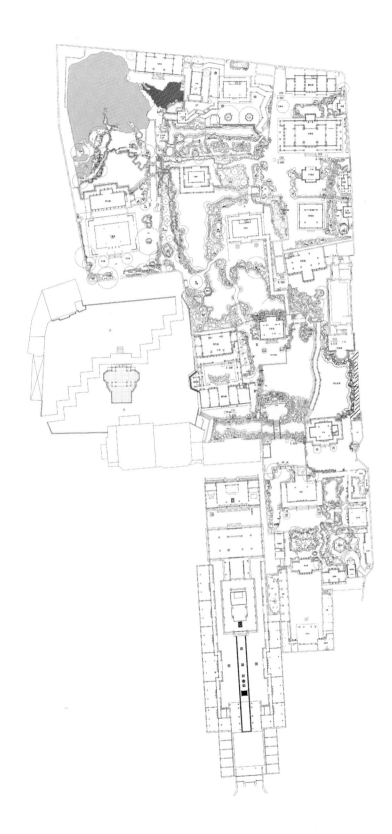

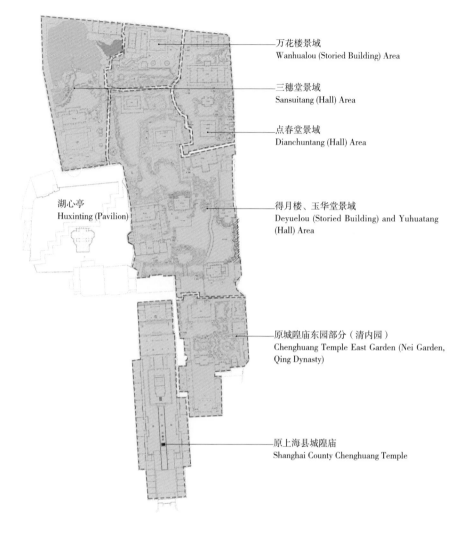

湖心亭
Huxinting (Pavilion)

万花楼景域
Wanhualou (Storied Building) Area

三穗堂景域
Sansuitang (Hall) Area

点春堂景域
Dianchuntang (Hall) Area

得月楼、玉华堂景域
Deyuelou (Storied Building) and Yuhuatang
(Hall) Area

原城隍庙东园部分（清内园）
Chenghuang Temple East Garden (Nei Garden,
Qing Dynasty)

原上海县城隍庙
Shanghai County Chenghuang Temple

豫园一层平面图
Ground floor plan of Yu Garden

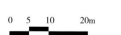
0    5   10      20m

N
E

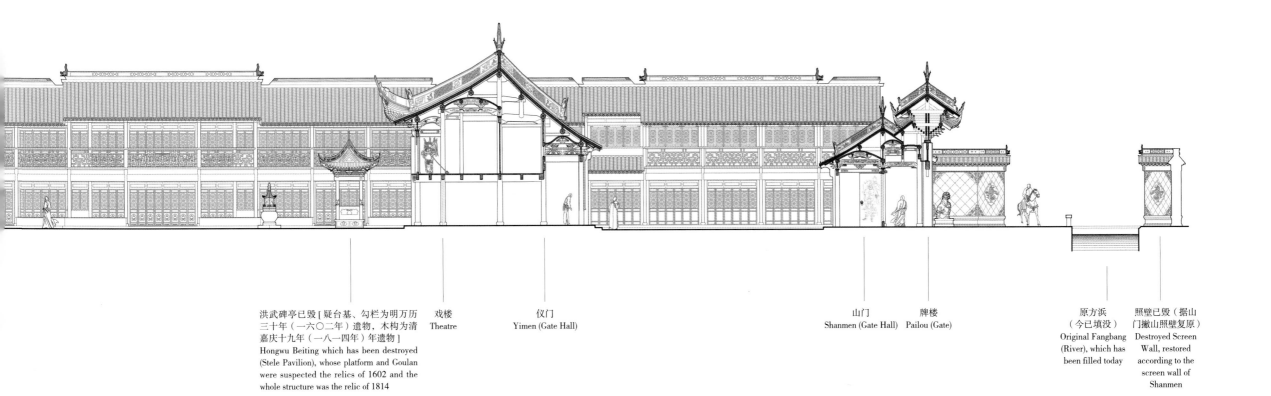

洪武碑亭已毁 [ 疑台基、勾栏为明万历
三十年（一六○二年）遗物，木构为清
嘉庆十九年（一八一四年）年遗物 ]
Hongwu Beiting which has been destroyed
(Stele Pavilion), whose platform and Goulan
were suspected the relics of 1602 and the
whole structure was the relic of 1814

戏楼
Theatre

仪门
Yimen (Gate Hall)

山门
Shanmen (Gate Hall)

牌楼
Pailou (Gate)

原方浜
（今已填没）
Original Fangbang
(River), which has
been filled today

照壁已毁（据山
门撇山照壁复原）
Destroyed Screen
Wall, restored
according to the
screen wall of
Shanmen

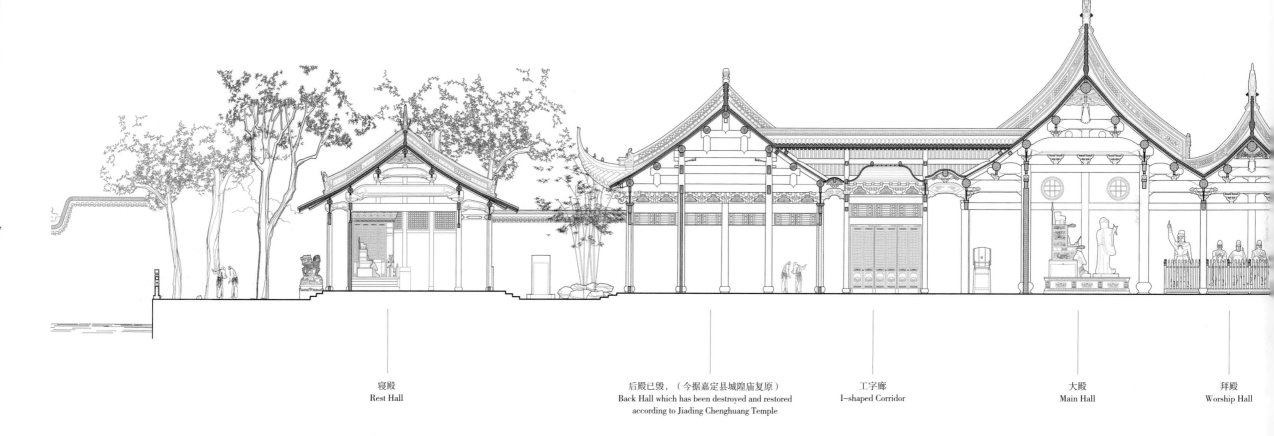

寝殿
Rest Hall

后殿已毁，（今据嘉定县城隍庙复原）
Back Hall which has been destroyed and restored
according to Jiading Chenghuang Temple

工字廊
I-shaped Corridor

大殿
Main Hall

拜殿
Worship Hall

上海县城隍庙与豫园三穗堂、大假山院落中轴剖面图（一）
Section of Shanghai Chenghuang Temple and Yu Garden' Sansuitang and North Rockery Courtyard (1)

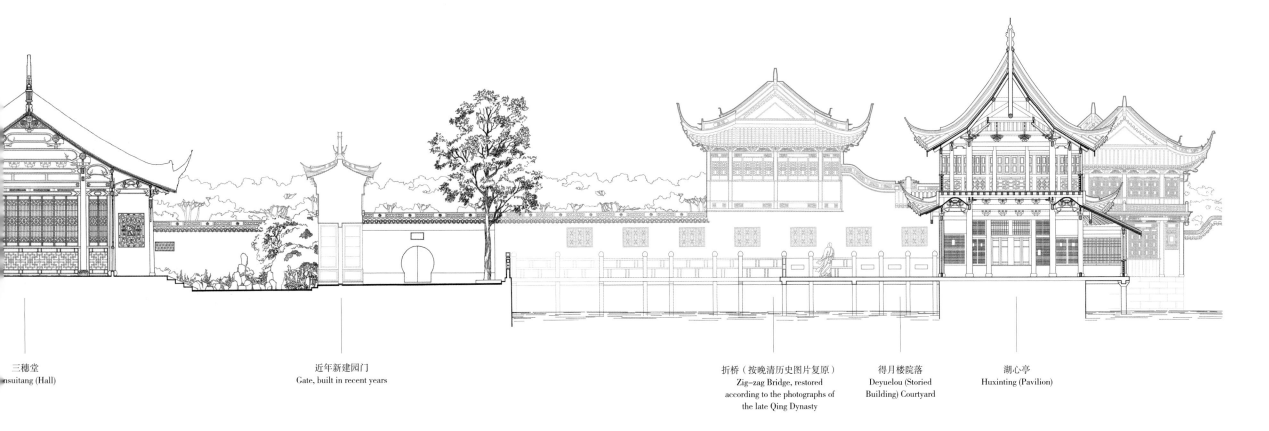

三穗堂
nsuitang (Hall)

近年新建园门
Gate, built in recent years

折桥（按晚清历史图片复原）
Zig-zag Bridge, restored
according to the photographs of
the late Qing Dynasty

得月楼院落
Deyuelou (Storied
Building) Courtyard

湖心亭
Huxinting (Pavilion)

今豫园北界
North Boundary of Yu
Garden today

望江亭
Wangjiangting
(Pavilion)

武康石北山，明张南阳所掇
Wukang Stone North Rockery, placed by ZHANG
Nanyang in the Ming Dynasty

挹秀亭
Yixiuting (Pavilion)

"渐入佳境"游廊
Jianru jiajing (Corridor)

卷雨楼（一层悬"仰山堂"匾）
Juanyulou (Storied Building), a plaque
named 'Yangshantang' is hang on the
first floor

上海县城隍庙与豫园三穗堂、大假山院落中轴剖面图（二）
Section of Shanghai Chenghuang Temple and Yu Garden' Sansuitang and North Rockery Courtyard (2)

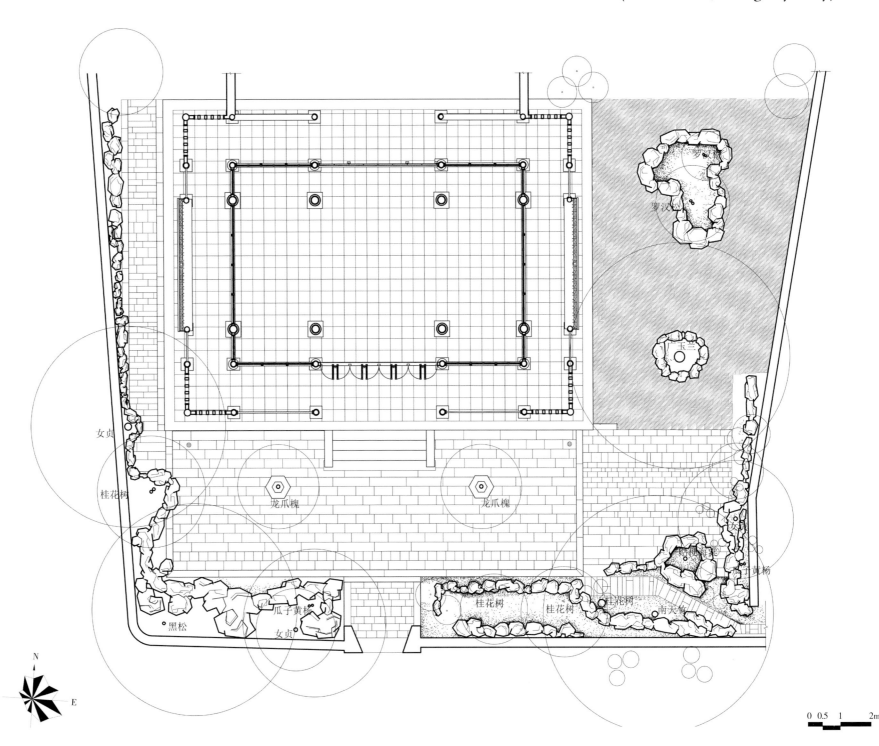

女贞

桂花树

龙爪槐

龙爪槐

黑松

瓜子黄杨

女贞

桂花树

桂花树

南天竹

罗汉

玉兰

三穗堂平面图
Plan of Sansuitang

N
E

0  0.5  1        2m

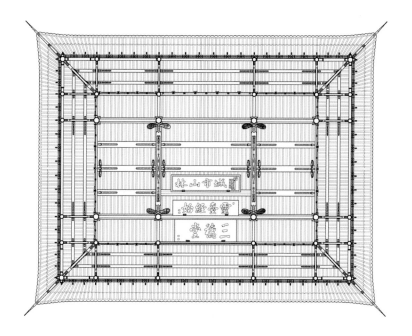

三穗堂屋架仰视图
Bottom view of Sansuitang's beams

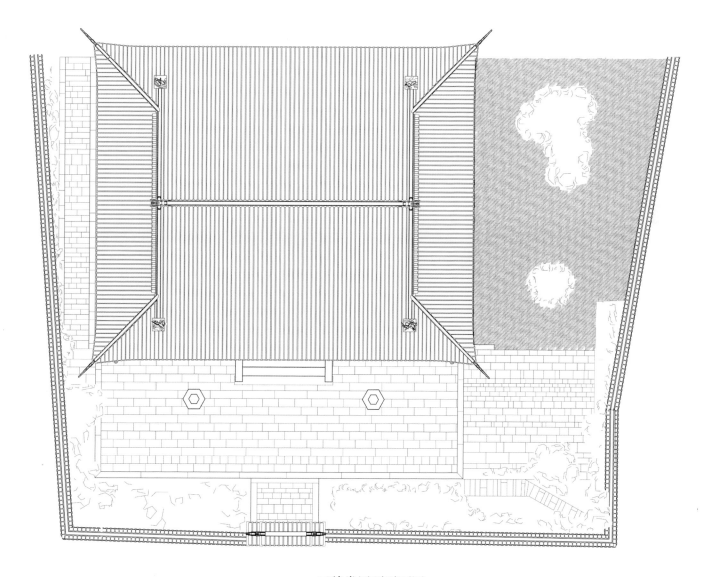

三穗堂屋顶平面图
Roof plan of Sansuitang

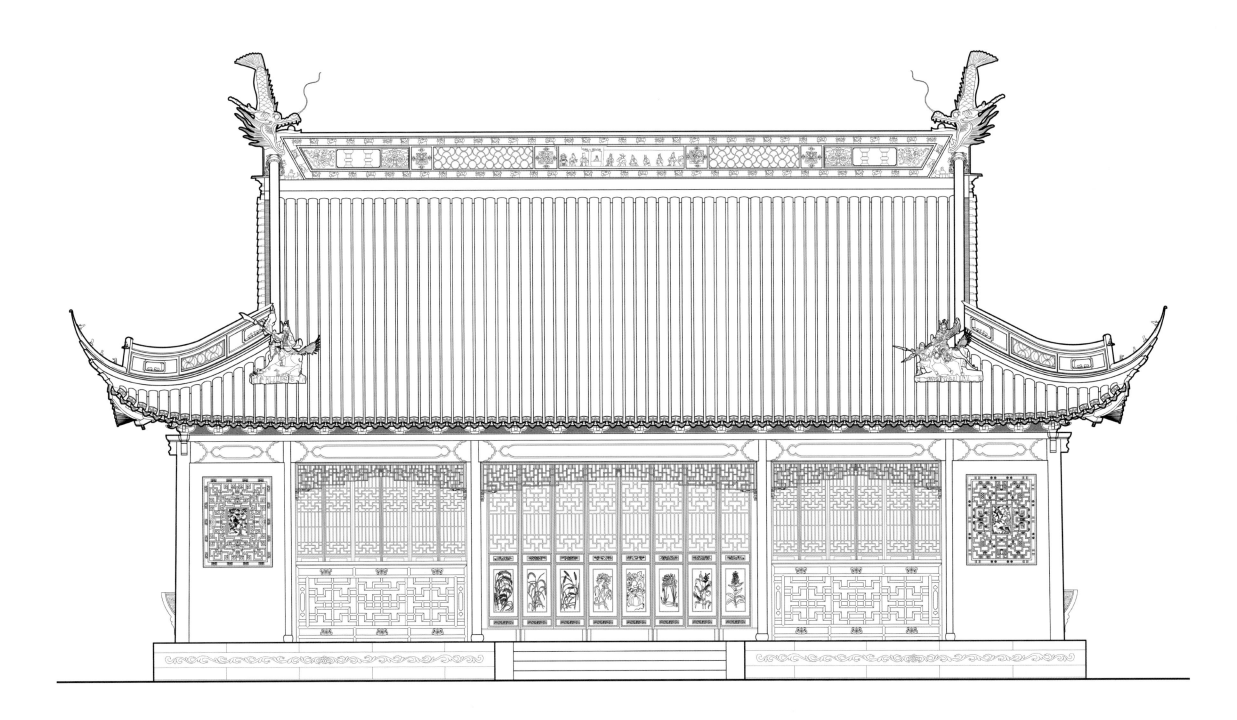

三穗堂正立面图
Front elevation of Sansuitang

0    0.5    1    2m

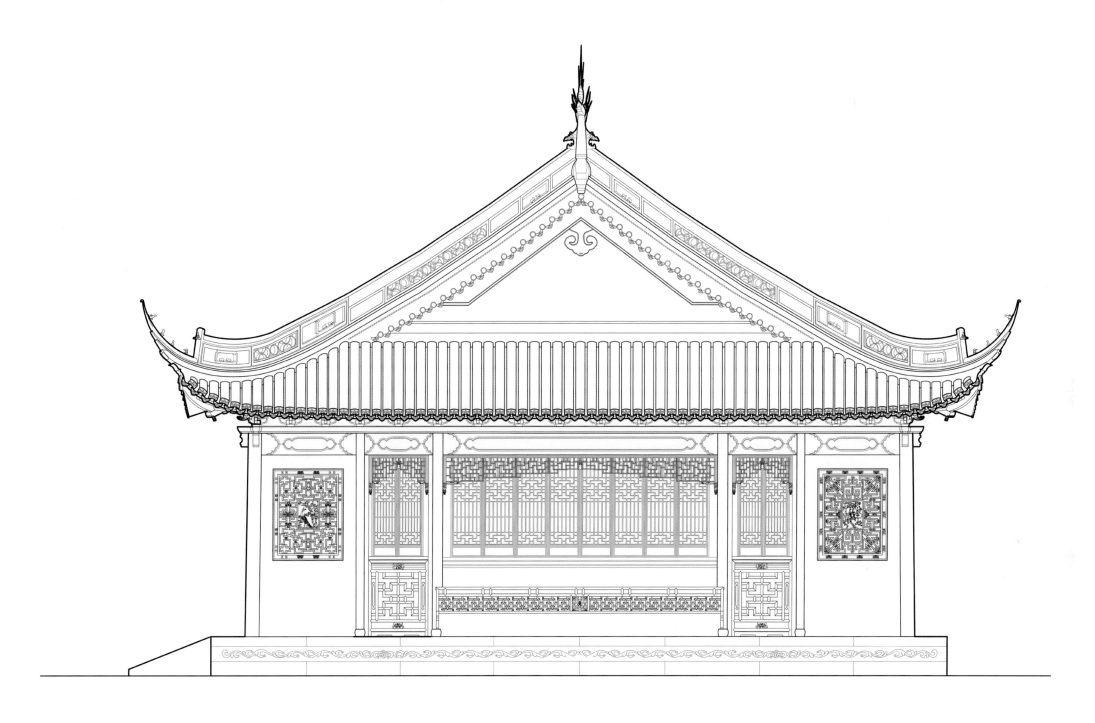

0  0.5  1  2m

三穗堂侧立面图
Side elevation of Sansuitang

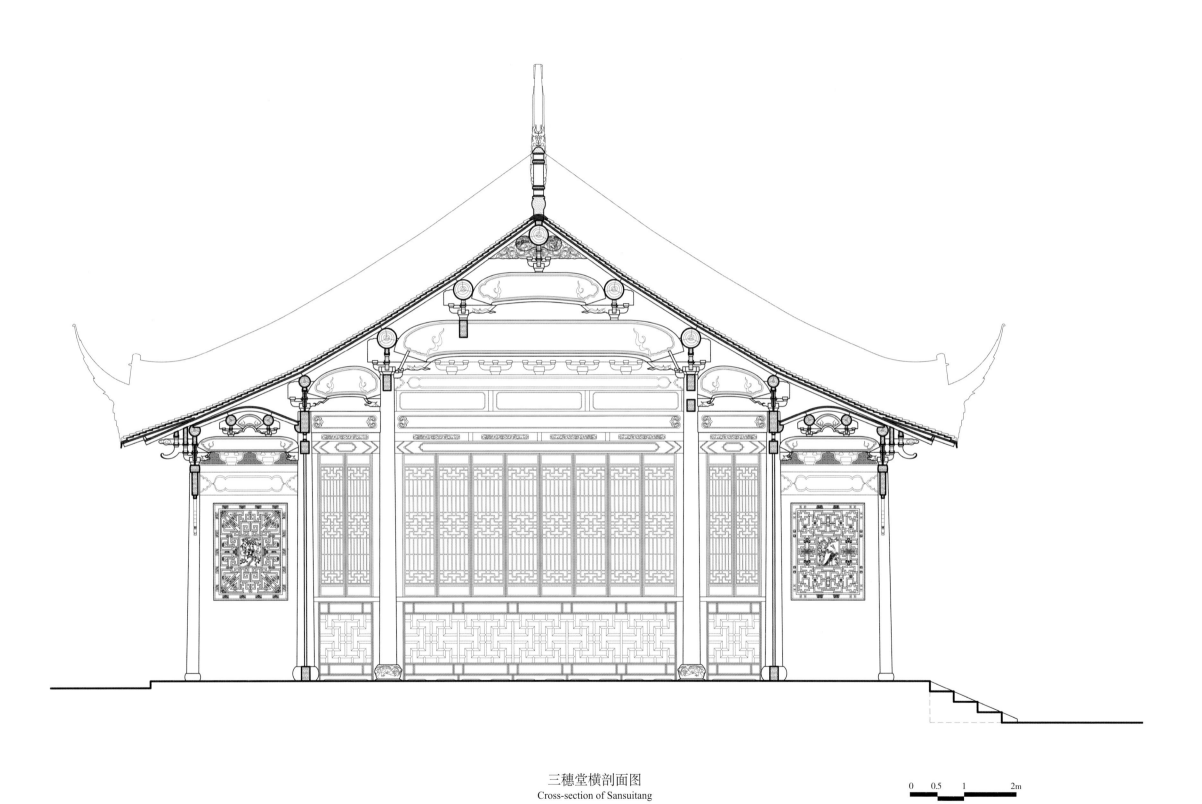

三穗堂横剖面图
Cross-section of Sansuitang

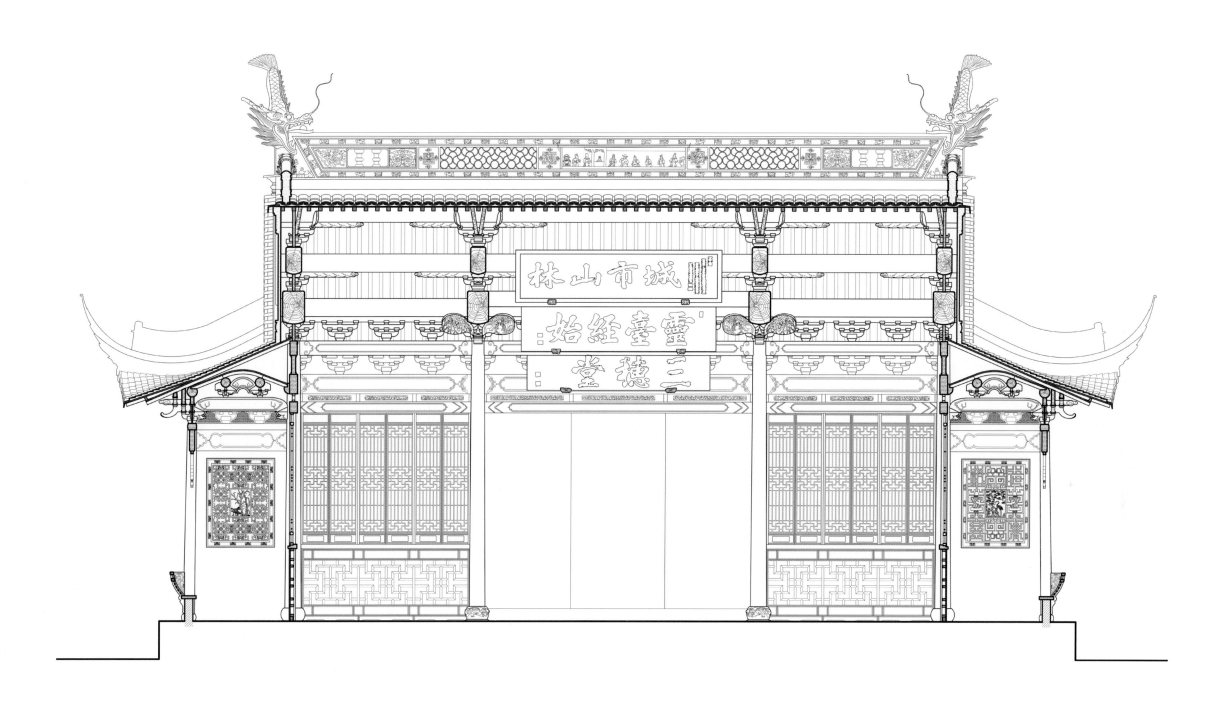

0    0.5    1    2m

三穗堂纵剖面图
Longitudinal section of Sansuitang

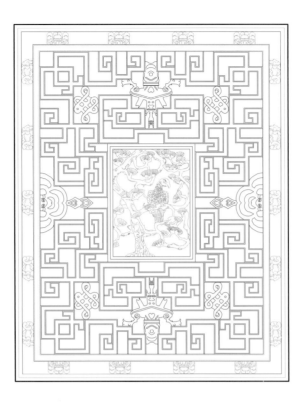
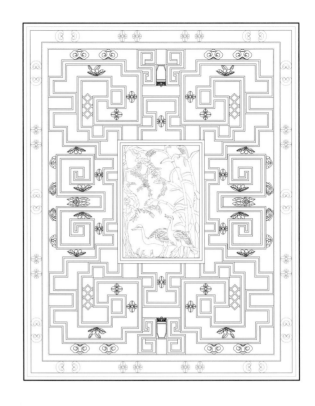

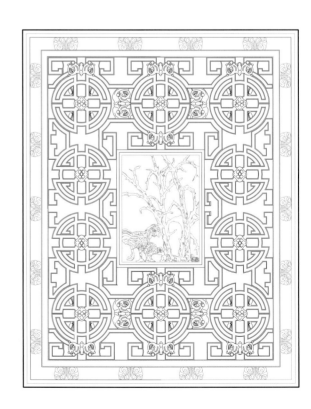

三穗堂漏窗大样图（一）
Window of Sansuitang (1)

0　0.1　0.2　　0.4m

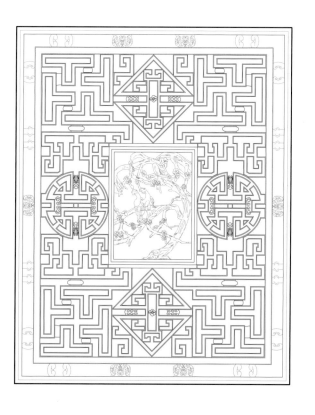

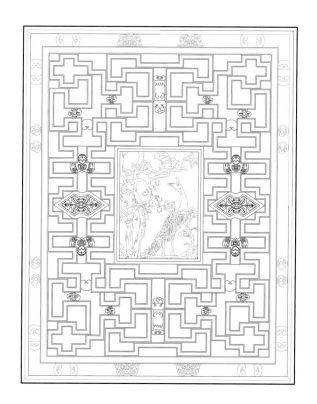

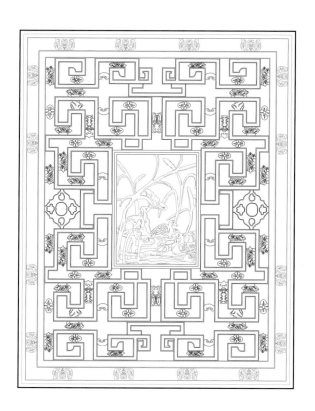

三穗堂漏窗大样图（二）
Window of Sansuitang (2)

0  0.1  0.2  0.4m

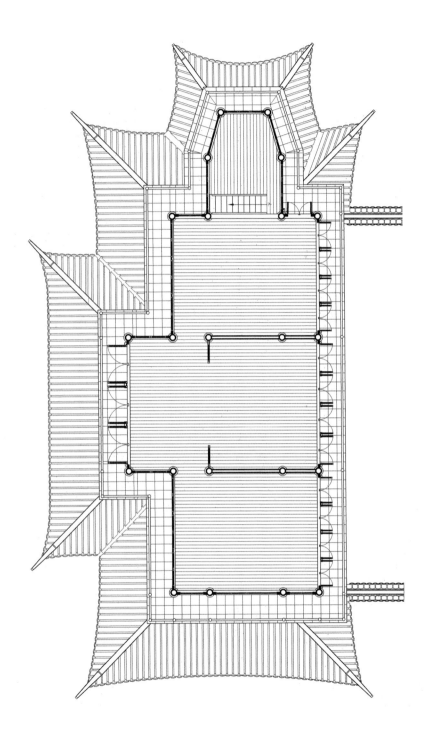

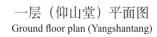

蜡梅

蜡梅

二层（卷雨楼）平面图
Second floor plan (Juanyulou)

一层（仰山堂）平面图
Ground floor plan (Yangshantang)

0　1　2　　　　5m

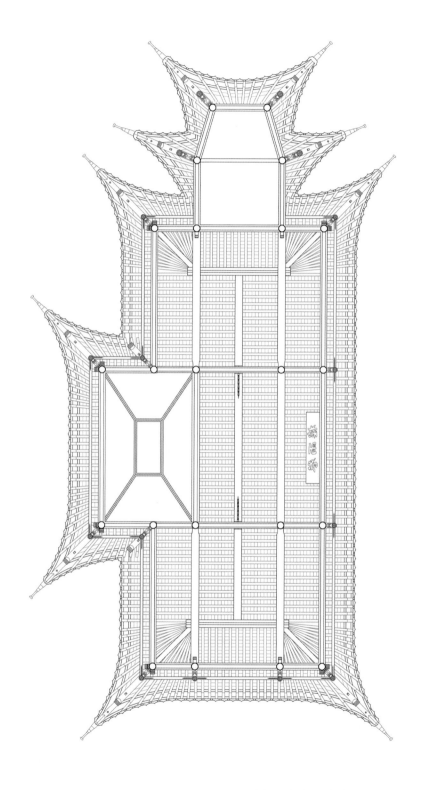

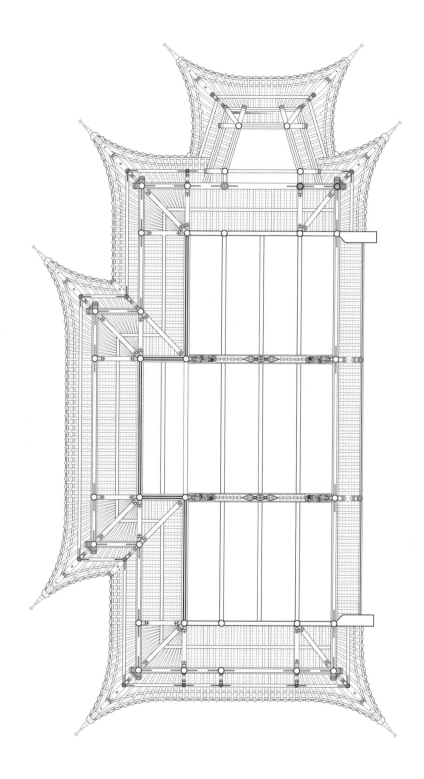

0 1 2 5m

二层（卷雨楼）屋架仰视图
Bottom view of Juanyulou's beams

一层（仰山堂）屋架仰视图
Bottom view of Yangshantang's beams

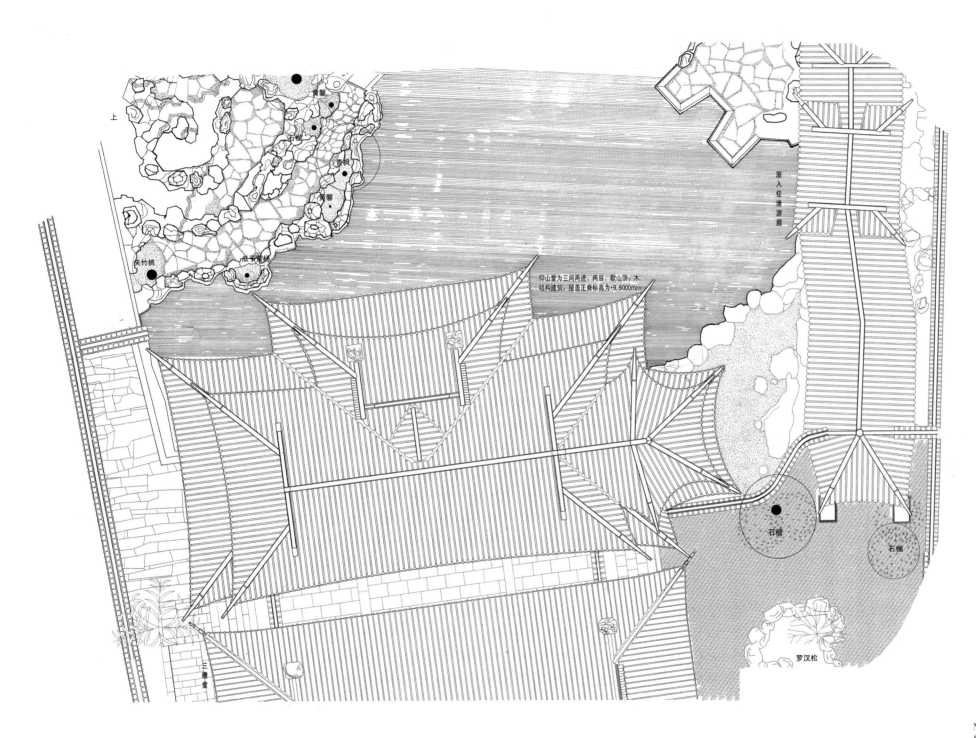

上

渐入佳境游廊

仰山堂为三间两进，两层，歇山顶，木结构建筑，屋盖正脊标高为+9.6000m。

矢竹桃

黄馨

柏

野块

筑馨

盘予海桃

三穗堂

石榴

石榴

罗汉松

卷雨楼屋顶平面图
Roof plan of Juanyulou

0 1 2 5m

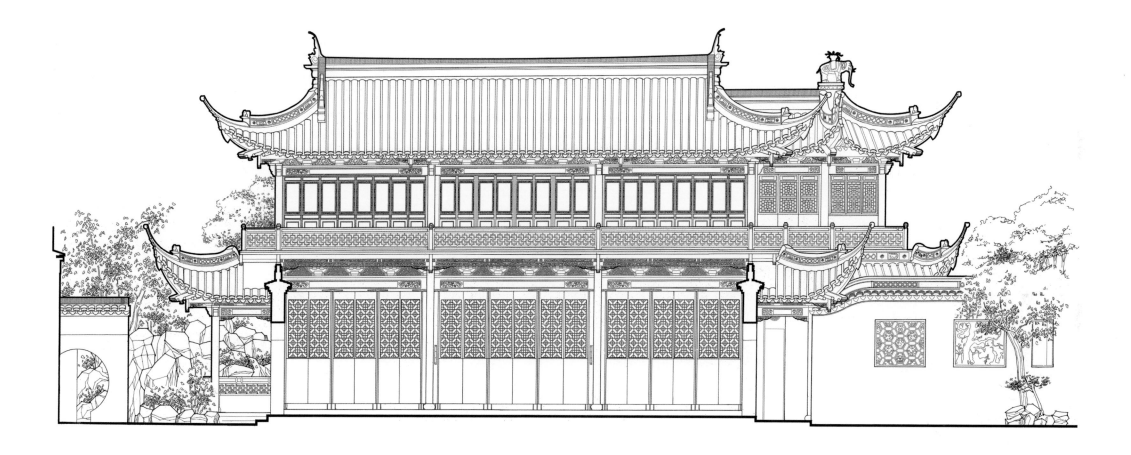

0　1　2　　　　5m

卷雨楼南立面图
South elevation of Juanyulou

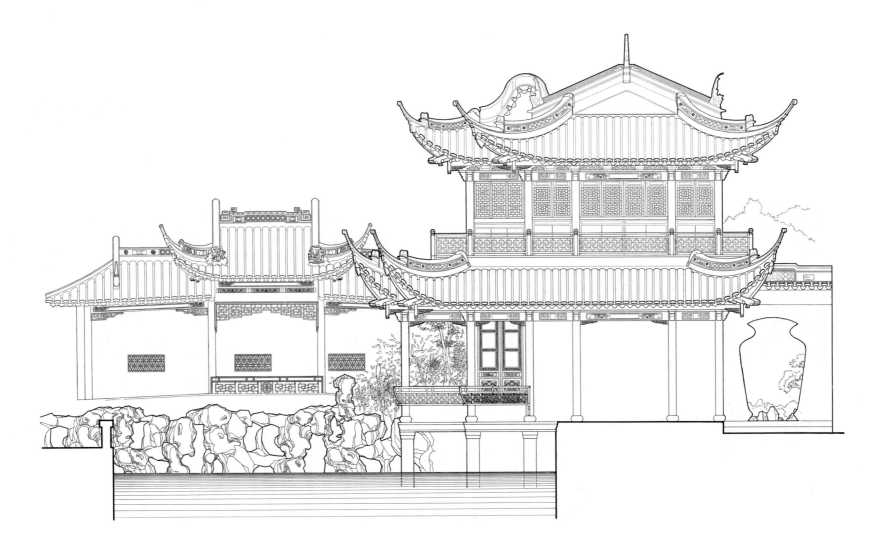

卷雨楼西立面图
West elevation of Juanyulou

0　1　2　　　　5m

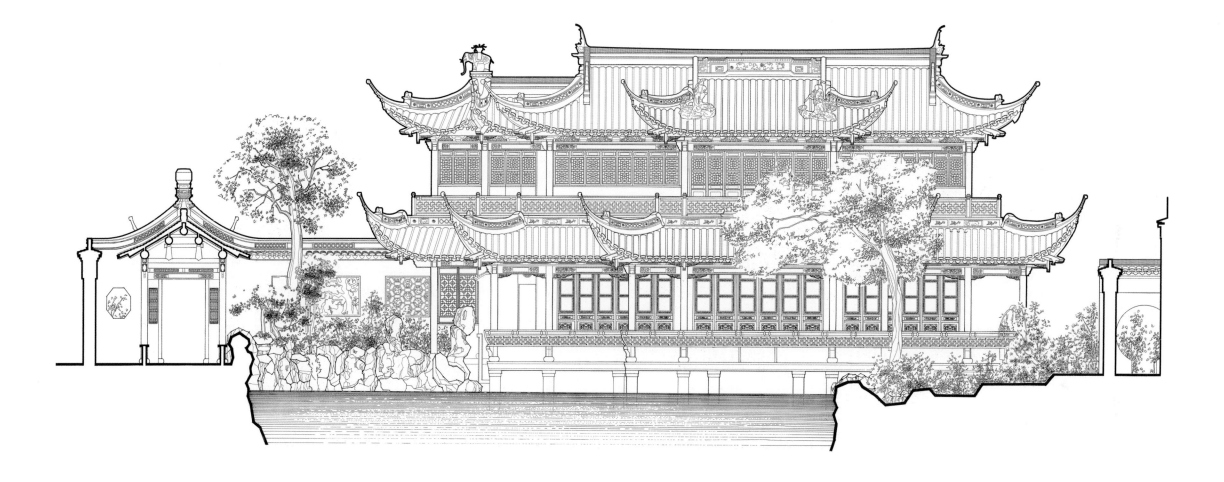

卷雨楼北立面图
North elevation of Juanyulou

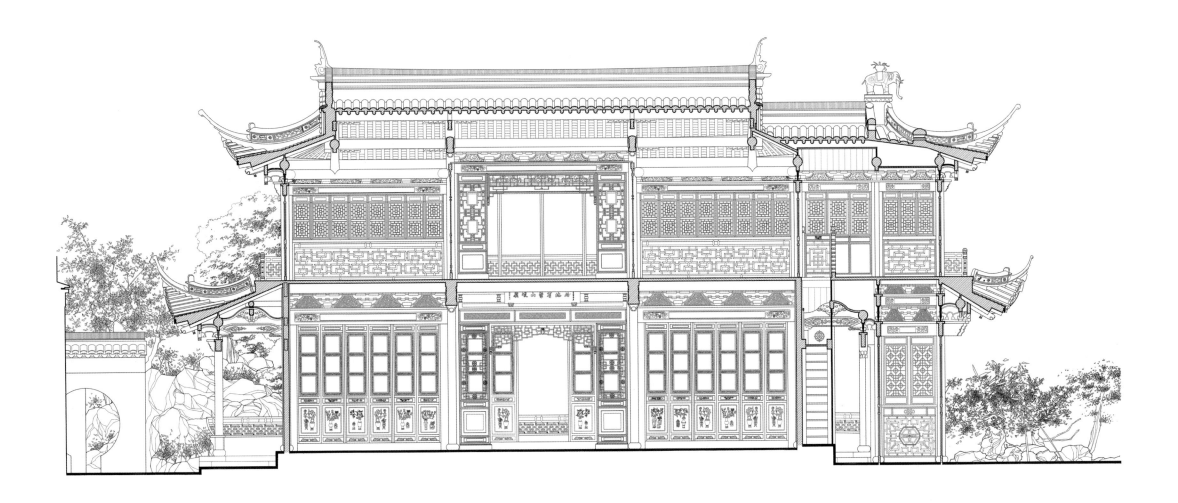

卷雨楼纵剖面图
Longitudinal section of Juanyulou

0 1 2 5m

卷雨楼轴测图

Axonometric drawing of Juanyulou

挹秀亭剖面图
Section of Yixiuting

挹秀亭屋架仰视图
Bottom view of Yixiuting's beams

挹秀亭屋顶平面图
Roof plan of Yixiuting

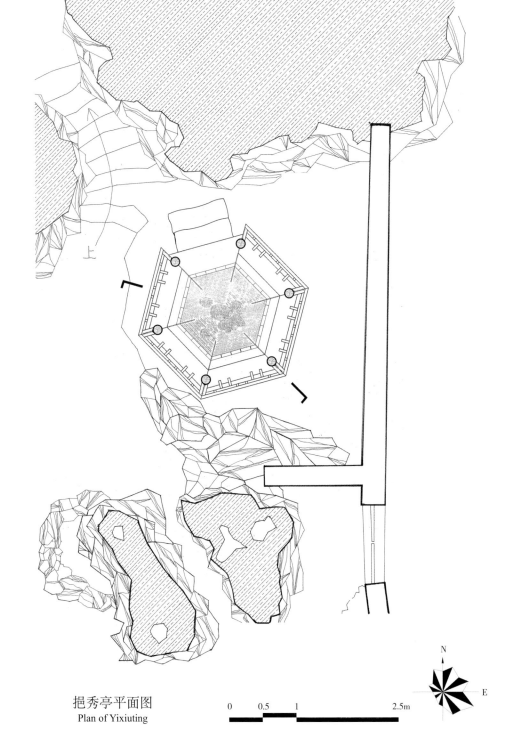

上

挹秀亭平面图
Plan of Yixiuting

N
E

0    0.5    1                    2.5m

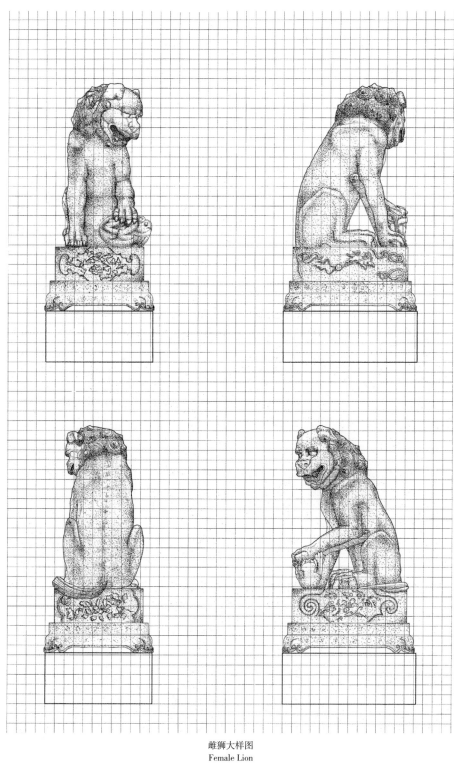

雌狮大样图
Female Lion

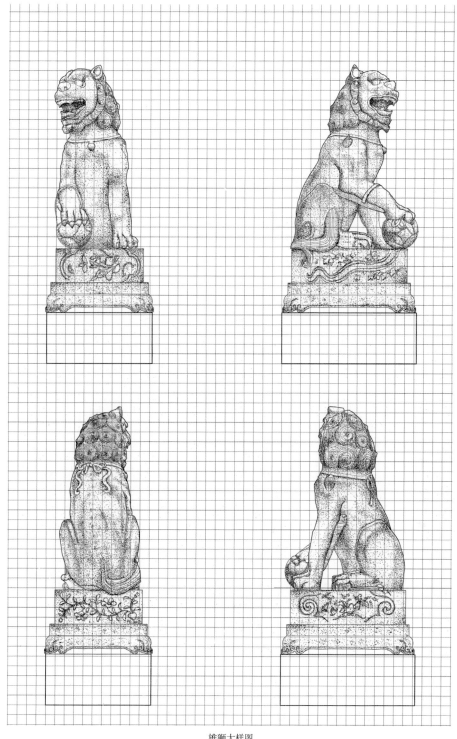

雄狮大样图
Male Lion

0    0.15    0.3    0.6m

原河南安阳"大元国至元二十九年"铁狮大样图
Iron lions of 1292 (year 29 of Zhiyuan Emperor of Yuan Dynasty) from Anyang, Henan Province

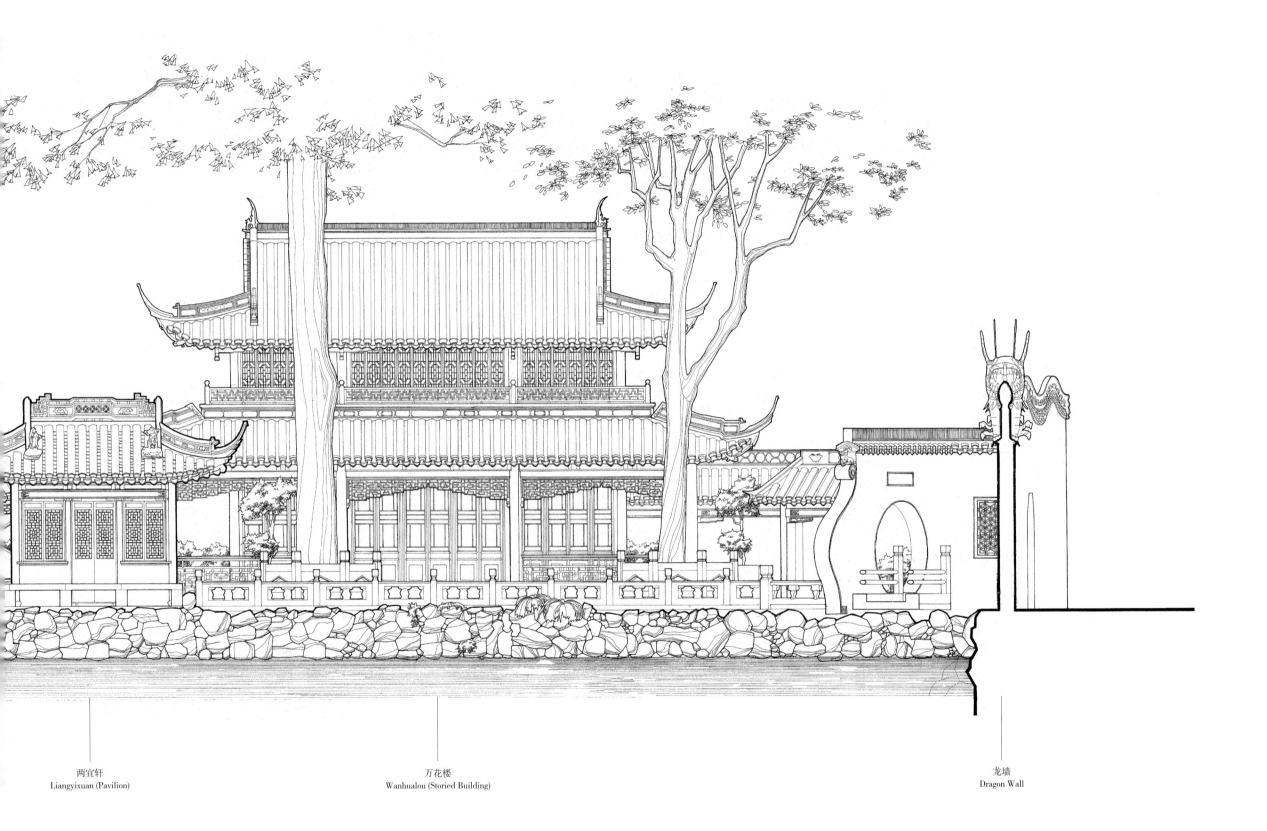

两宜轩
Liangyixuan (Pavilion)

万花楼
Wanhualou (Storied Building)

龙墙
Dragon Wall

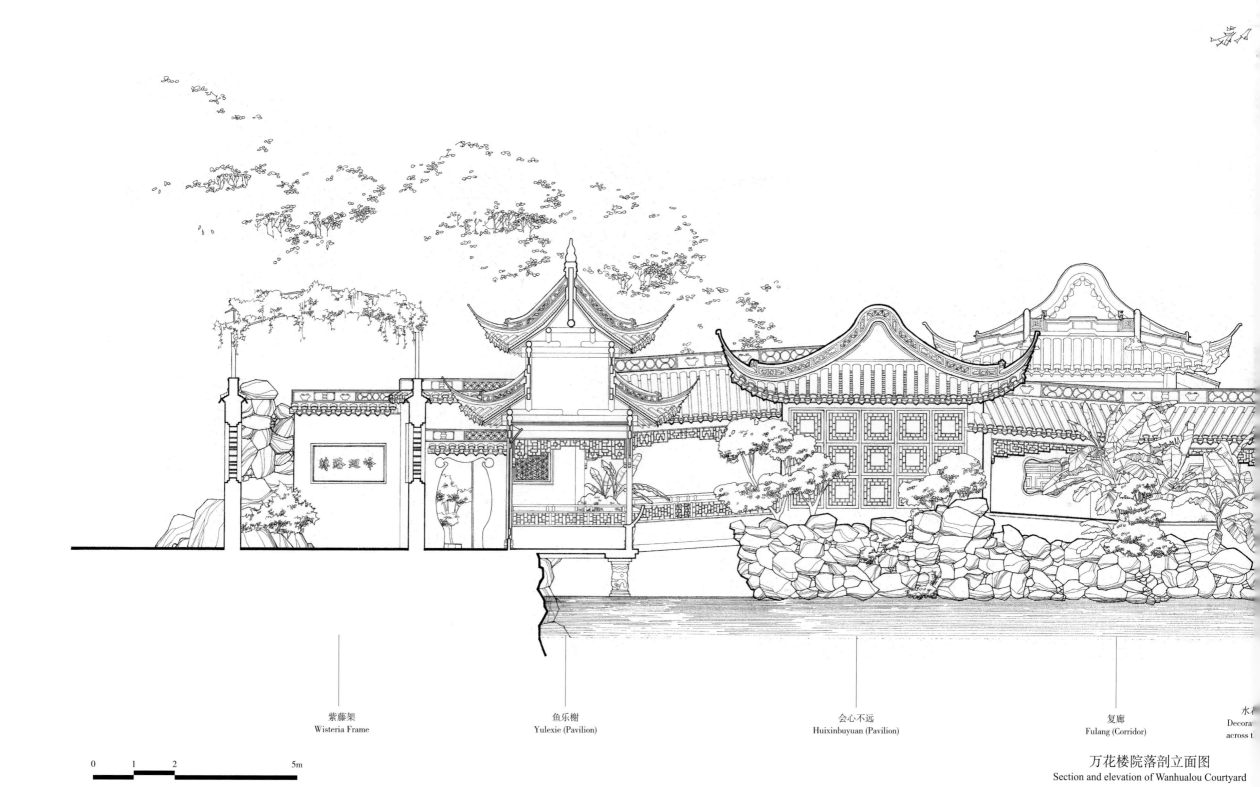

峰迴路转

| 紫藤架 | 鱼乐榭 | 会心不远 | 复廊 | 水石 |
|---|---|---|---|---|
| Wisteria Frame | Yulexie (Pavilion) | Huixinbuyuan (Pavilion) | Fulang (Corridor) | Decora... across t... |

0　1　2　　　　5m

万花楼院落剖立面图
Section and elevation of Wanhualou Courtyard

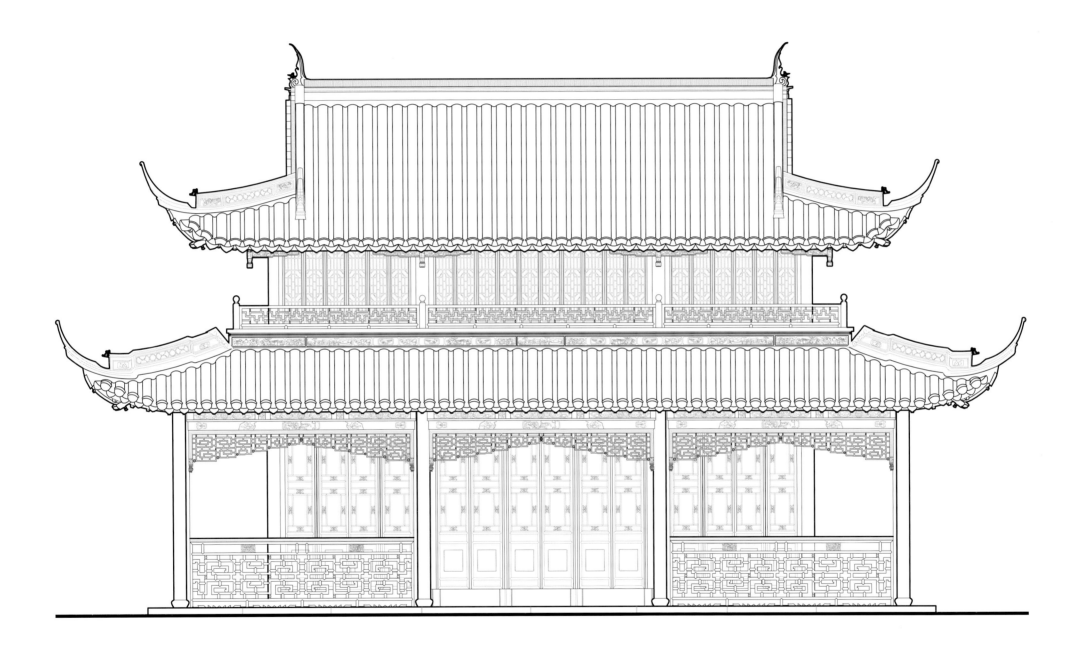

万花楼正立面图
Front elevation of Wanhualou

0    0.5    1m

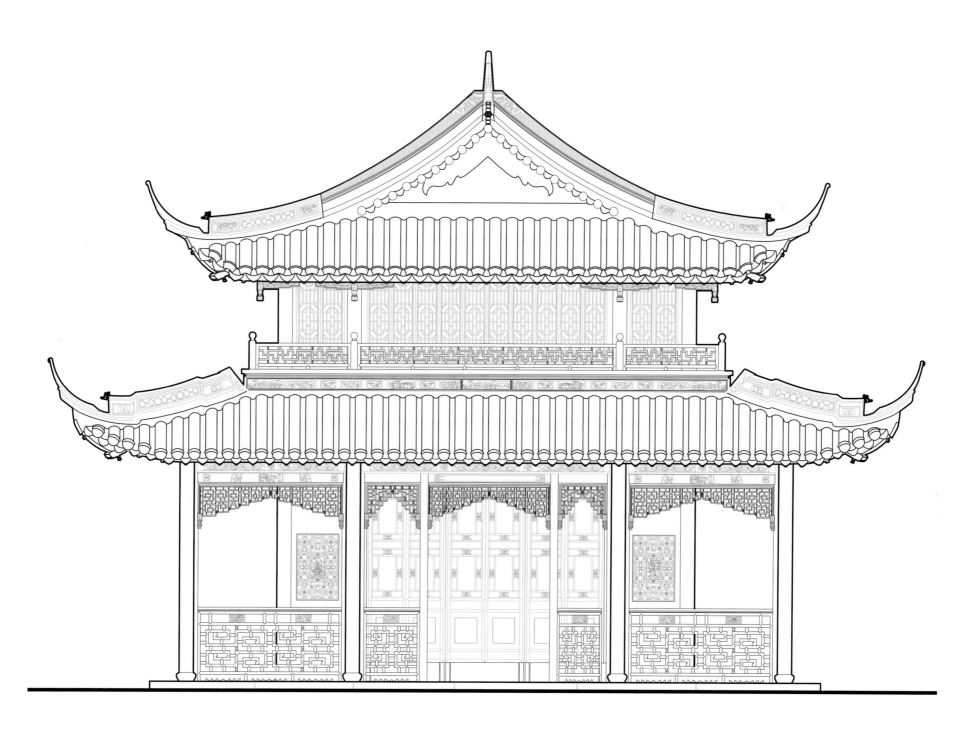

0    0.5    1m

万花楼侧立面图
Side elevation of Wanhualou

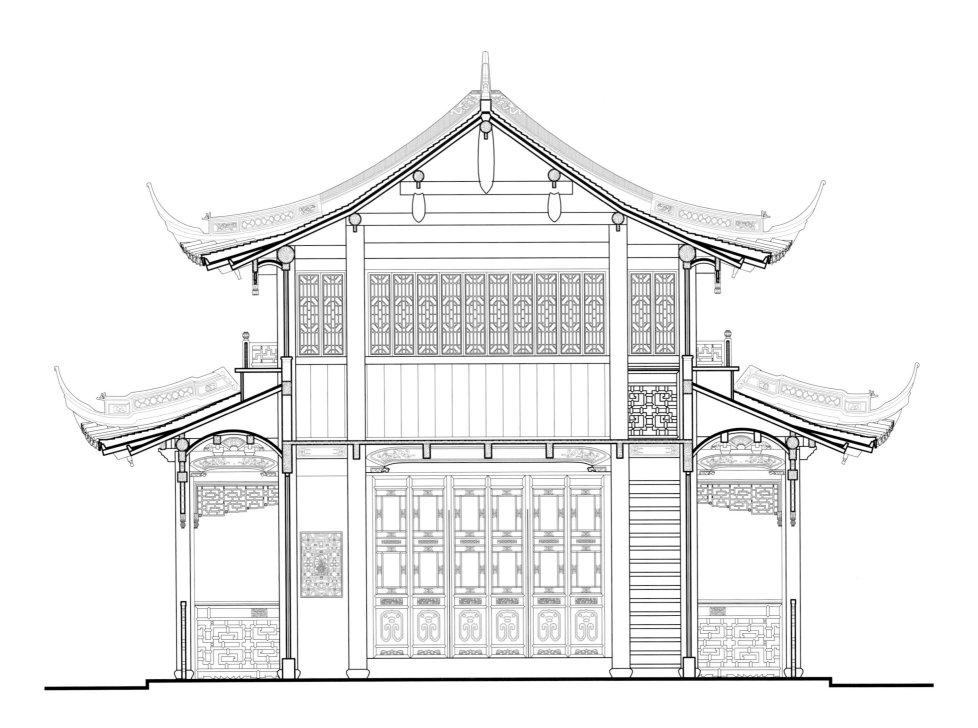

万花楼横剖面图
Cross-section of Wanhualou

0    0.5    1m

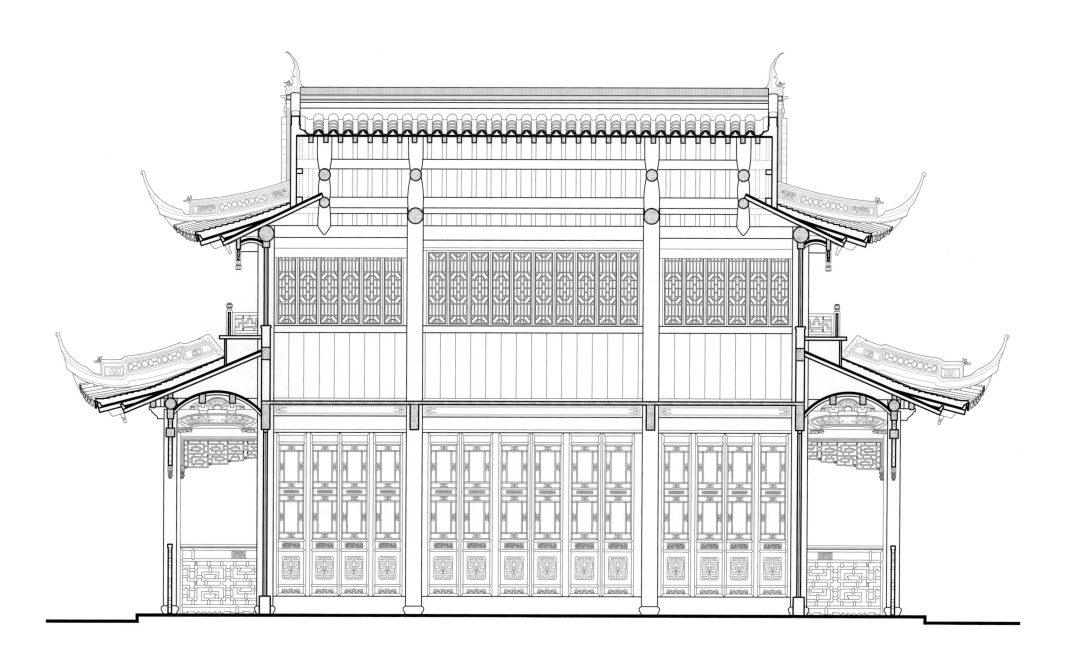

0    0.5    1m

万花楼纵剖面图
Longitudinal section of Wanhualou

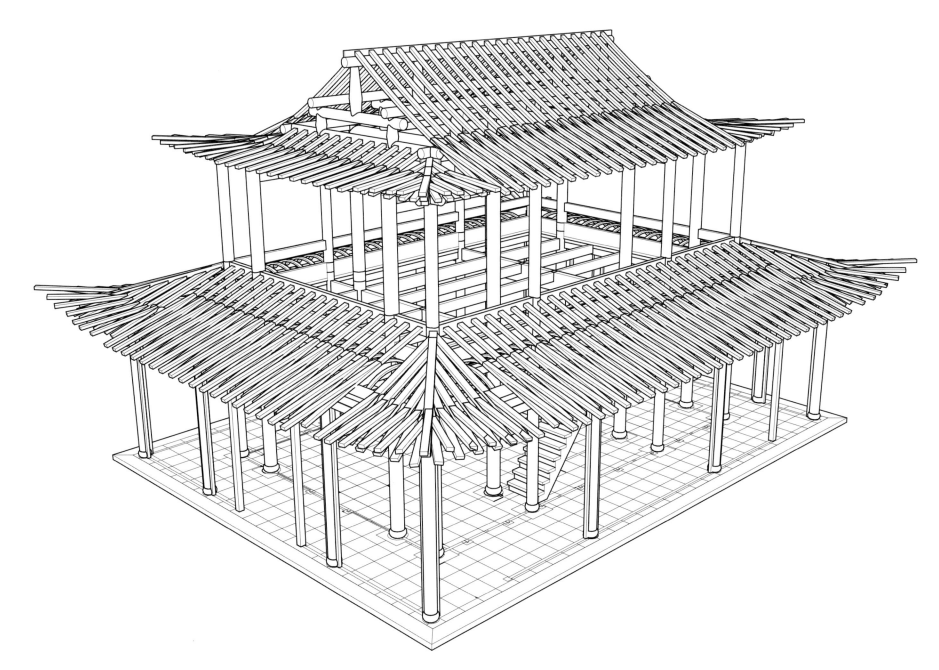

万花楼构架透视图（一）
Perspective view of Wanhualou's structure I

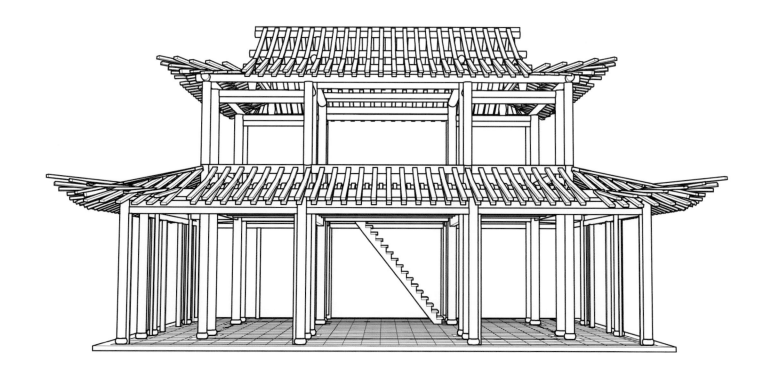

万花楼构架透视图（二）
Perspective view of Wanhualou's structure II

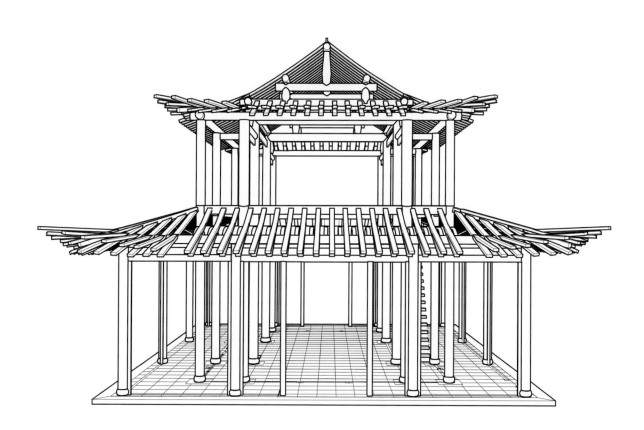

万花楼构架透视图（三）
Perspective view of Wanhualou's structure III

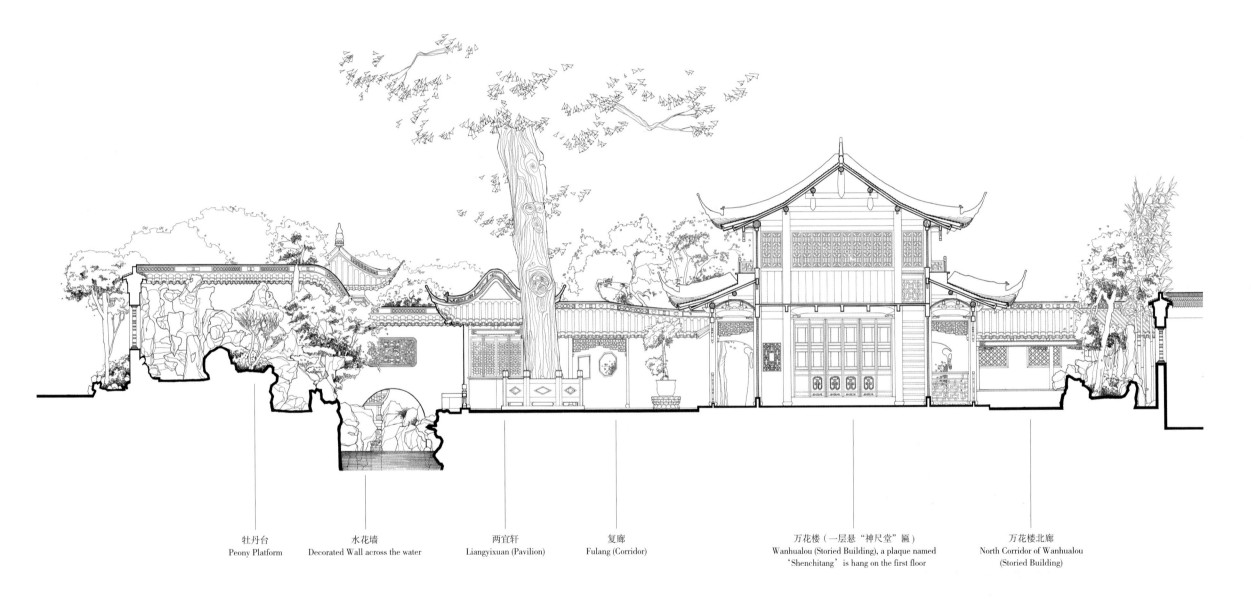

牡丹台
Peony Platform

水花墙
Decorated Wall across the water

两宜轩
Liangyixuan (Pavilion)

复廊
Fulang (Corridor)

万花楼（一层悬"神尺堂"匾）
Wanhualou (Storied Building), a plaque named
'Shenchitang' is hang on the first floor

万花楼北廊
North Corridor of Wanhualou
(Storied Building)

万花楼院落南北剖面图
South-north section of Wanhualou Courtyard

0    1    2         5m

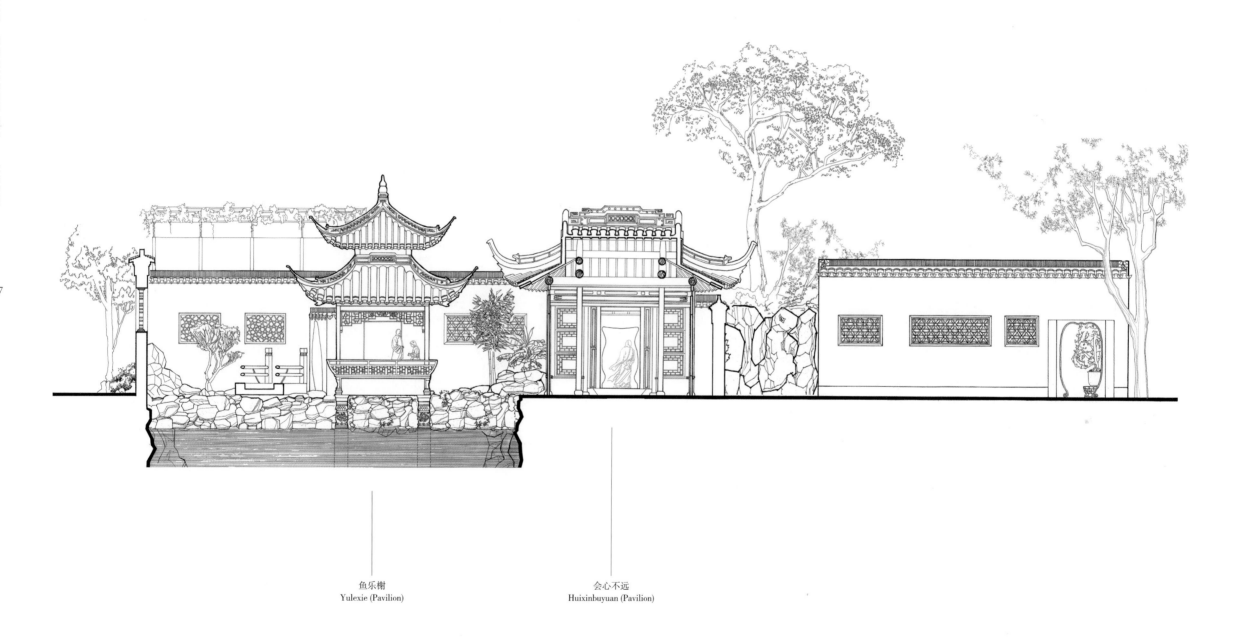

鱼乐榭
Yulexie (Pavilion)

会心不远
Huixinbuyuan (Pavilion)

0　1　2　　　　5m

鱼乐榭院落南北剖面图
South-north section of Yulexie Courtyard

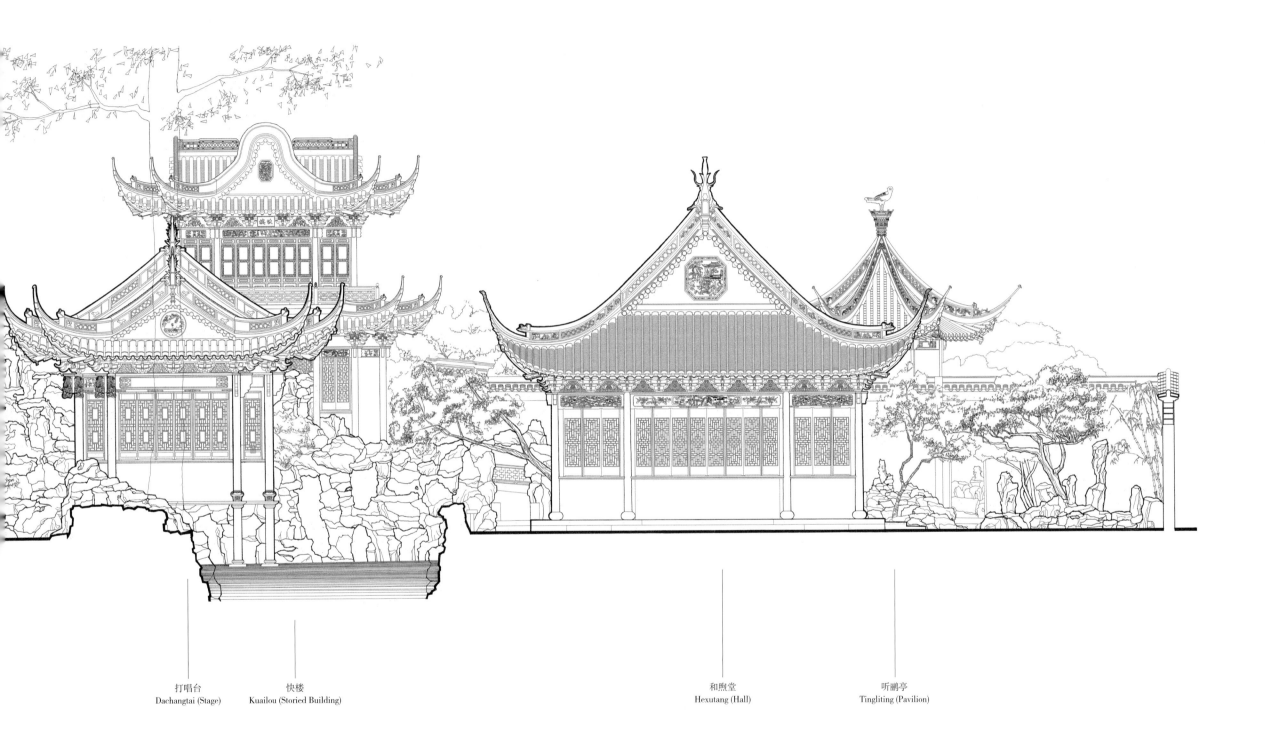

打唱台
Dachangtai (Stage)

快楼
Kuailou (Storied Building)

和煦堂
Hexutang (Hall)

听鹂亭
Tingliting (Pavilion)

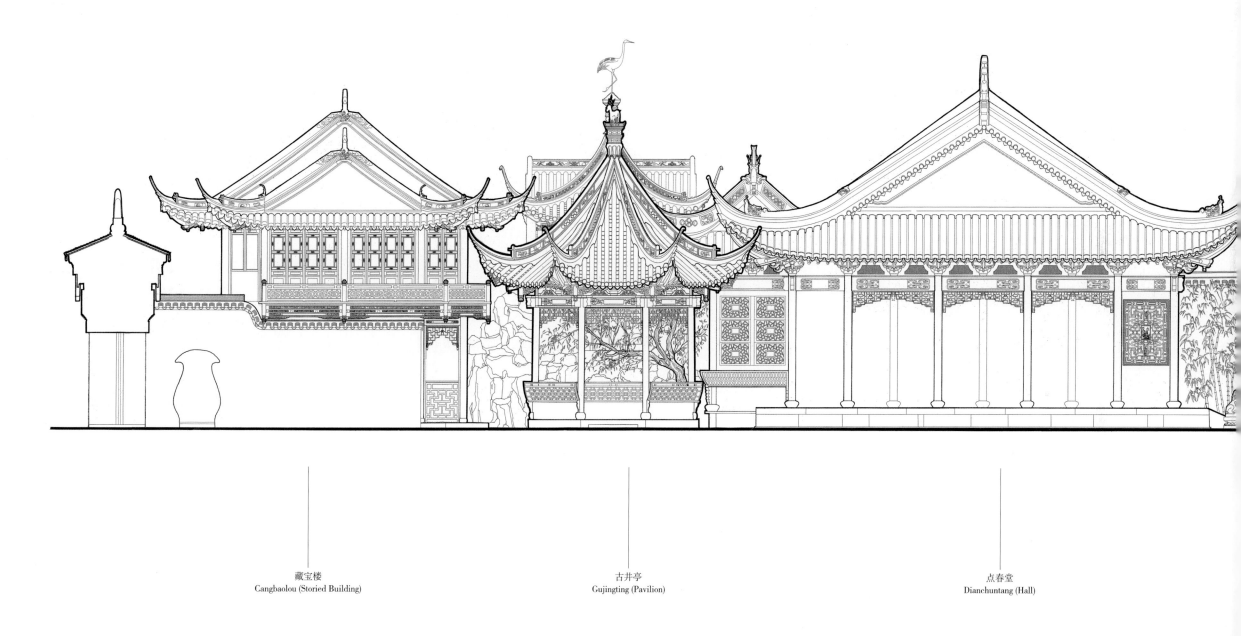

藏宝楼
Cangbaolou (Storied Building)

古井亭
Gujingting (Pavilion)

点春堂
Dianchuntang (Hall)

0    1    2        5m

点春堂院落中轴建筑西立面图
West elevation of the architectures on the main axis of Dianchuntang Courtyard

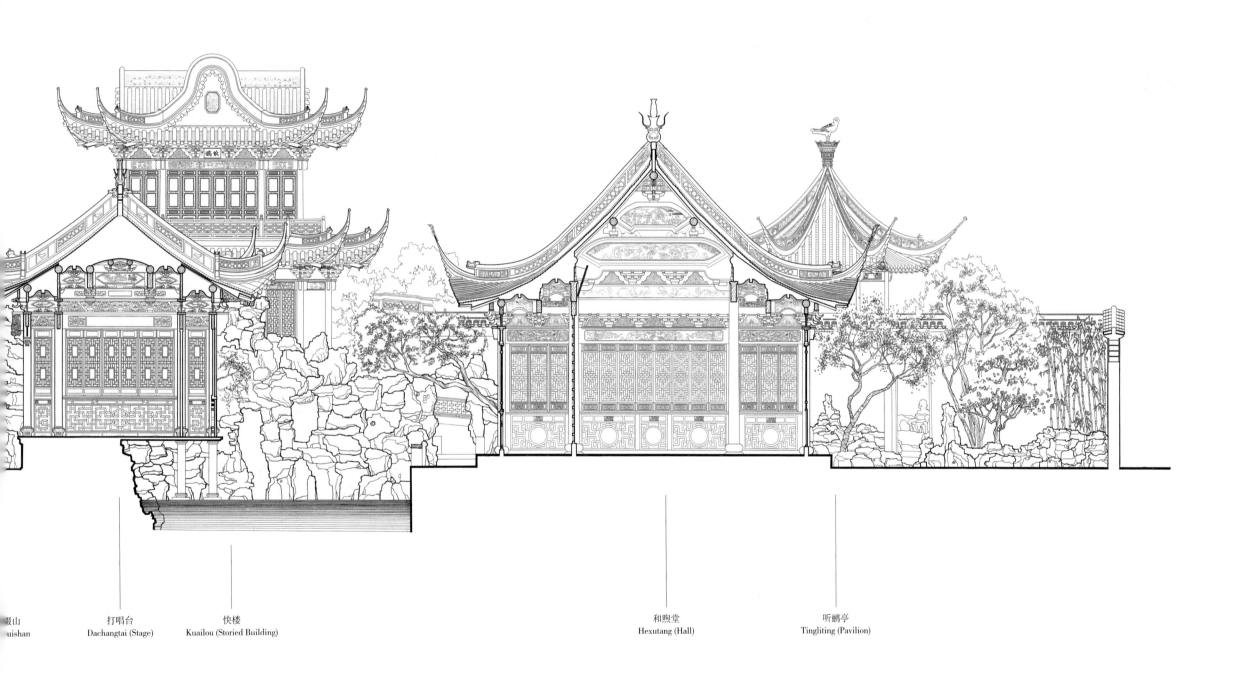

叠山
uishan

打唱台
Dachangtai (Stage)

快楼
Kuailou (Storied Building)

和煦堂
Hexutang (Hall)

听鹂亭
Tinglíting (Pavilion)

藏宝楼
Cangbaolou (Storied Building)

学圃
Xuepu

点春堂
Dianchuntang (Hall)

0  1  2  5m

点春堂院落中轴剖面图
Section of the architectures on the main axis of Dianchuntang Courtyard

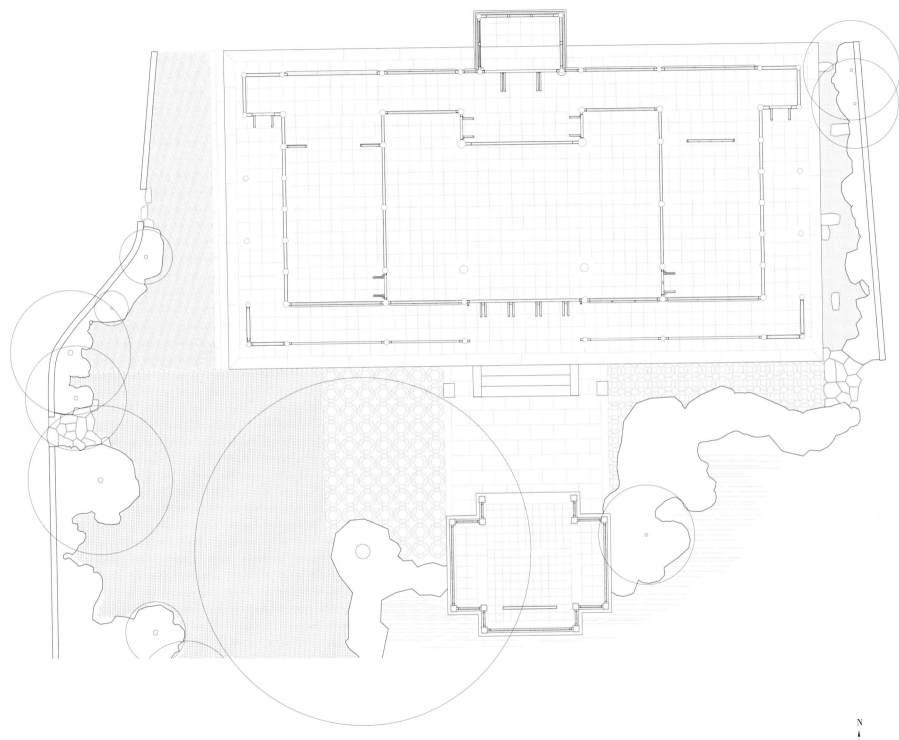

点春堂、打唱台一层平面图
Ground floor plan of Dianchuntang and Dachangtai

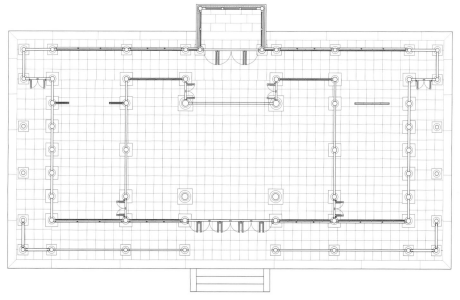

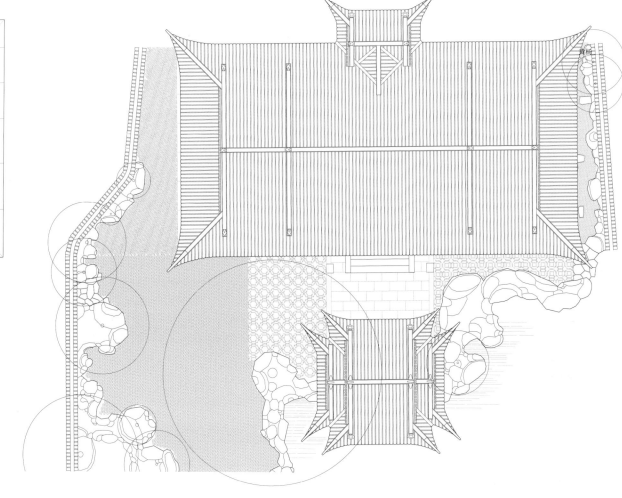

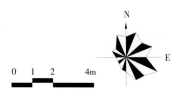

N

E

0 1 2 4m

点春堂一层平面图
Ground floor plan of Dianchuntang

点春堂、打唱台屋顶平面图
Roof plan of Dianchuntang and Dachangtai

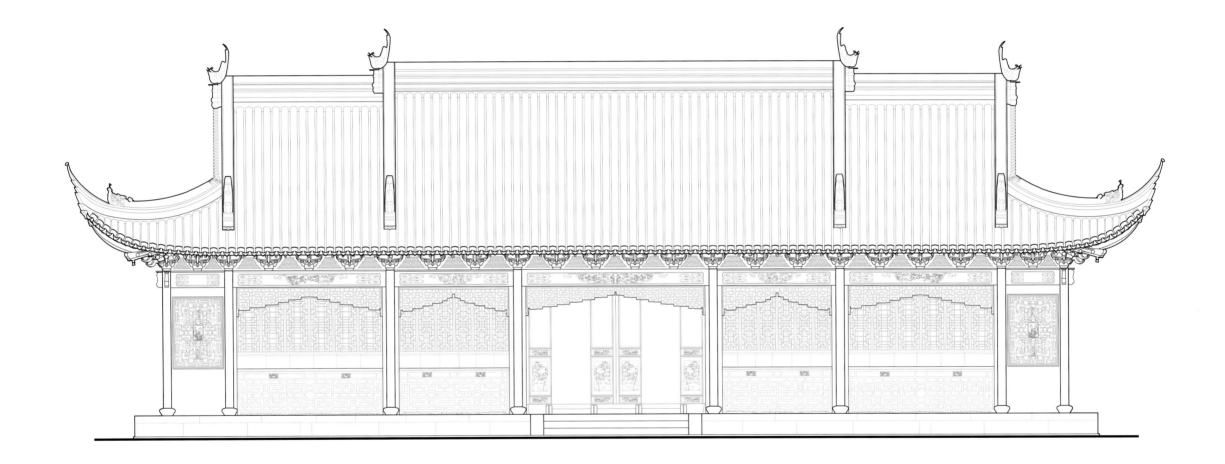

点春堂主立面图
Front elevation of Dianchuntang

0      1.6      3.2m

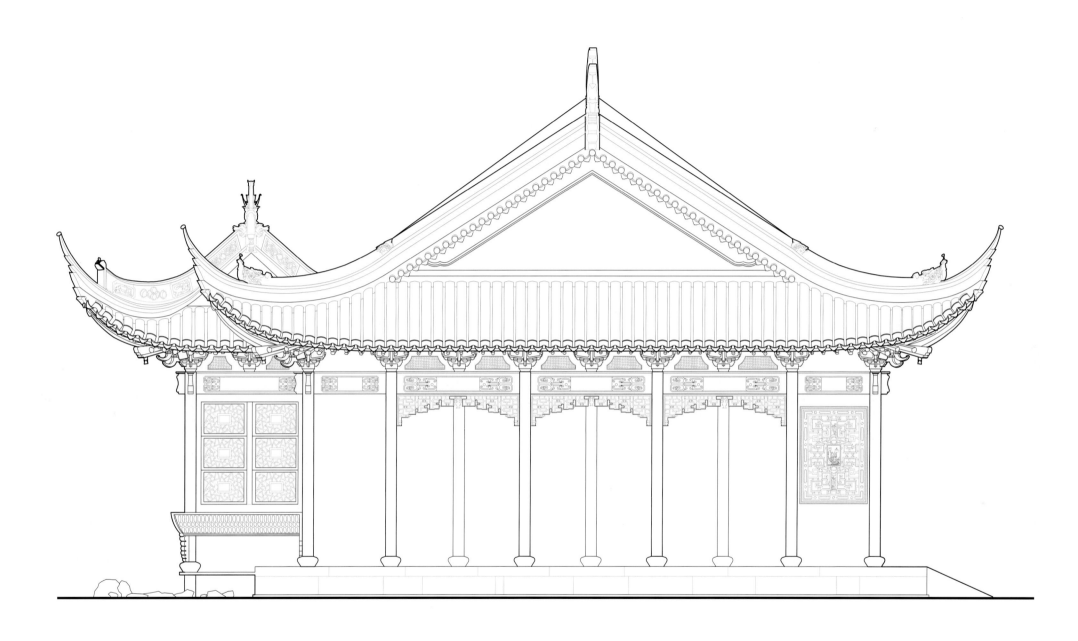

0  0.5  1  2m

点春堂侧立面图
Side elevation of Dianchuntang

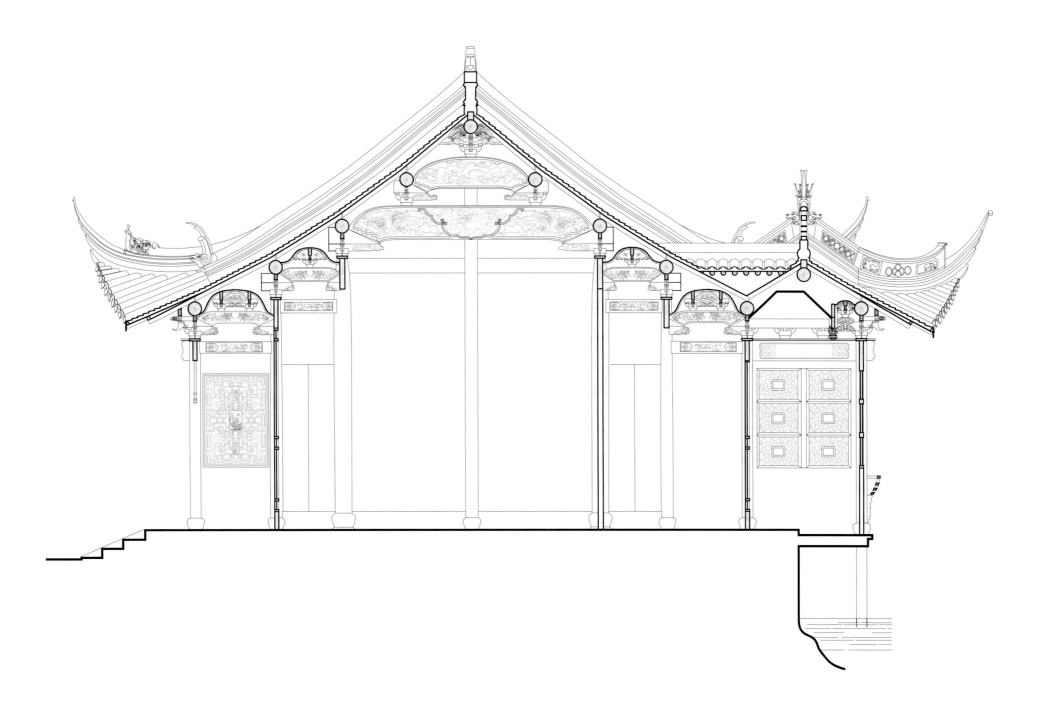

点春堂横剖面图
Cross-section of Dianchuntang

0　0.5　1　2m

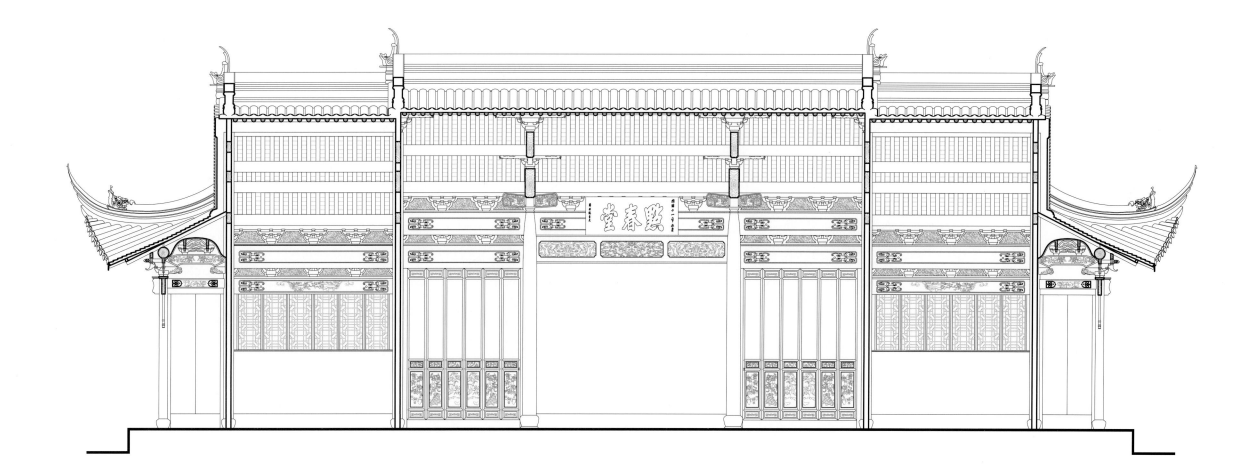

点春堂纵剖面图
Longitudinal section of Dianchuntang

穿云龙墙立面展开图

Expanded elevation of Chuanyun Dragon Wall

0　0.15　0.3　　0.6m

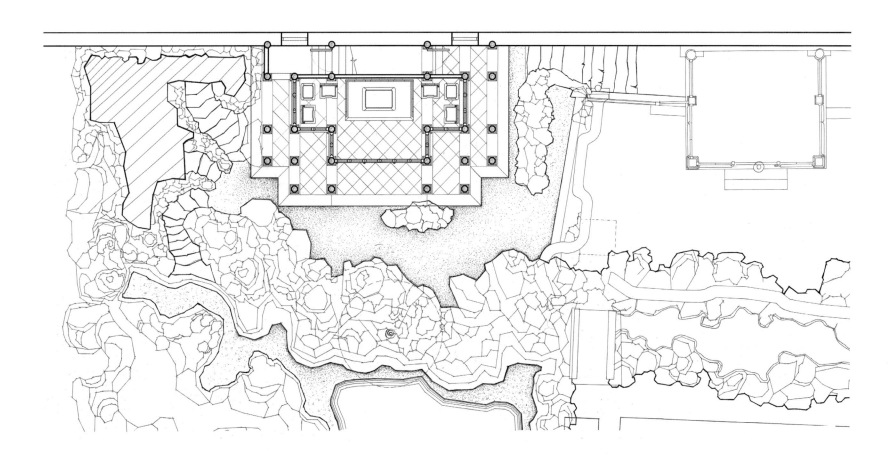

快楼一层平面图
Ground floor plan of Kuailou

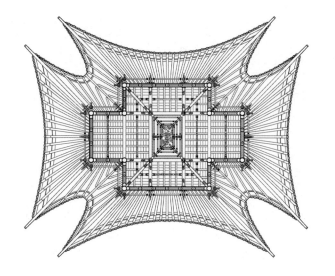

快楼屋架仰视图
Bottom view of Kuailou's beams

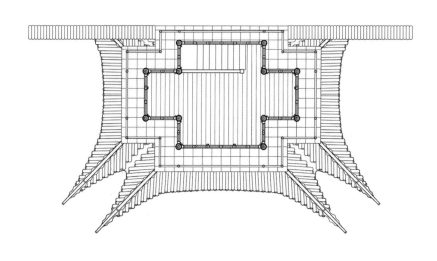

快楼二层平面图
Second floor plan of Kuailou

0 1 2 5m

快楼南立面图
South elevation of Kuailou

快楼西立面图
West elevation of Kuailou

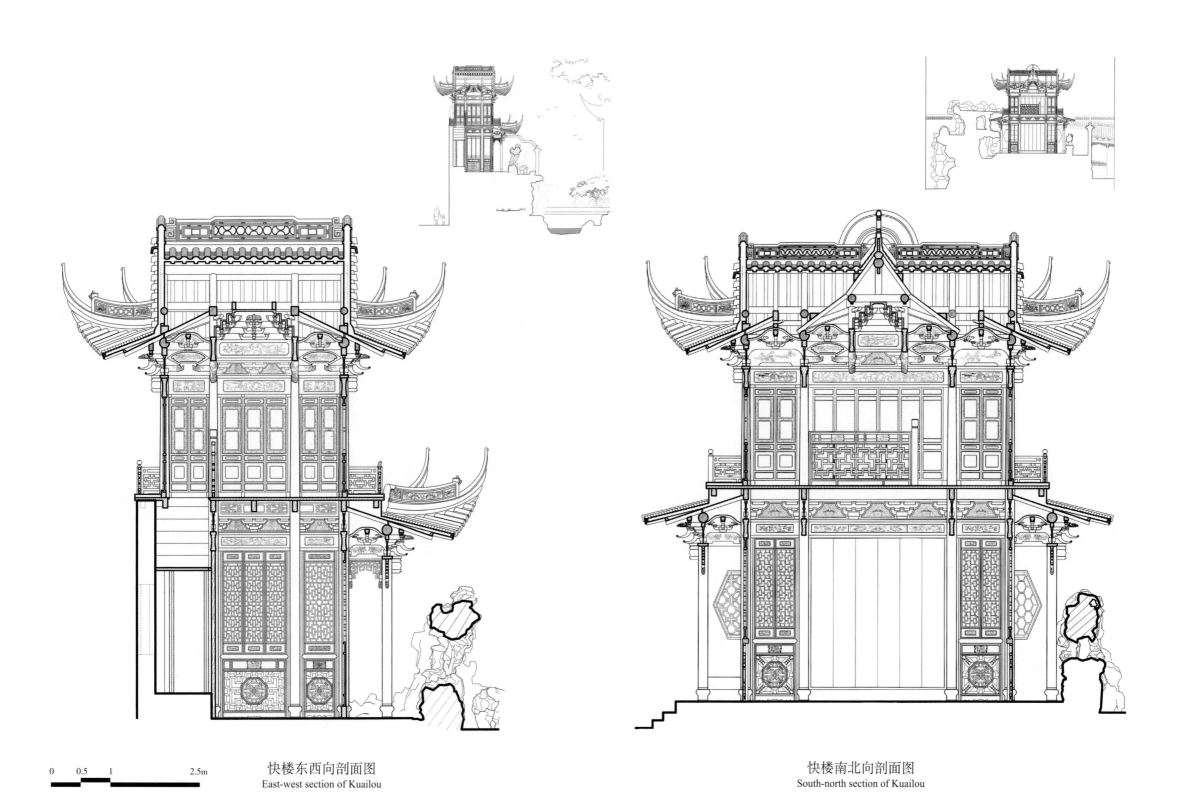

0    0.5    1                    2.5m

快楼东西向剖面图
East-west section of Kuailou

快楼南北向剖面图
South-north section of Kuailou

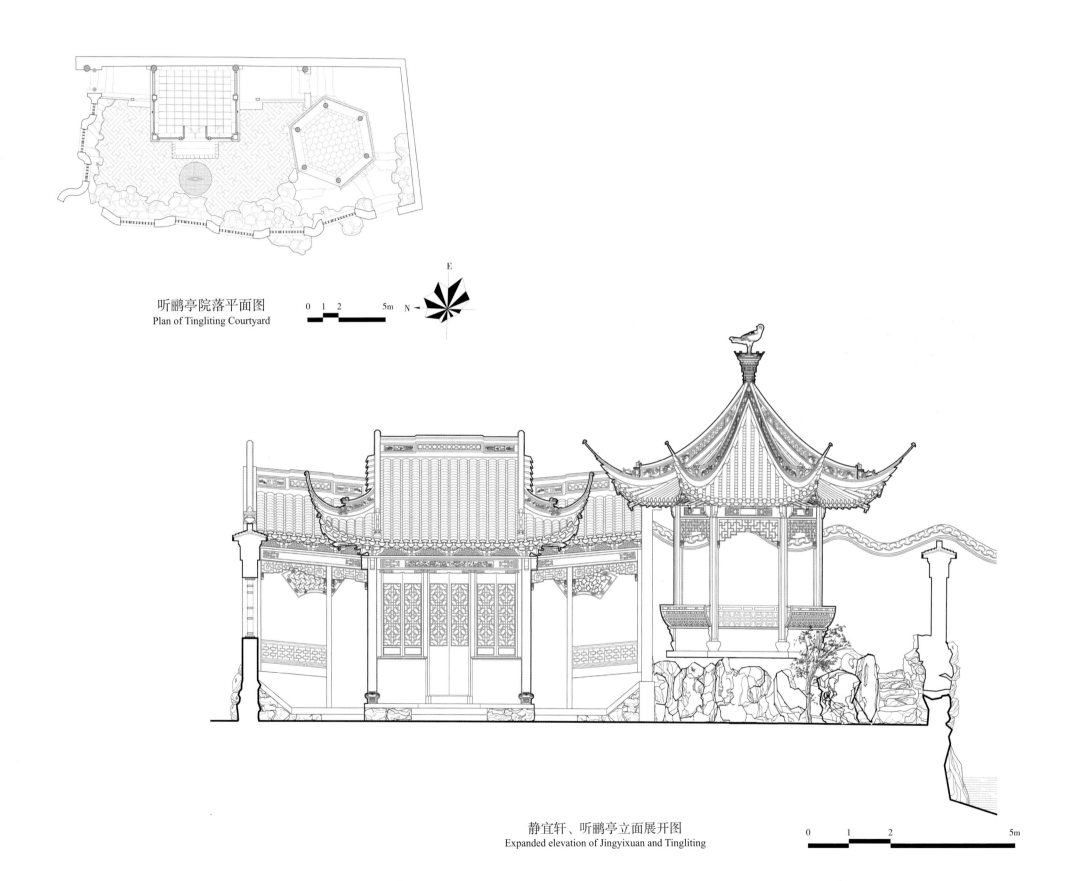

听鹂亭院落平面图
Plan of Tingliting Courtyard

0 1 2 5m N

E

静宜轩、听鹂亭立面展开图
Expanded elevation of Jingyixuan and Tingliting

0 1 2 5m

听鹂亭院落轴测解析图
Axonometric and exploded drawing of Tingliting Courtyard

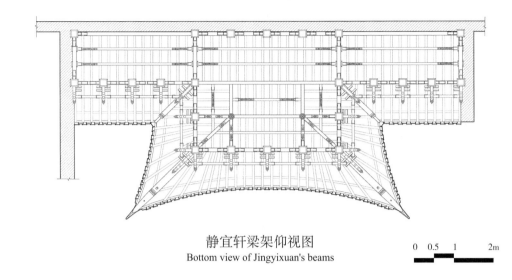

静宜轩梁架仰视图
Bottom view of Jingyixuan's beams

0  0.5  1  2m

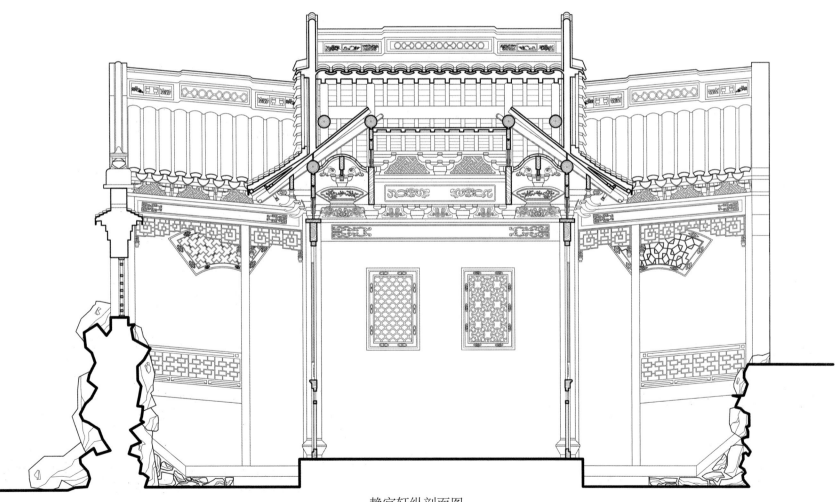

静宜轩纵剖面图
Longitudinal section of Jingyixuan

0  0.5  1  2m

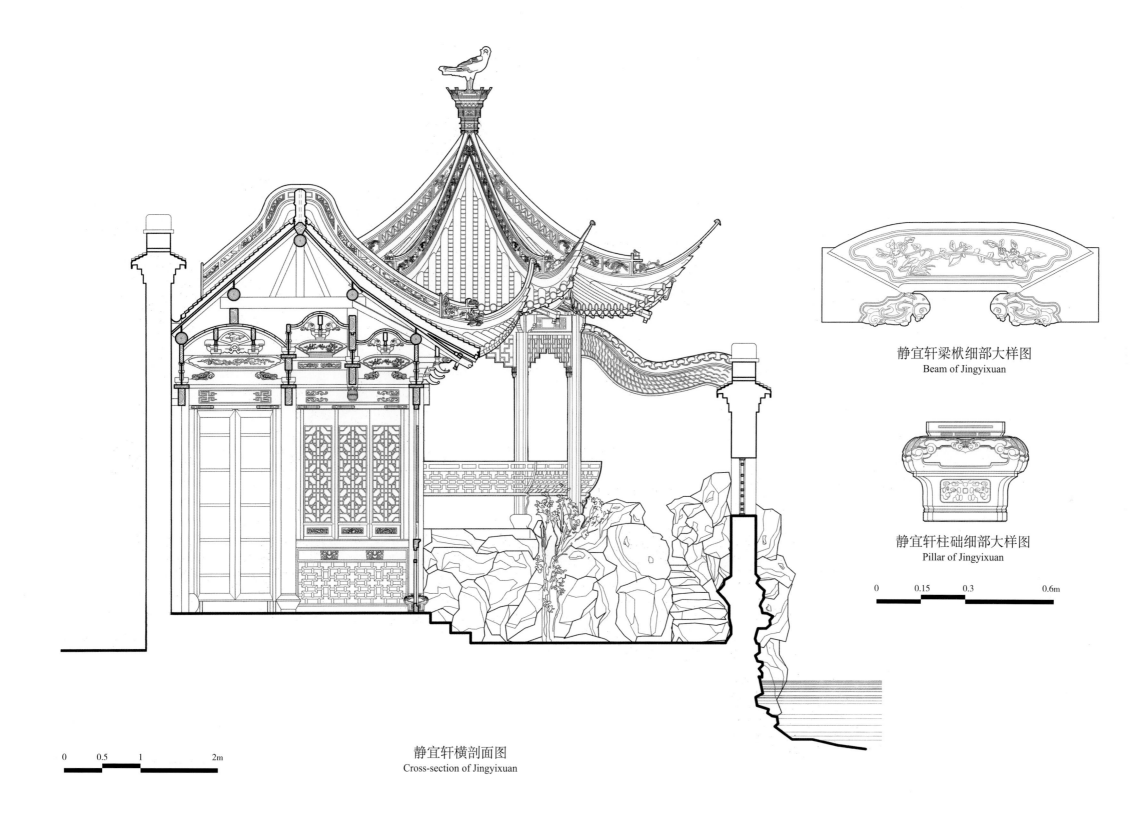

静宜轩梁枋细部大样图
Beam of Jingyixuan

静宜轩柱础细部大样图
Pillar of Jingyixuan

0    0.15    0.3         0.6m

静宜轩横剖面图
Cross-section of Jingyixuan

0    0.5    1         2m

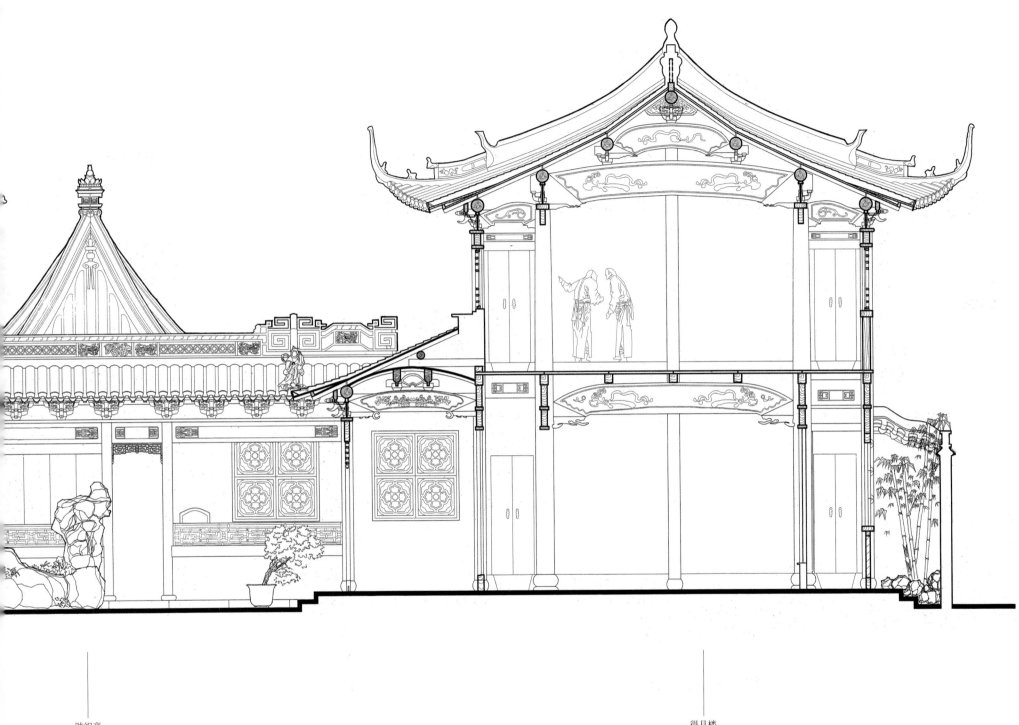

跂织亭
Qizhiting (Pavilion)

得月楼
Deyuelou (Storied Building)

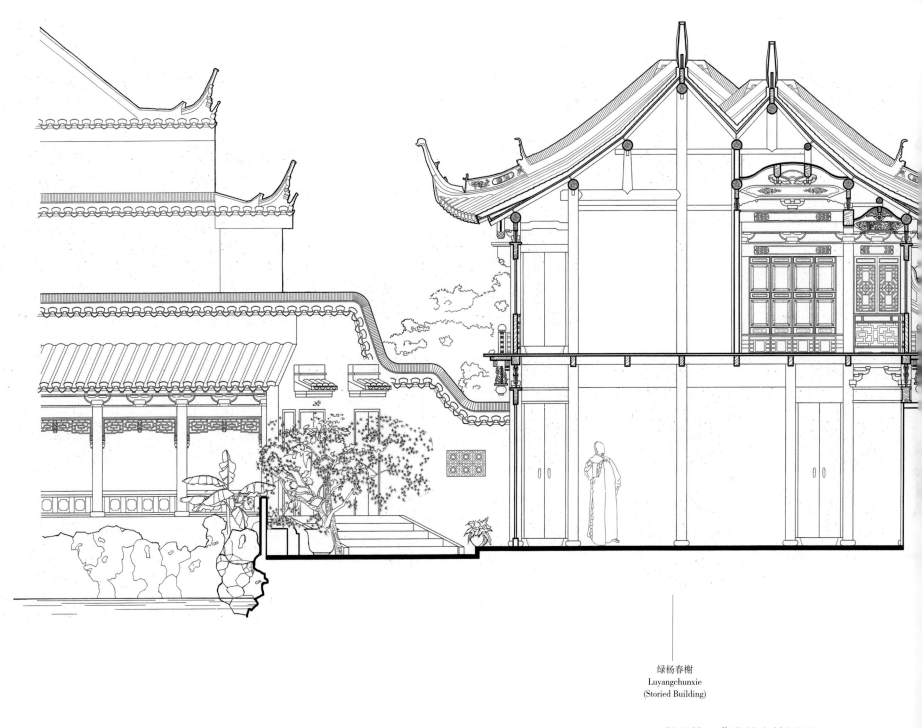

绿杨春榭
Luyangchunxie
(Storied Building)

得月楼、藏书楼中轴剖面图
Section of Deyuelou and Cangshulou

0 1 2 5m

得月楼
Deyuelou (Storied Building)

原城隍庙西园

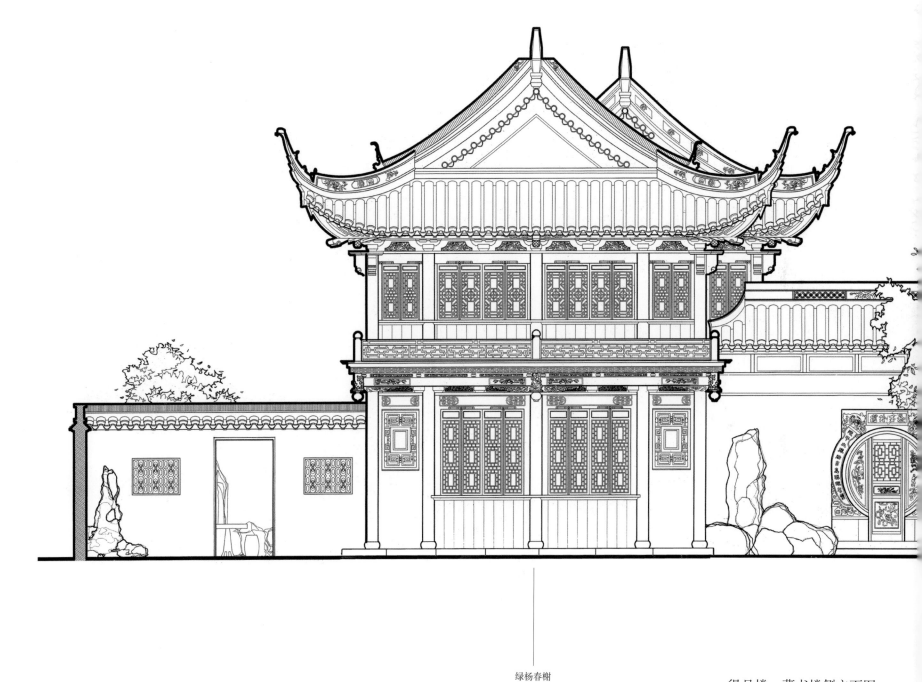

0 1 2 5m

绿杨春榭
Luyangchunxie (Storied Building)

得月楼、藏书楼侧立面图
Side elevation of Deyuelou and Cangshulou

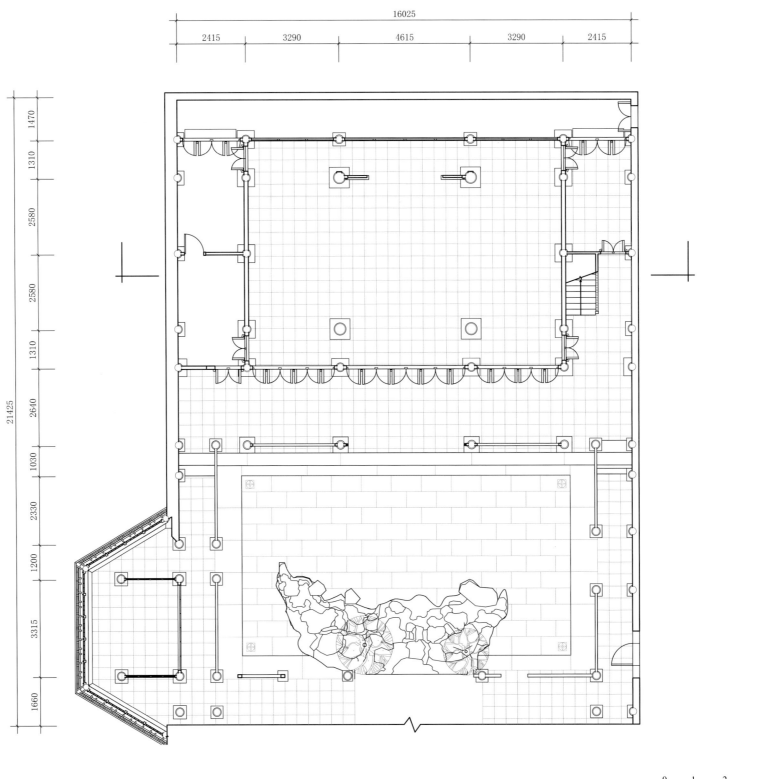

得月楼一层平面图
Ground floor plan of Deyuelou

0  1  2  4m

N
E

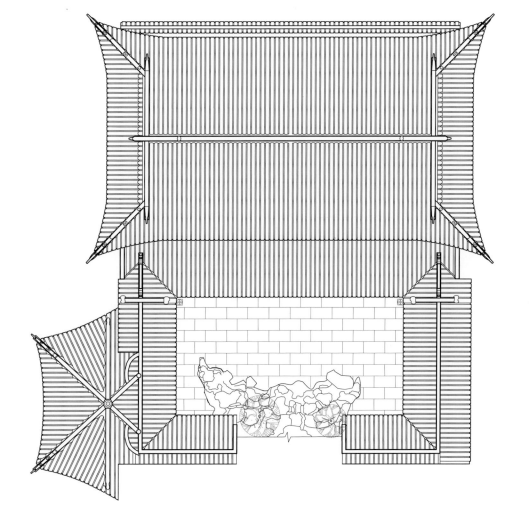

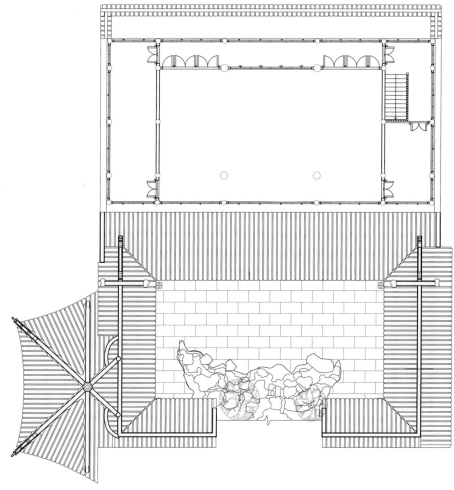

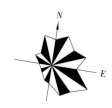

0　1　2　　4m

得月楼屋顶平面图
Roof plan of Deyuelou

得月楼二层平面图
Second floor plan of Deyuelou

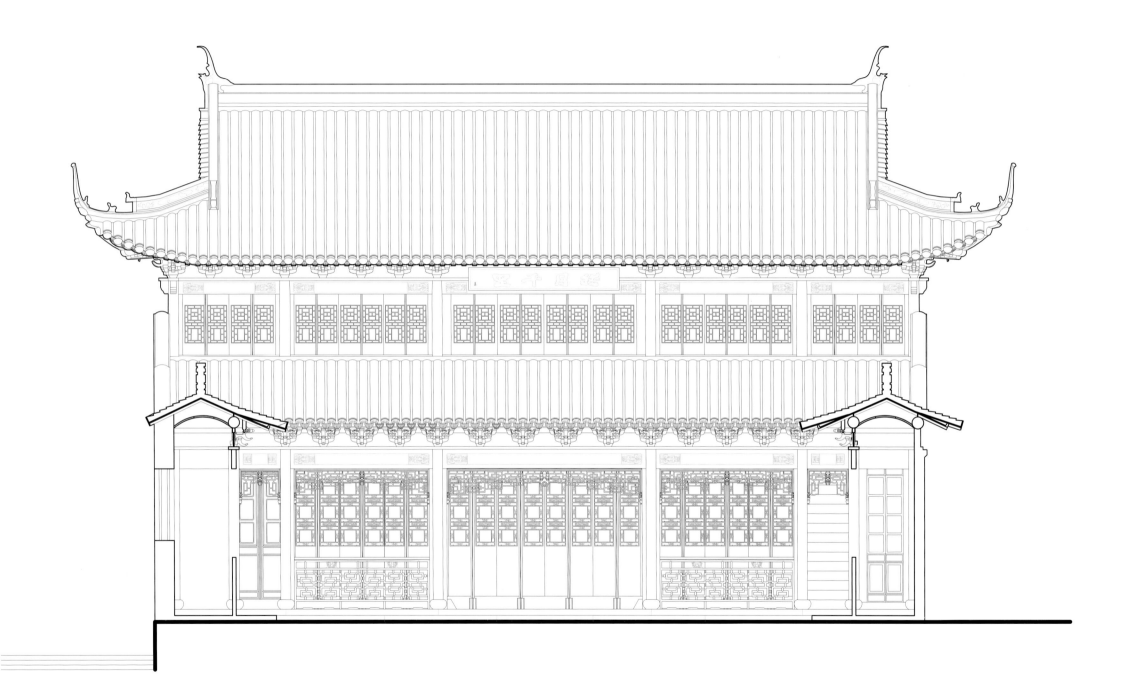

得月楼正立面图
Front elevation of Deyuelou

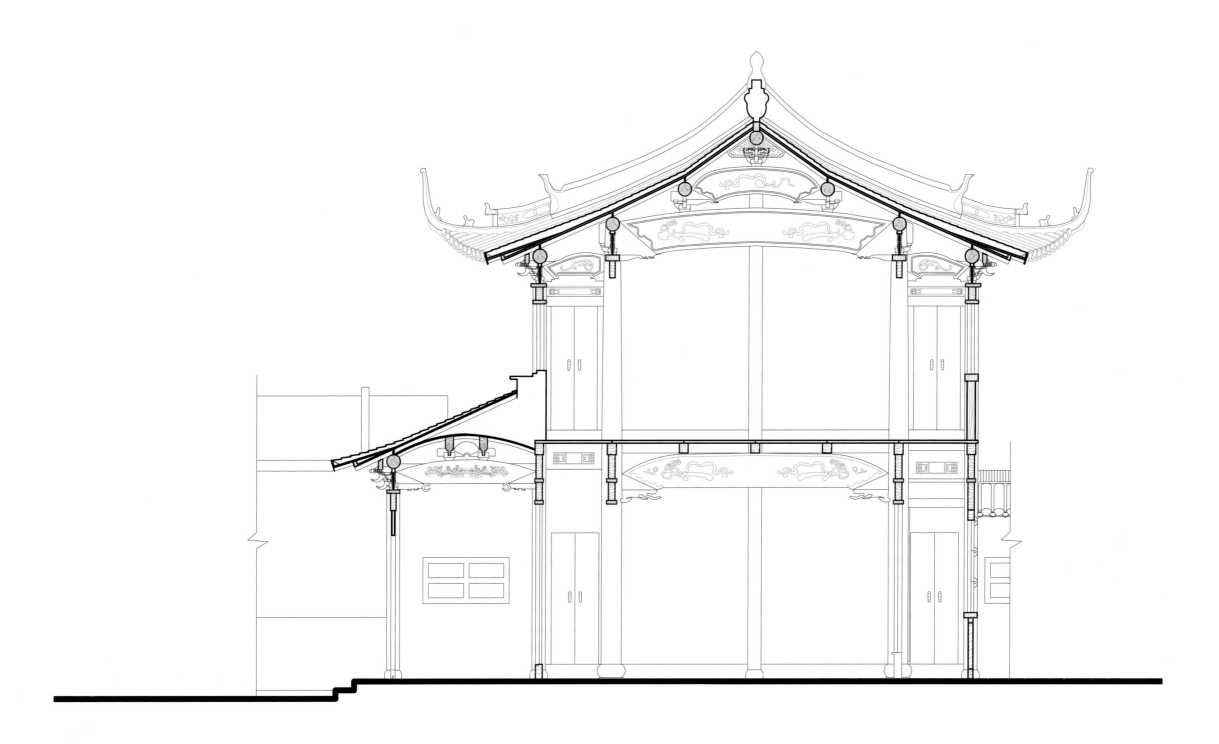

得月楼横剖面图
Cross-section of Deyuelou

0 1 2 4m

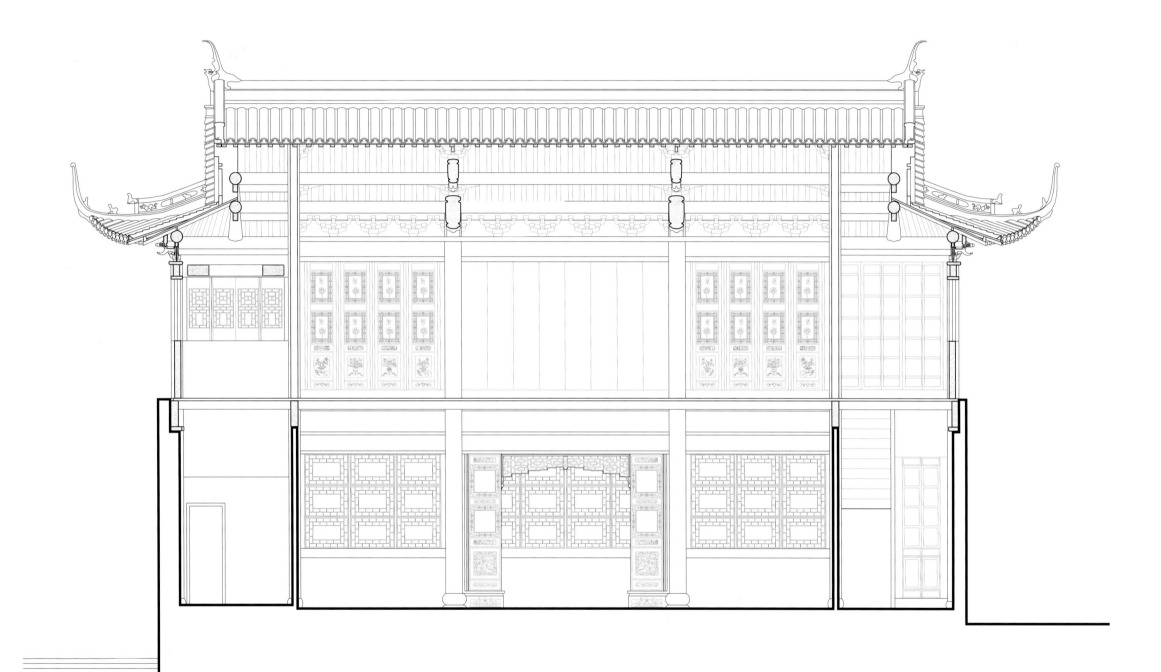

得月楼纵剖面图
Longitudinal section of Deyuelou

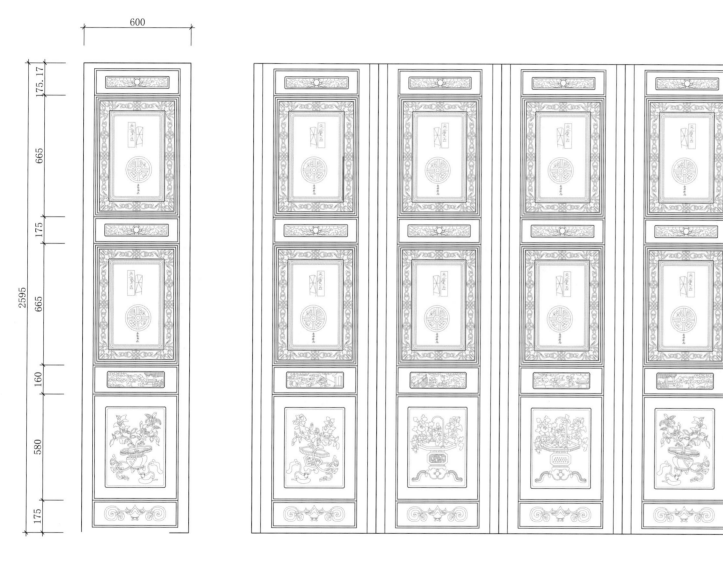

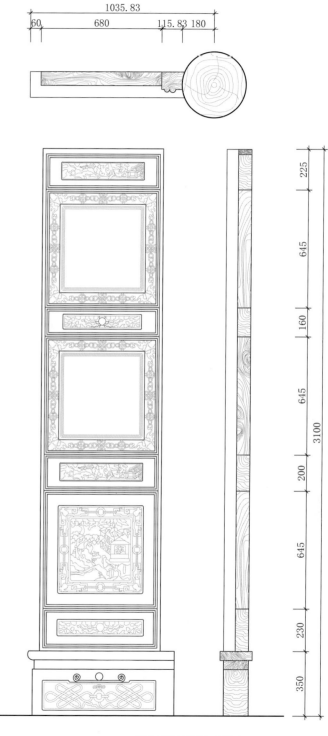

得月楼长窗大样图
Long windows of Deyuelou

得月楼纱隔大样图
Shage of Deyuelou

0   0.1  0.2   0.4m

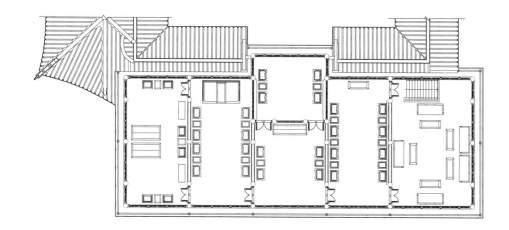

绿杨春榭二层平面图
Second floor plan of Luyangchunxie

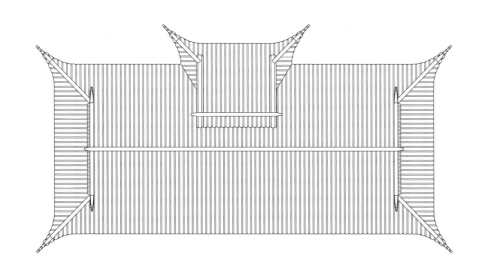

绿杨春榭屋顶平面图
Roof plan of Luyangchunxie

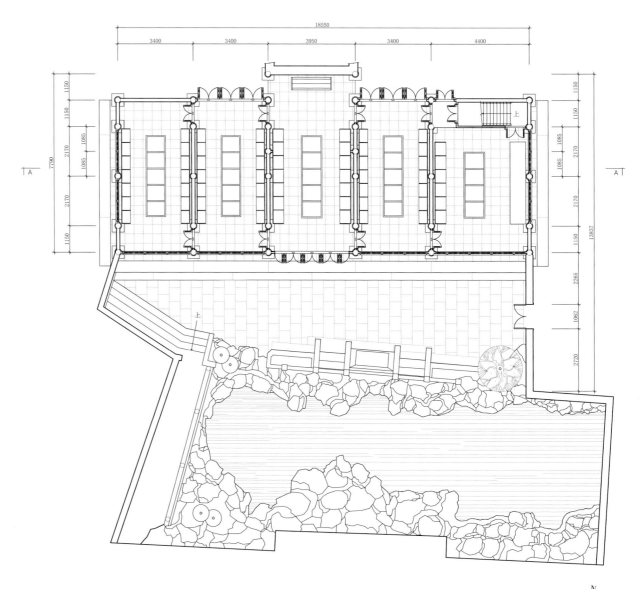

绿杨春榭一层平面图
Ground floor plan of Luyangchunxie

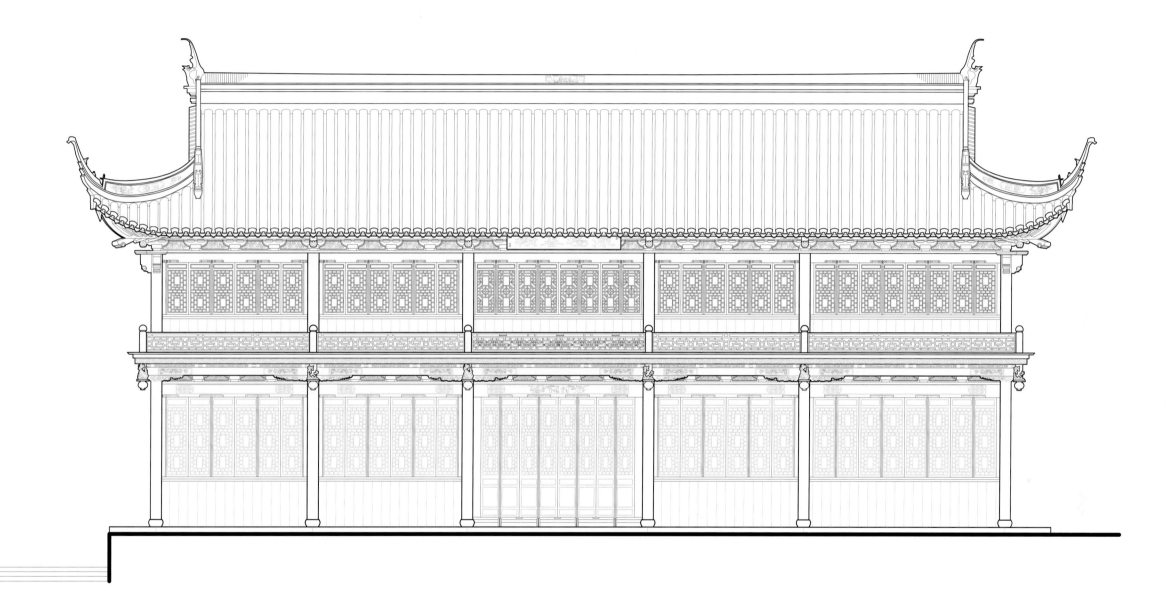

绿杨春榭正立面图
Front elevation of Luyangchunxie

0　1　2　　　　　5m

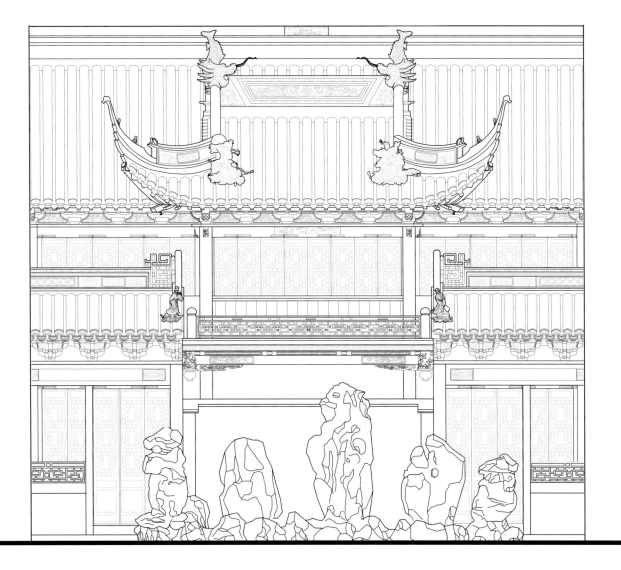

绿杨春榭抱厦（原应为戏楼）立面图
Elevation of Luyangchunxie's affiliated house（the original stage）

0　　1　　2　　　　5m

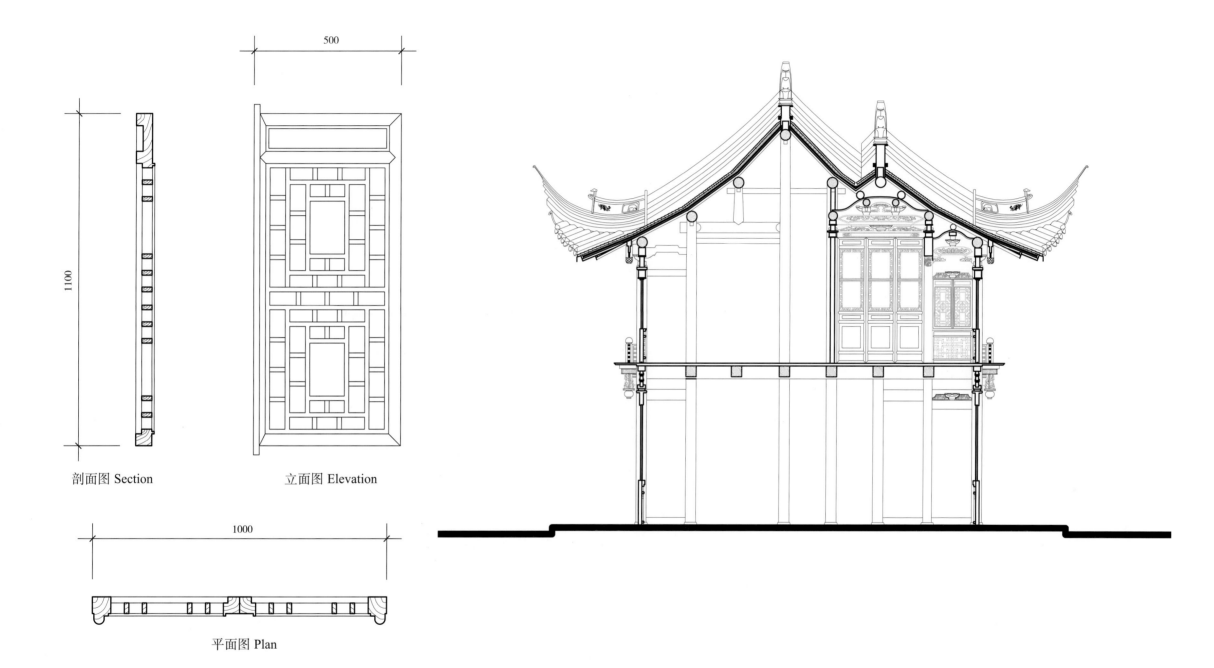

500

1100

1000

剖面图 Section

立面图 Elevation

平面图 Plan

宫式地坪窗大样图
Gongshi Dipingchuang of Luyangchunxie

0   0.1   0.2      0.5m

绿杨春榭横剖面图
Cross-section of Luyangchunxie

0   1   2      5m

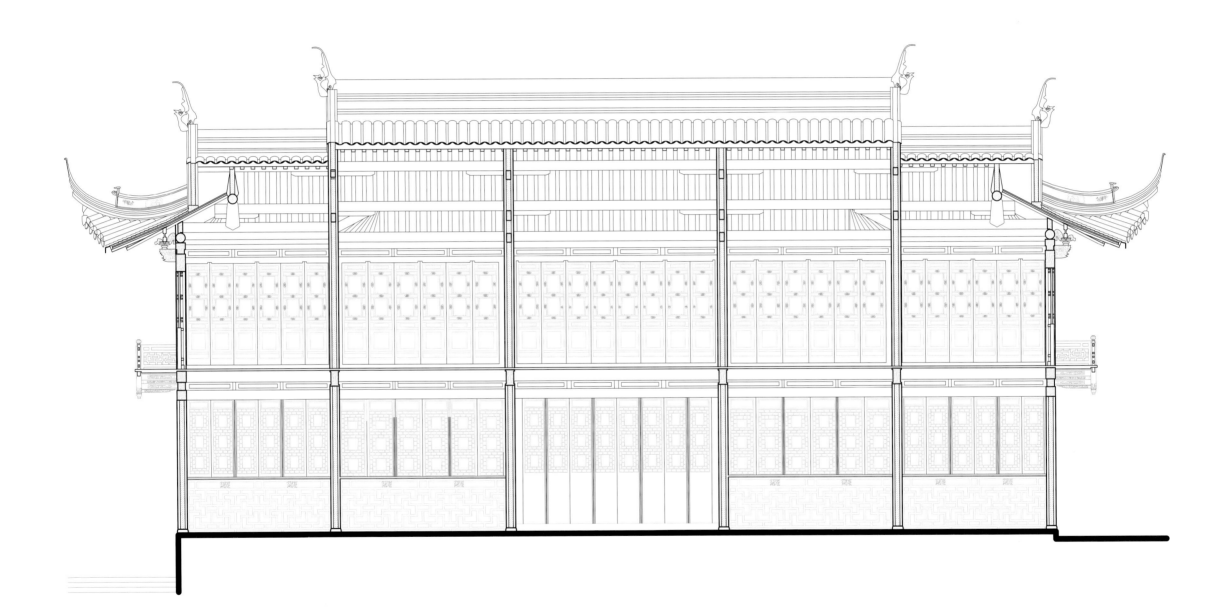

绿杨春榭纵剖面图
Longitudinal section of Luyangchunxie

0  1  2  5m

绿杨春榭山墙戏文人物装饰大样图
Brick decoration gable wall of Luyangchunxie

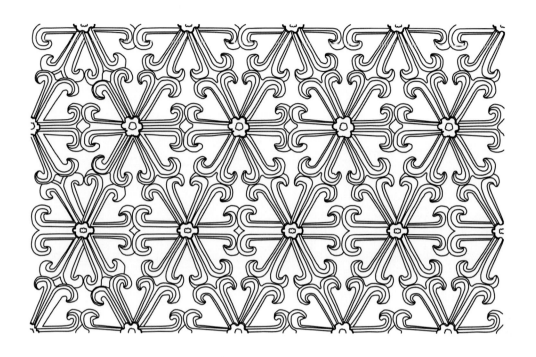

绿杨春榭窗格大样图
Chuangge of Luyangchunxie

绿杨春榭垂柱大样图
Chuizhu of Luyangchunxie

0　　0.1　　0.2　　　　　0.5m

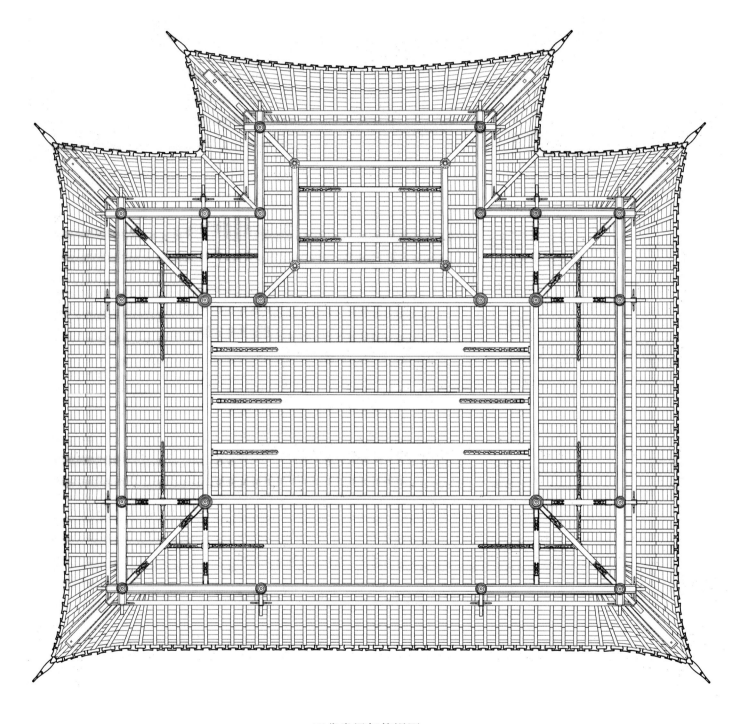

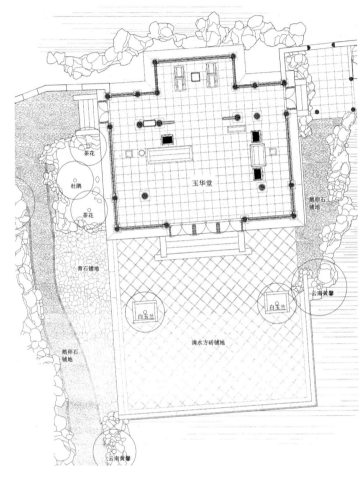

玉华堂平面图
Plan of Yuhuatang

玉华堂梁架仰视图
Bottom view of Yuhuatang's beams

0 0.5 1 2.5m

玉华堂北立面图
North elevation of Yuhuatang

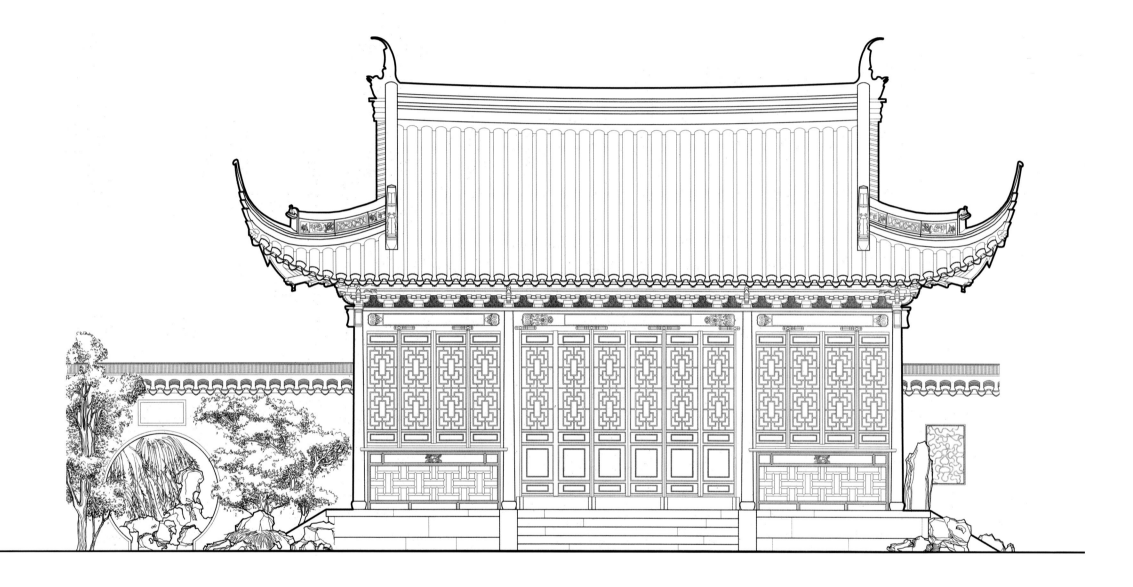

玉华堂南立面图
South elevation of Yuhuatang

0　0.5　1　　　　2.5m

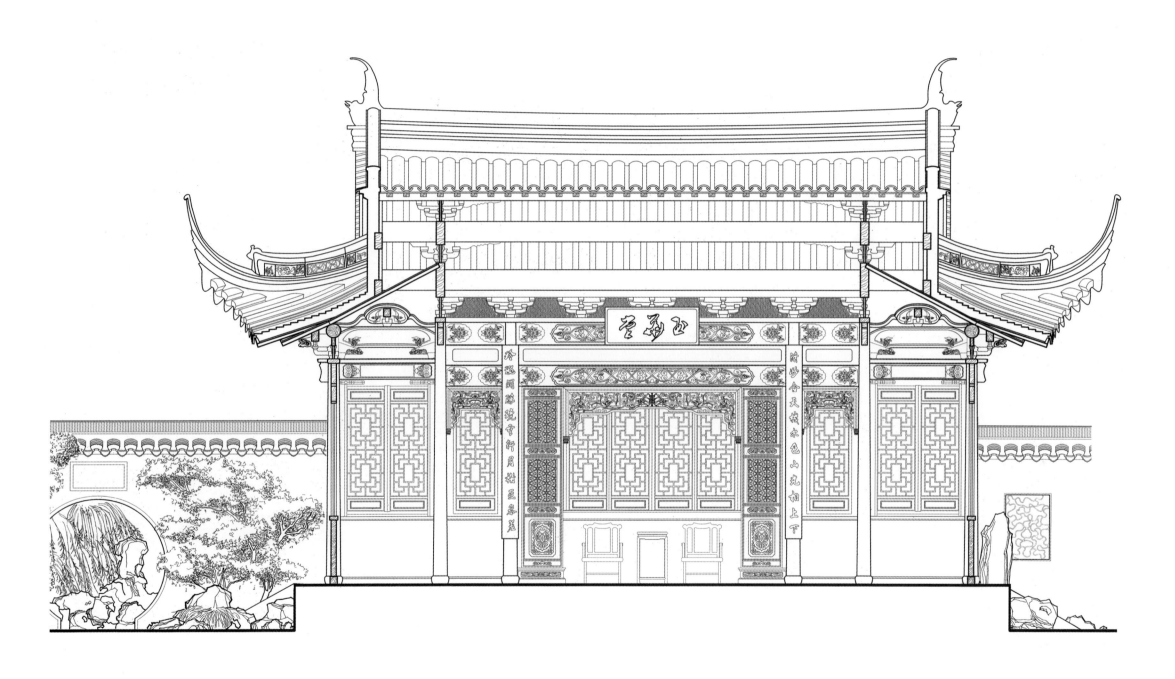

0    0.5    1         2.5m

玉华堂纵剖面图
Longitudinal section of Yuhuatang

玉华堂西立面图

West elevation of Yuhuatang

0    0.5    1              2.5m

玉华堂横剖面图
Cross-section of Yuhuatang

0 0.5 1 2.5m

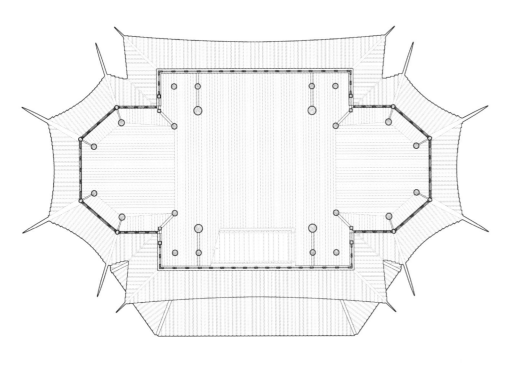

注：试按清中叶原貌复原

湖心亭二层平面图
Second floor plan of Huxin Pavilion

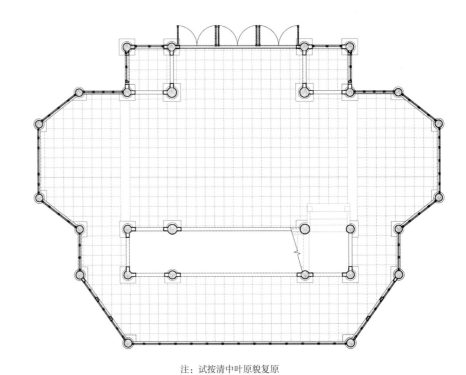

注：试按清中叶原貌复原

湖心亭一层平面图
Ground floor plan of Huxin Pavilion

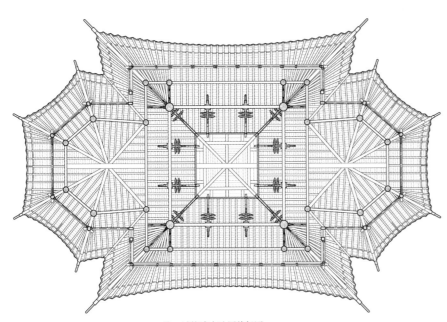

注：试按清中叶原貌复原

湖心亭二层梁架仰视图
Bottom view of Huxin Pavilion's second floor beams

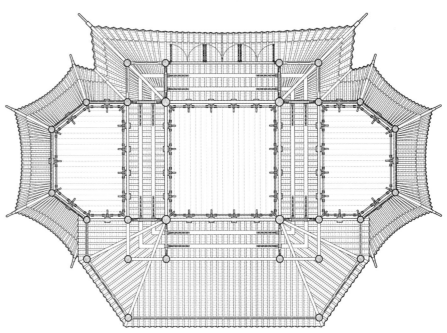

注：试按清中叶原貌复原

湖心亭一层梁架仰视图
Bottom view of Huxin Pavilion's ground floor beams

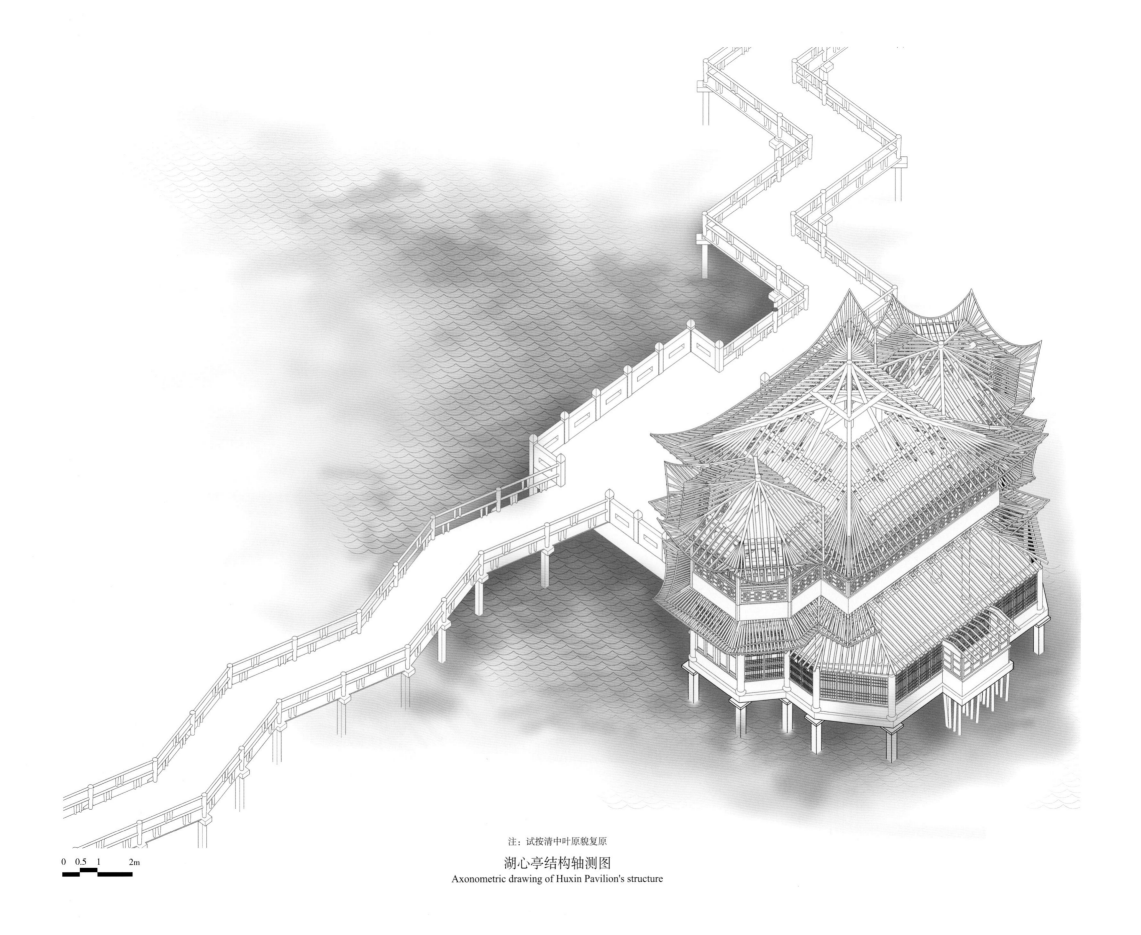

注：试按清中叶原貌复原

湖心亭结构轴测图

Axonometric drawing of Huxin Pavilion's structure

0  0.5  1    2m

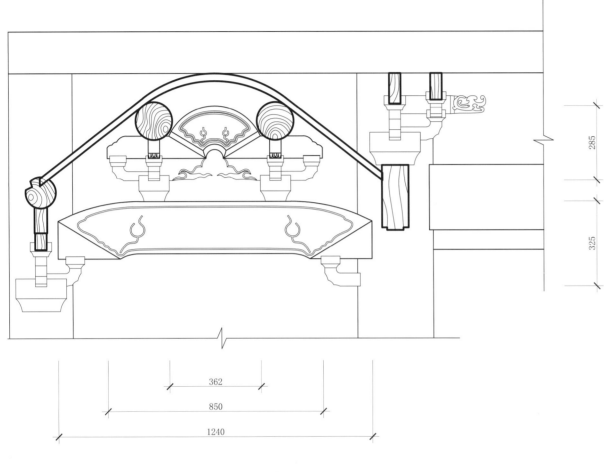

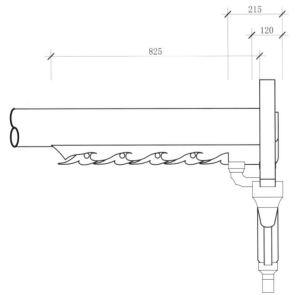

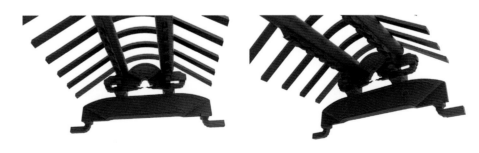

湖心亭一层船篷轩大样图
Chuanpengxuan on the ground floor of Huxin Pavilion

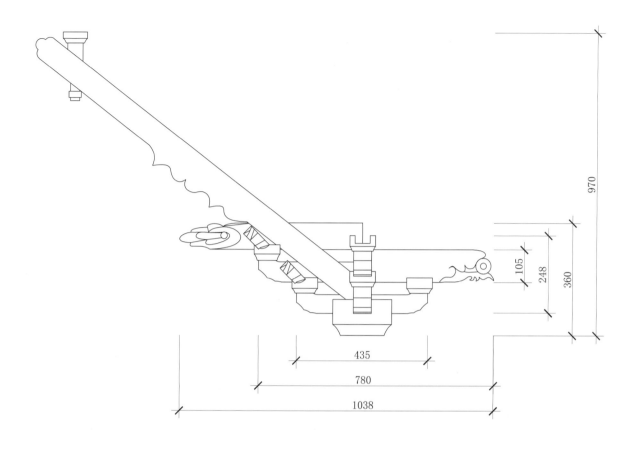

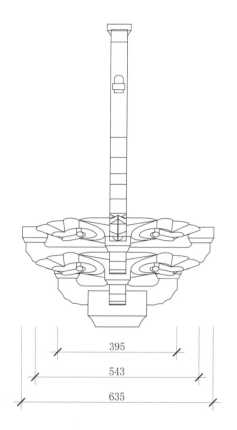

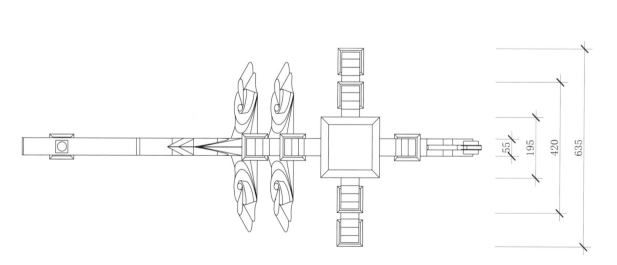

湖心亭二层藻井斗栱大样图

Zaojing Dougong on the second floor of Huxin Pavilion

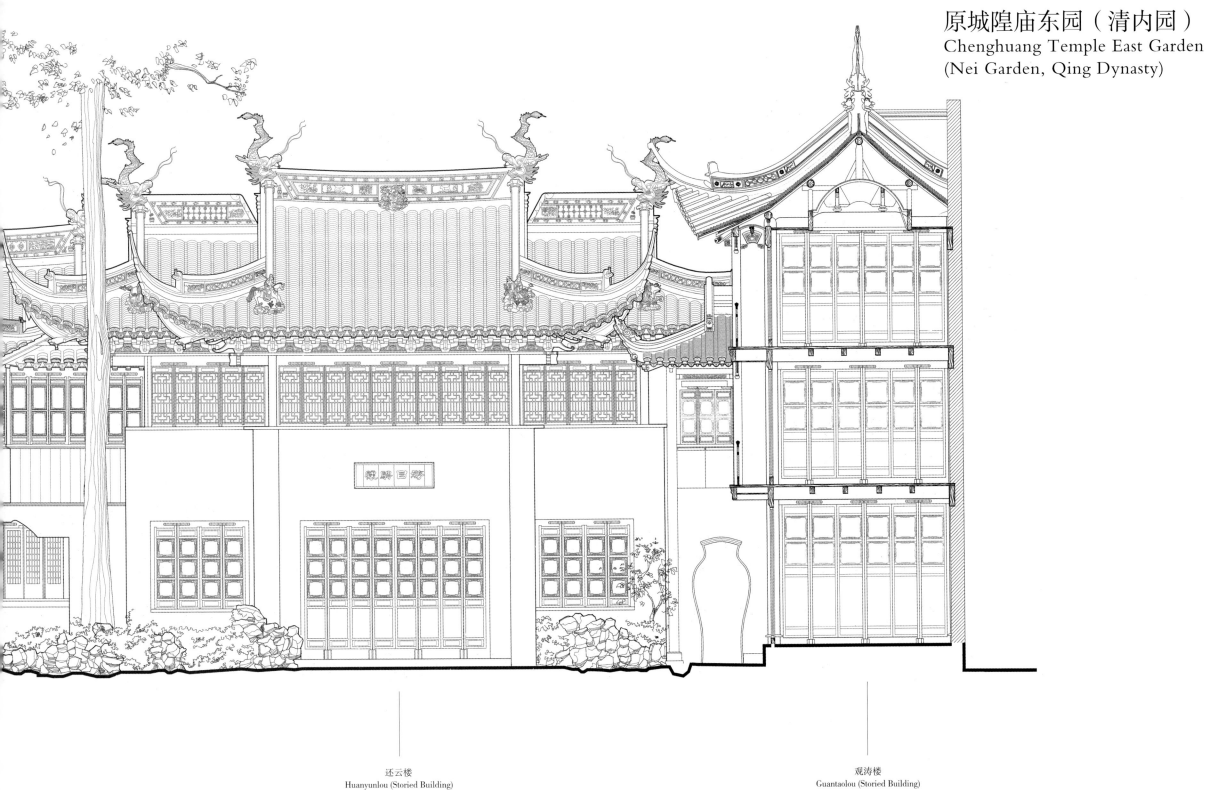

还云楼
Huanyunlou (Storied Building)

观涛楼
Guantaolou (Storied Building)

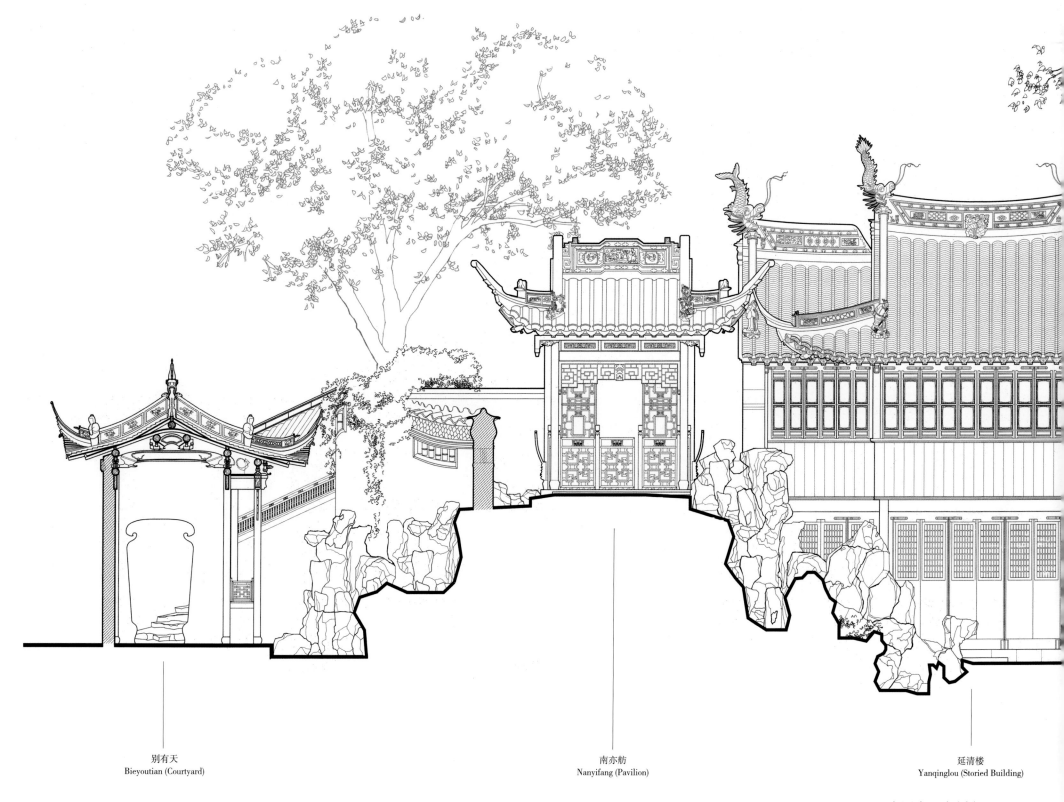

别有天
Bieyoutian (Courtyard)

南亦舫
Nanyifang (Pavilion)

延清楼
Yanqinglou (Storied Building)

0 0.5 1 2.5m

内园东西向剖立面图
East-west section and elevation of Nei Garden

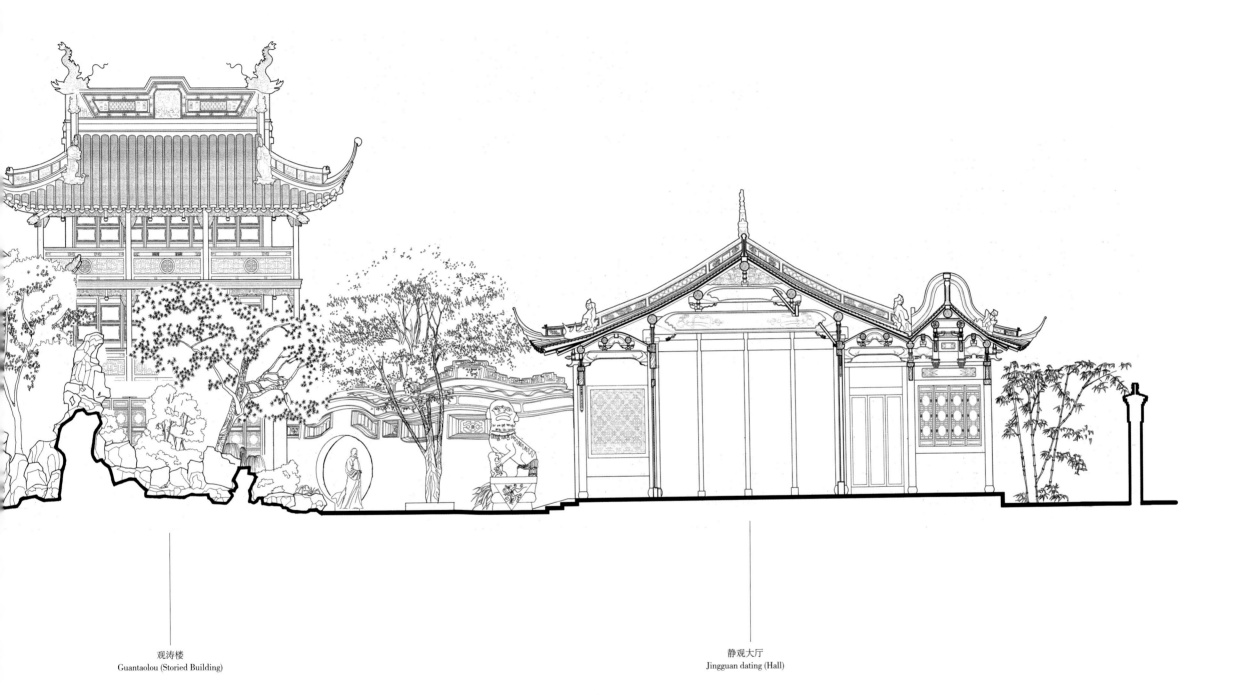

观涛楼
Guantaolou (Storied Building)

静观大厅
Jingguan dating (Hall)

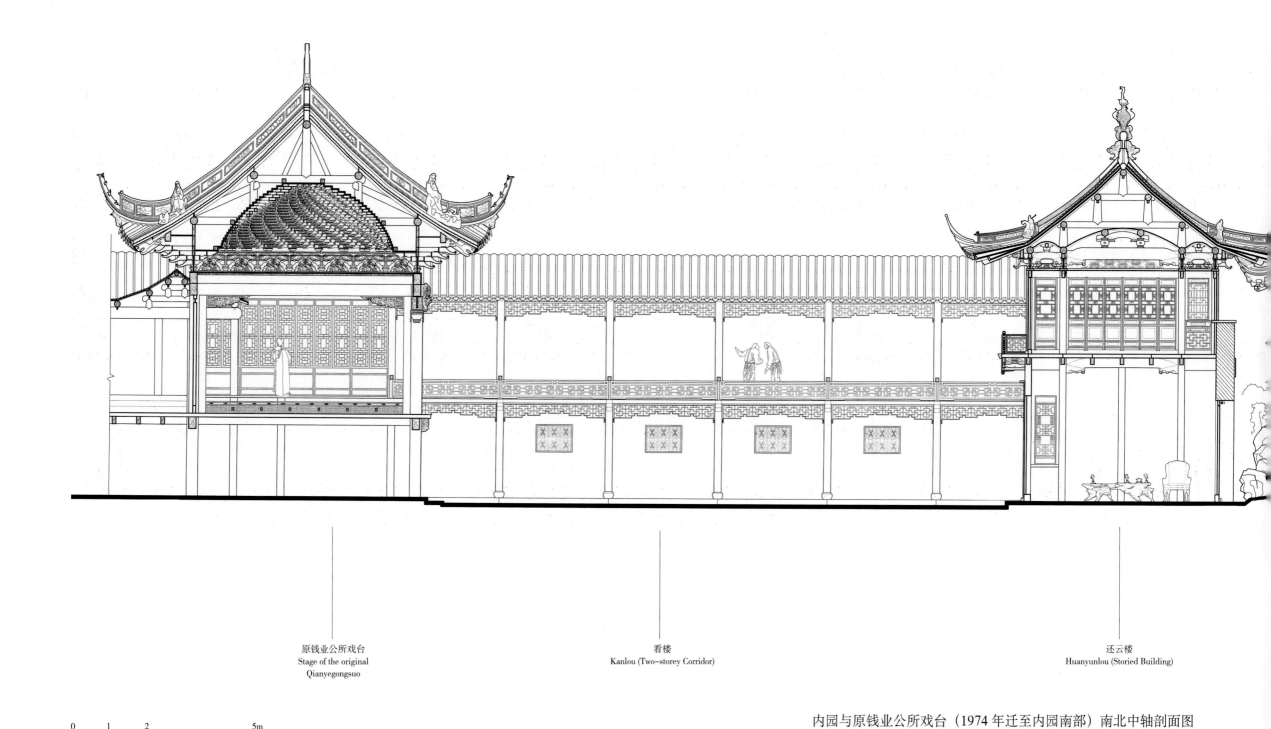

原钱业公所戏台
Stage of the original
Qianyegongsuo

看楼
Kanlou (Two-storey Corridor)

还云楼
Huanyunlou (Storied Building)

0　1　2　　　　5m

内园与原钱业公所戏台（1974年迁至内园南部）南北中轴剖面图
South-north section of Nei Garden and the original Qianyegongsuo's Xitai (which was transferred to the south of Nei Garden)

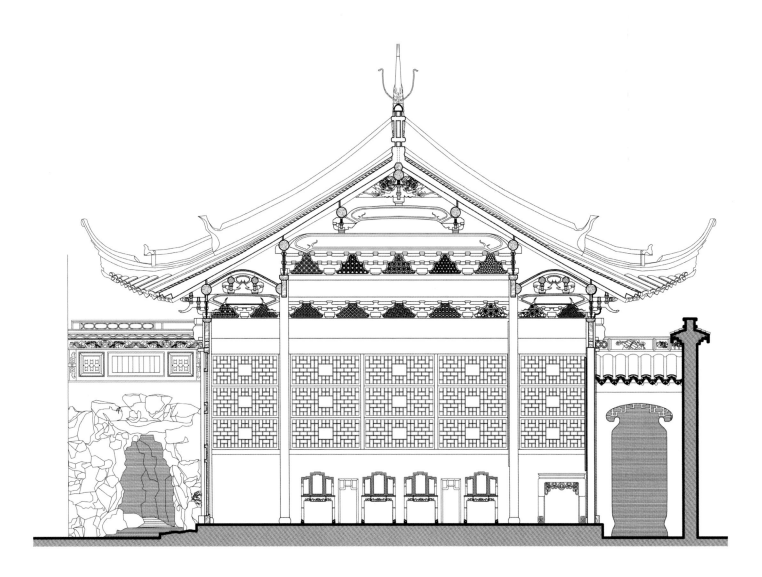

内园可以观横剖面图
Cross-section of Nei Garden's Keyiguan

0    0.5    1                    2m

靜觀

內園靜觀大廳與九龍潭東西向剖面圖
East-west section of Nei Garden's Jingguandating and Jiulongtan

0  1  2      5m

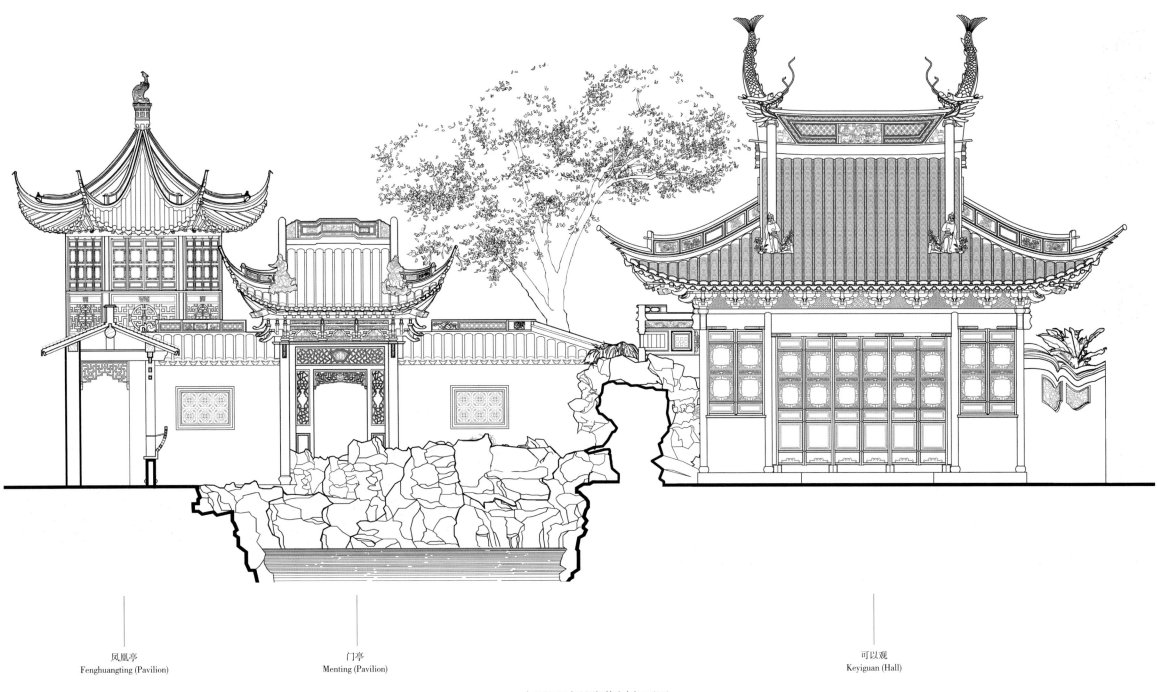

凤凰亭
Fenghuangting (Pavilion)

门亭
Menting (Pavilion)

可以观
Keyiguan (Hall)

内园可以观院落剖立面图
Section and elevation of Nei Garden's Keyiguan Courtyard

0　0.5　1　　　　2.5m

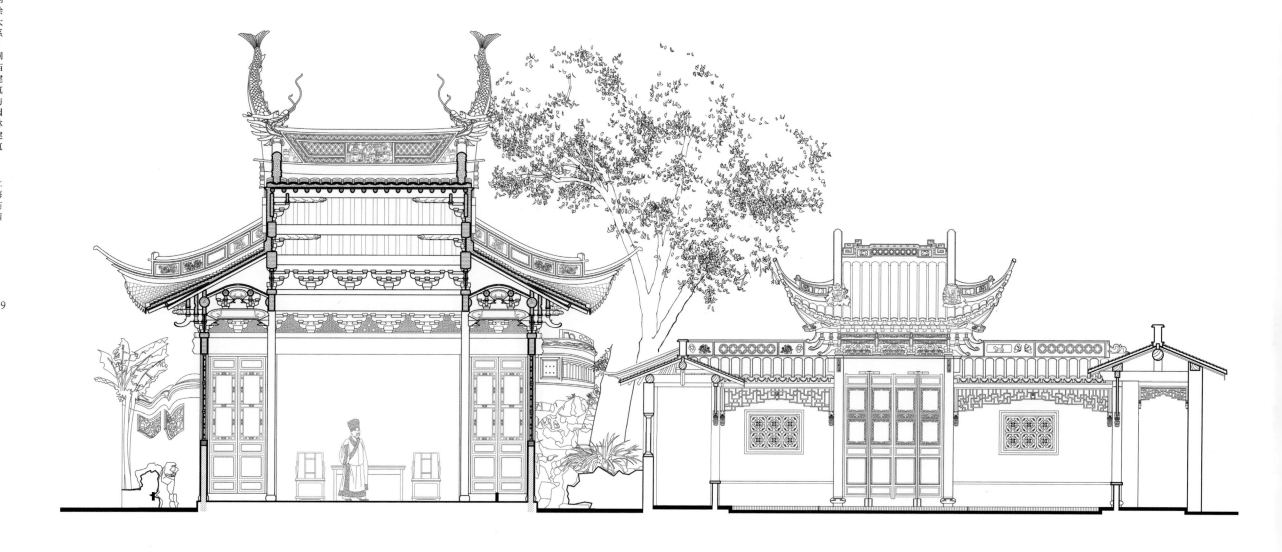

可以观
Keyiguan (Hall)

门亭
Menting (Pavilion)

0  0.5  1      2.5m

内园可以观院落剖立面图
Section and elevation of Nei Garden's Keyiguan Courtyard

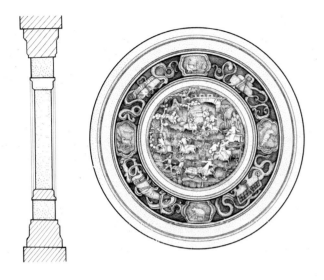

内园砖雕壁饰大样图（一）
Brick-carved wall decoration of Nei Garden（1）

内园砖雕壁饰大样图（二）
Brick-carved wall decoration of Nei Garden（2）

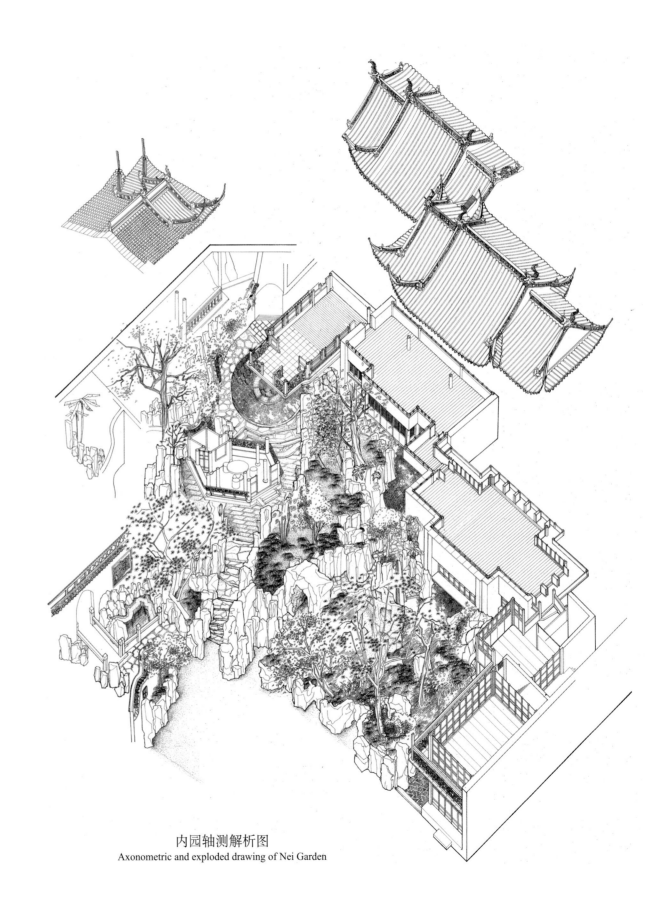

内园轴测解析图
Axonometric and exploded drawing of Nei Garden

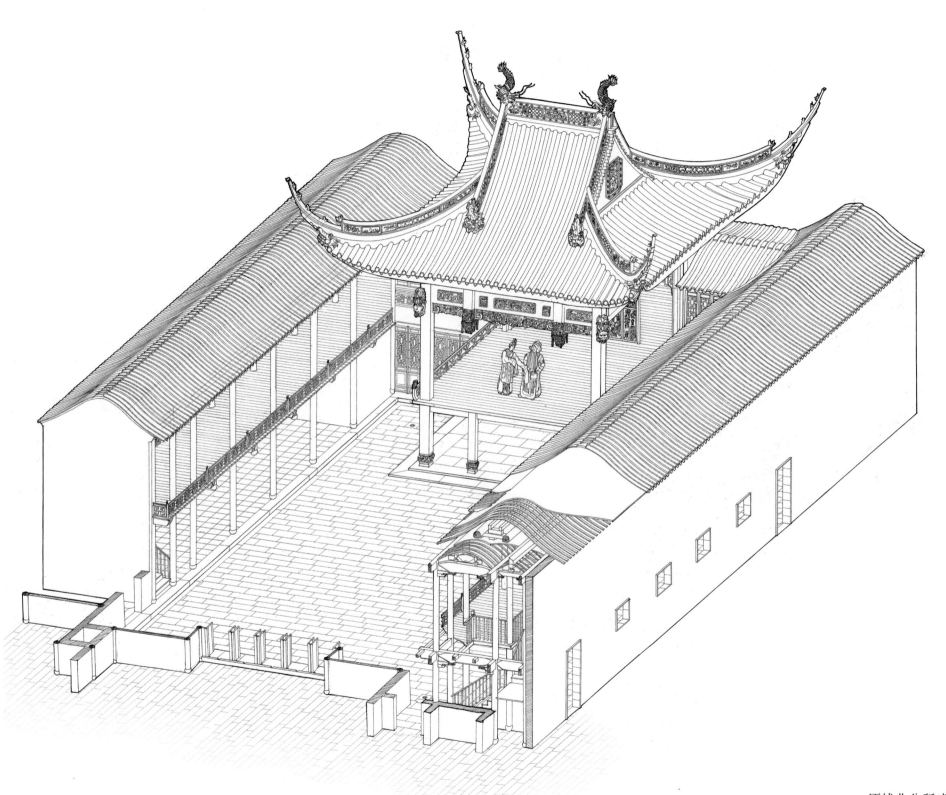

原钱业公所戏台轴测图

Axonometric drawing of the original Qianyegongsuo's Xitai

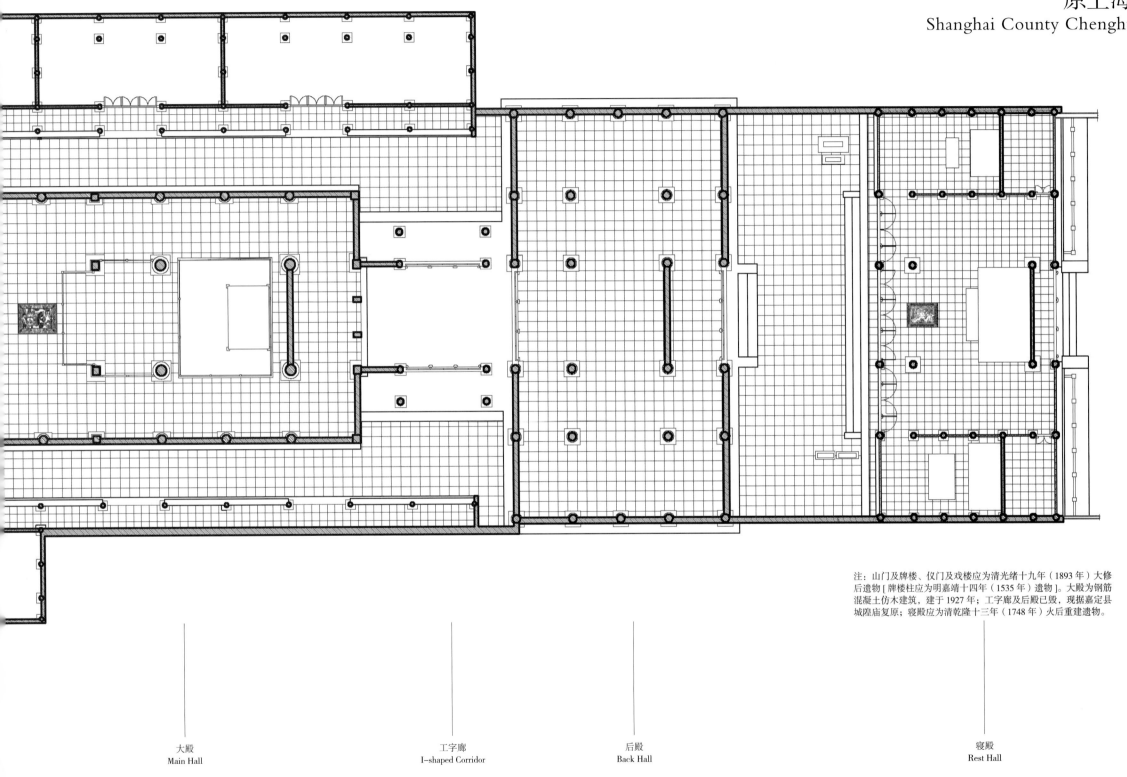

注：山门及牌楼、仪门及戏楼应为清光绪十九年（1893年）大修后遗物［牌楼柱应为明嘉靖十四年（1535年）遗物］。大殿为钢筋混凝土仿木建筑，建于1927年；工字廊及后殿已毁，现据嘉定县城隍庙复原；寝殿应为清乾隆十三年（1748年）火后重建遗物。

大殿
Main Hall

工字廊
I-shaped Corridor

后殿
Back Hall

寝殿
Rest Hall

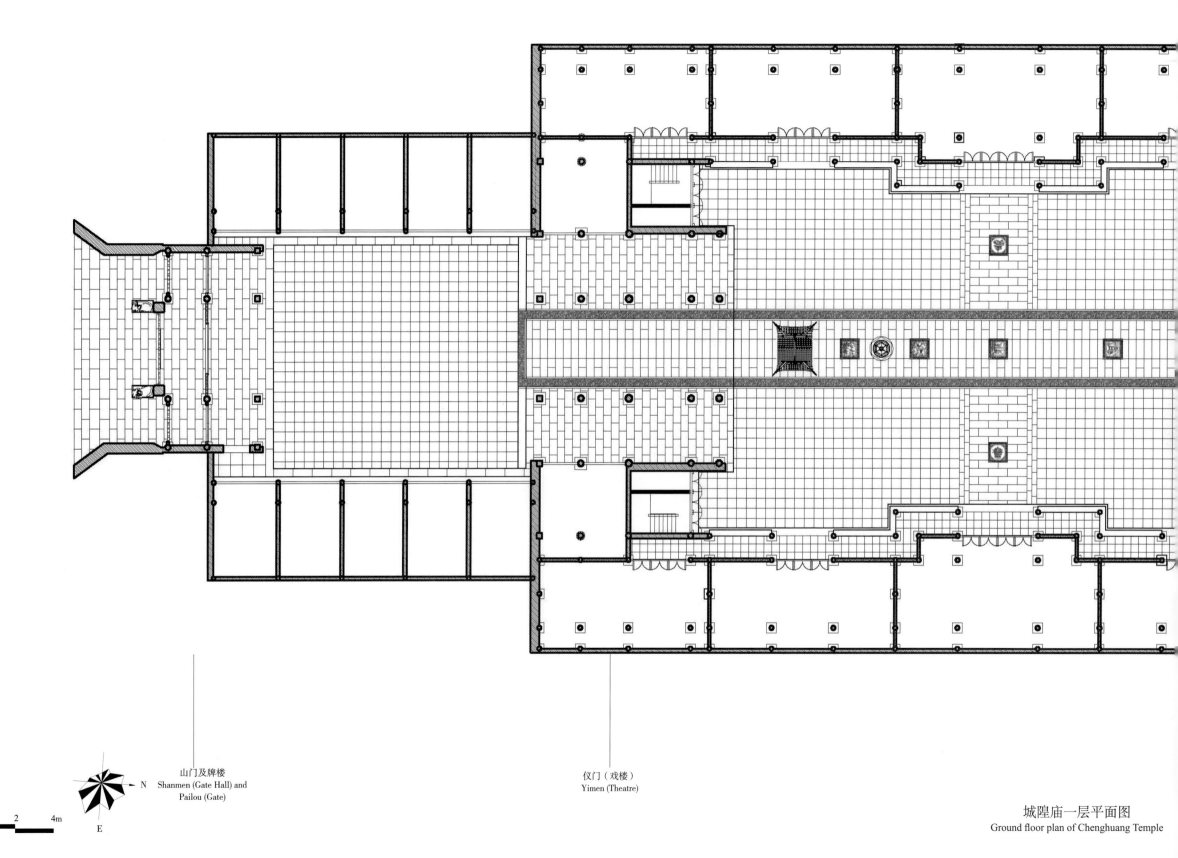

山门及牌楼
Shanmen (Gate Hall) and
Pailou (Gate)

仪门（戏楼）
Yimen (Theatre)

N

E

0 1 2 4m

城隍庙一层平面图
Ground floor plan of Chenghuang Temple

城隍庙山门、牌楼南立面图
South elevation of Chenghuang Temple's Shanmen and Pailou

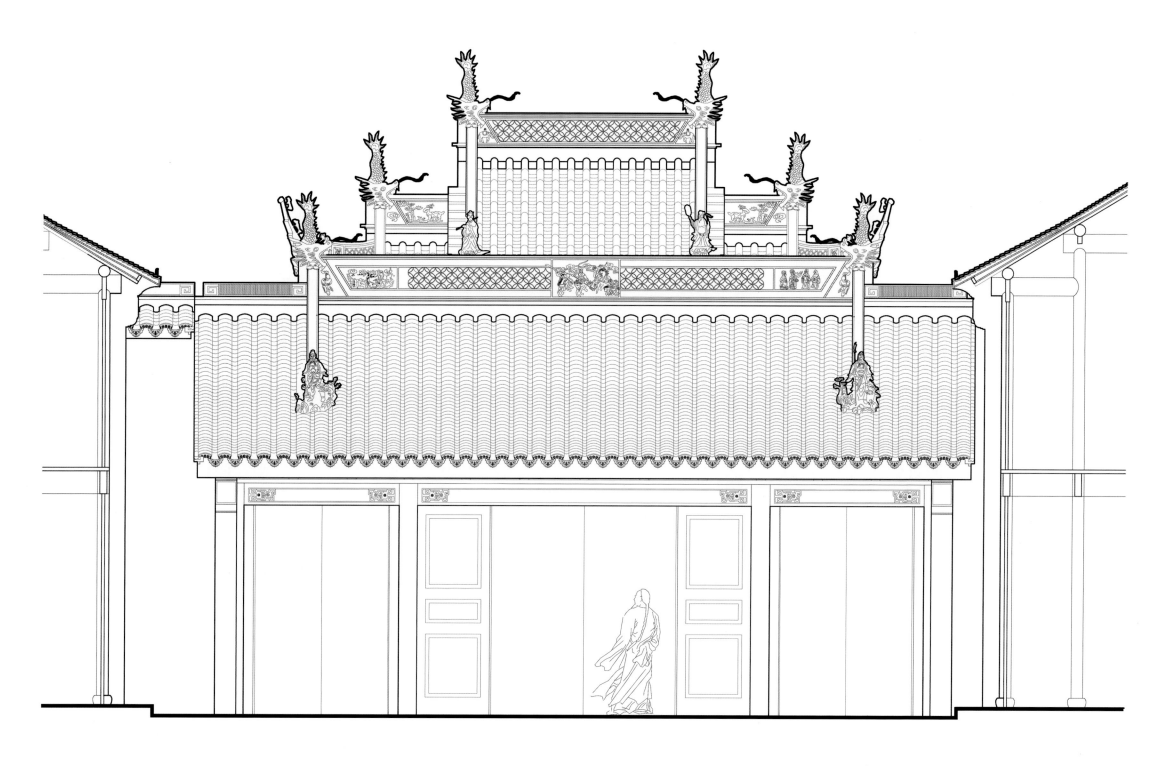

城隍庙山门、牌楼北立面图
North elevation of Chenghuang Temple's Shanmen and Pailou

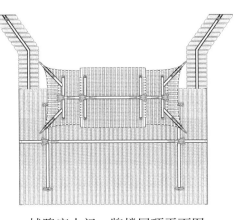

城隍庙山门、牌楼屋顶平面图
Roof plan of Chenghuang Temple's Shanmen and Pailou

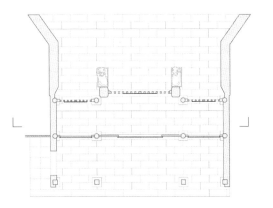

城隍庙山门、牌楼平面图
Plan of Chenghuang Temple's Shanmen and Pailou

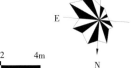

城隍庙山门、牌楼剖面图
Section of Chenghuang Temple's Shanmen and Pailou

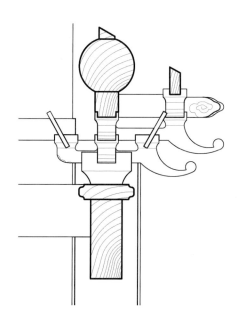

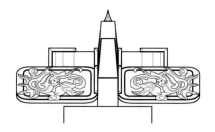

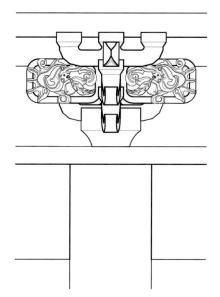

城隍庙山门柱头斗栱大样图
Zhutou Dougong of Chenghuang Temple's Shanmen

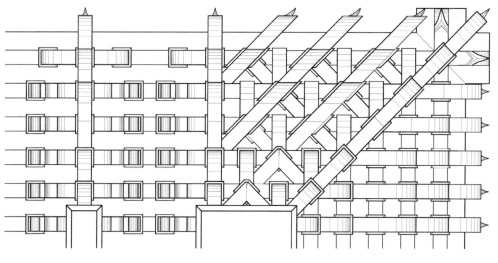

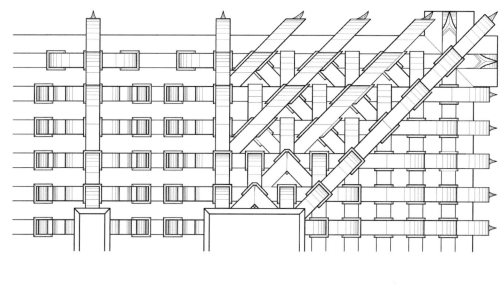

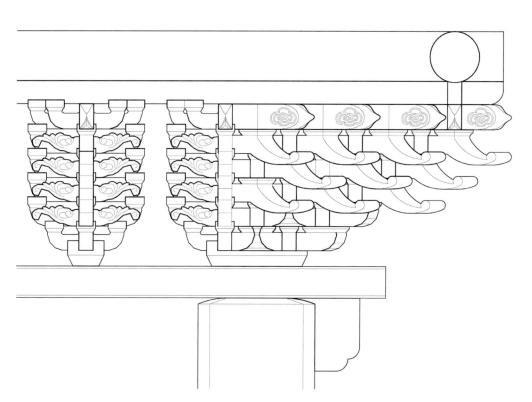

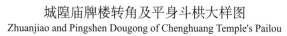

城隍庙牌楼转角及平身斗栱大样图
Zhuanjiao and Pingshen Dougong of Chenghuang Temple's Pailou

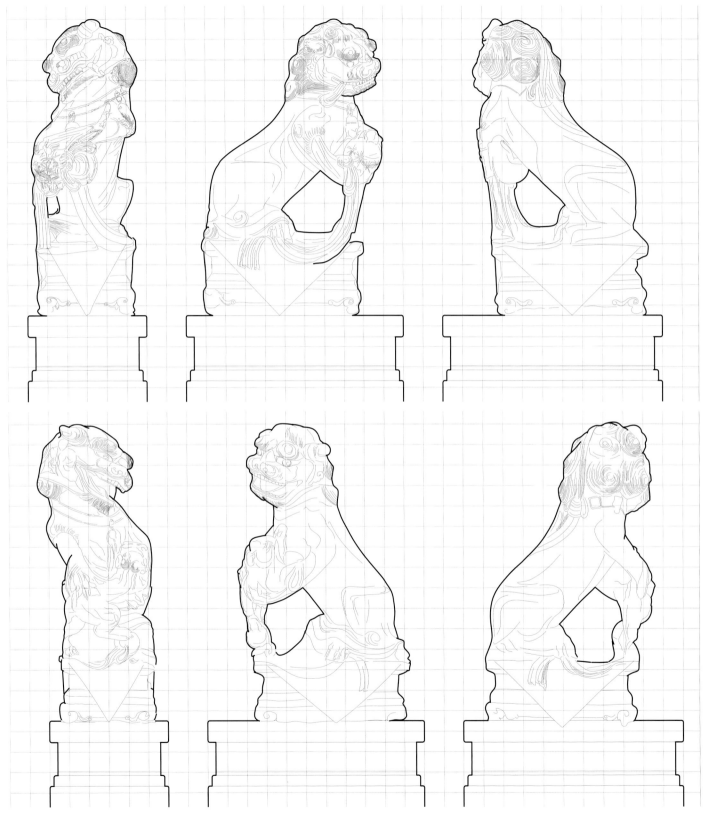

城隍庙山门、牌楼明风青石狮雄狮大样图
Ming style Qing Stone male lion in front of Chenghuang Temple's Shanmen and Pailou

城隍庙山门、牌楼明风青石狮雌狮大样图
Ming style Qing Stone female lion in front of Chenghuang Temple's Shanmen and Pailou

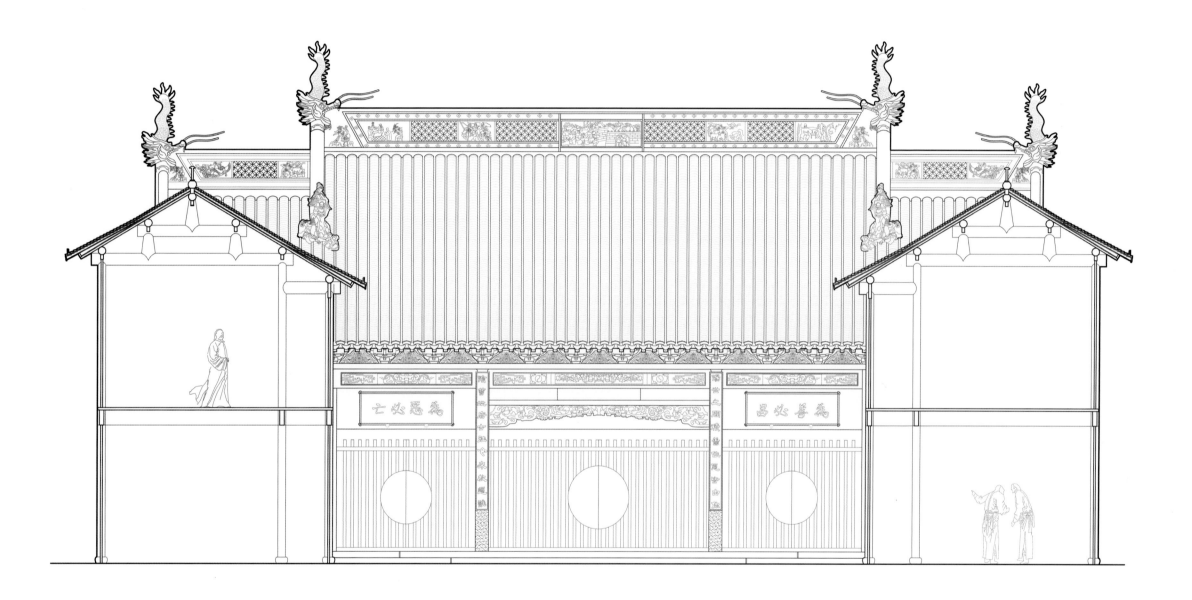

0    0.5    1        2m

城隍庙仪门（北面为戏楼）南立面现状图
Current south elevation of Chenghuang Temple's Yimen (Xilou in the north)

城隍庙戏楼（南面为仪门）北立面现状图
Current north elevation of Chenghuang Temple's Xilou (Yimen in the south)

0    0.5    1    2m

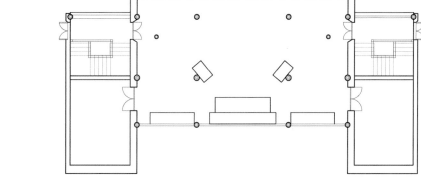

城隍庙戏楼二层平面图
Second floor plan of Chenghuang Temple's Xilou

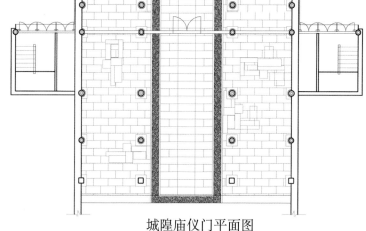

城隍庙仪门平面图
Plan of Chenghuang Temple's Yimen

城隍庙戏楼（南面为仪门）北立面复原图
Recovery north elevation of Chenghuang Temple's Xilou (Yimen in the south)

0    0.5    1    2m

0    0.5    1    2m

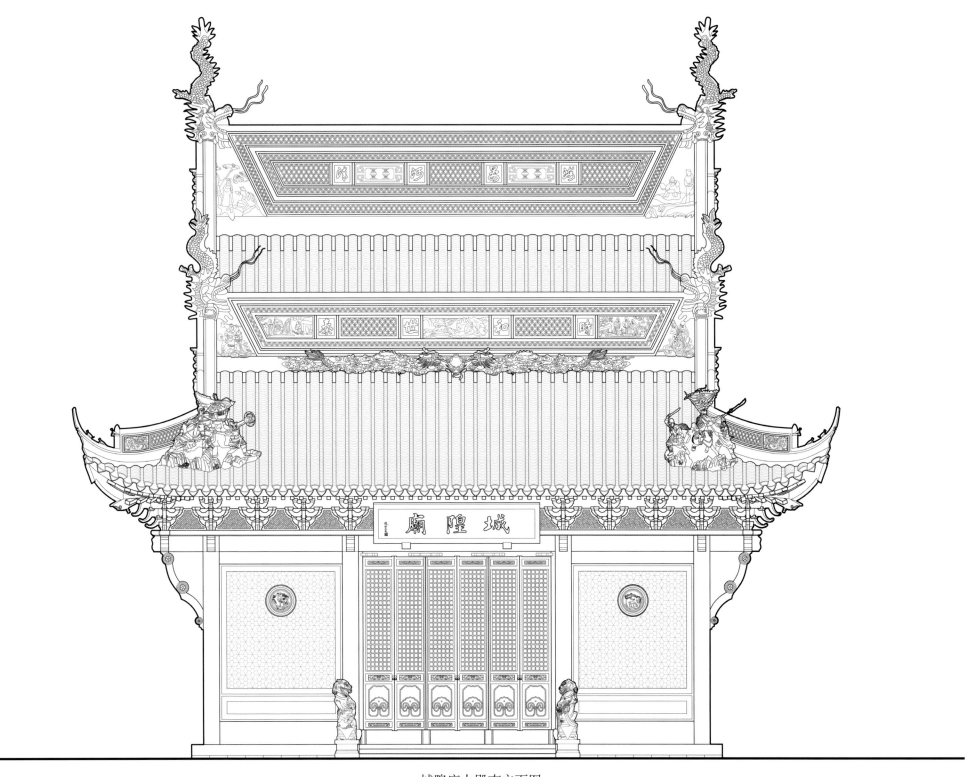

城隍庙大殿南立面图
South elevation of Chenghuang Temple's main hall

0  1  2  4m

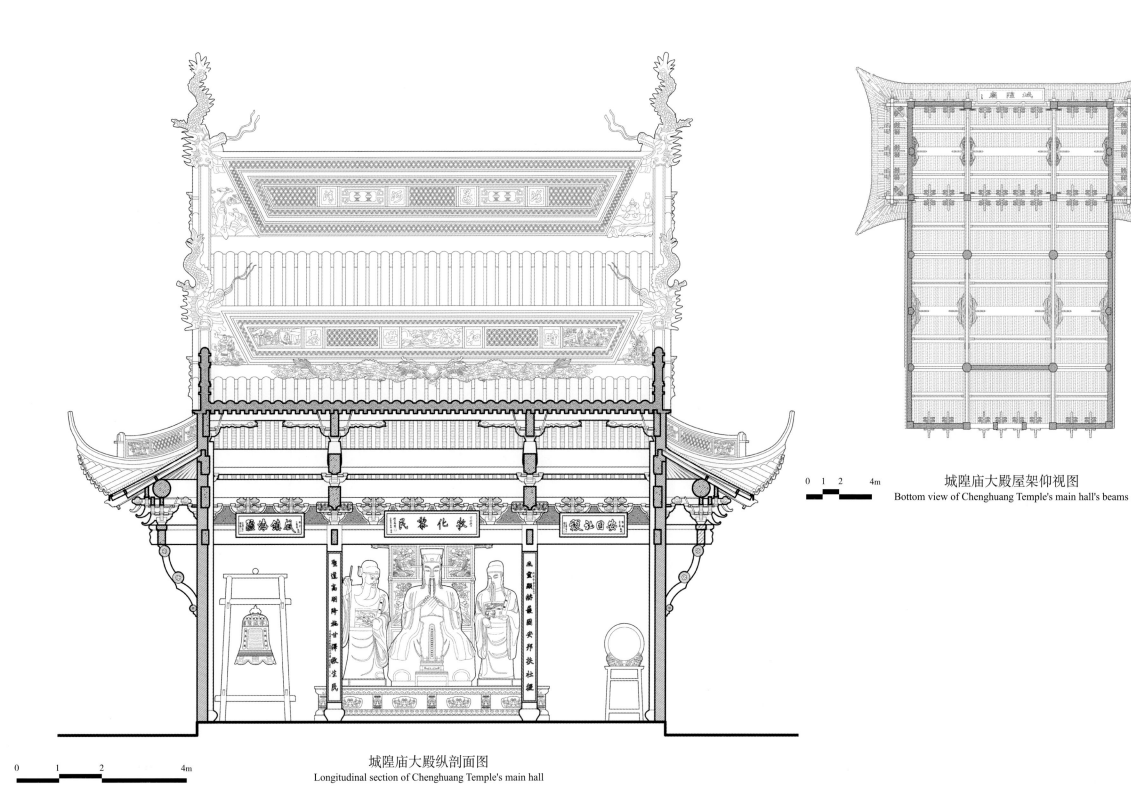

城隍庙大殿屋架仰视图
Bottom view of Chenghuang Temple's main hall's beams

0　1　2　　　　4m

城隍庙大殿纵剖面图
Longitudinal section of Chenghuang Temple's main hall

0　1　2　　　　4m

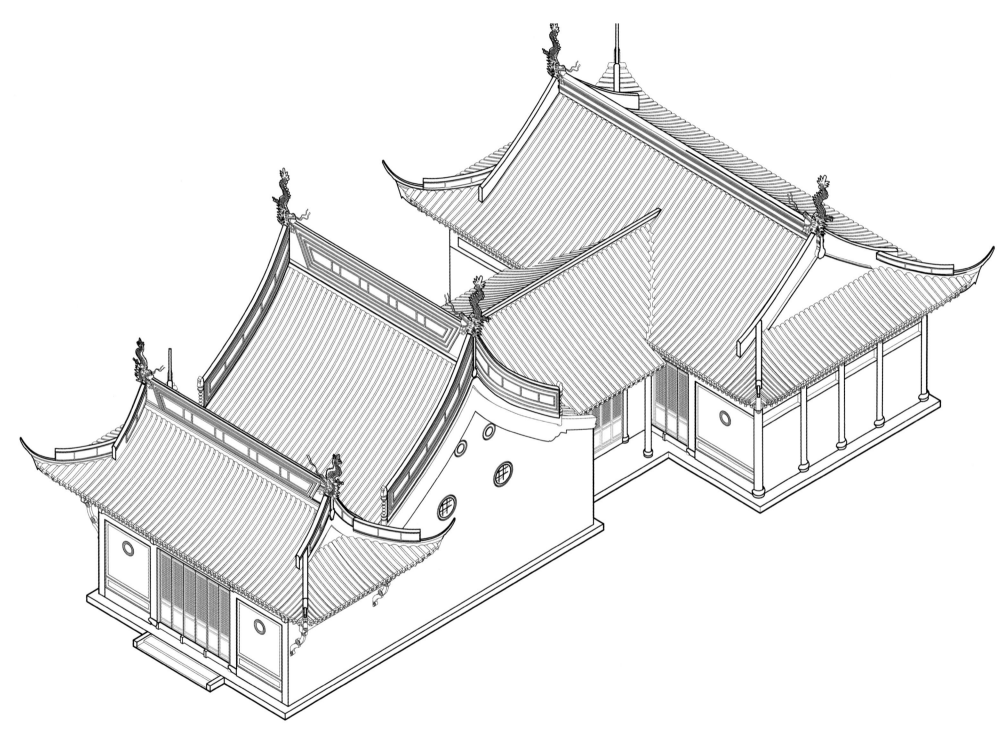

注：工字殿及后殿已毁，现据方志图及嘉定县城隍庙后殿复原

城隍庙大殿轴测图
Axonometric drawing of Chenghuang Temple's main hall

0 　 2 　 4 　 　 8m

注：工字殿及后殿为假想复原图

城隍庙大殿结构轴测渲染图
Rendering of the structure of Chenghuang Temple's main hall

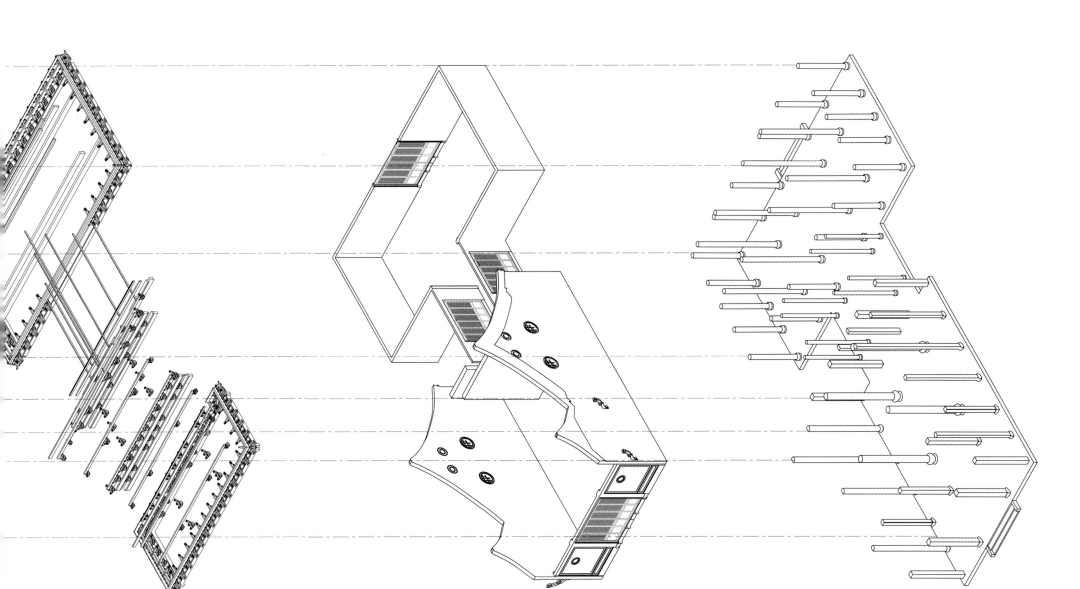

注：工字殿及后殿为假想恢复原图

城隍庙大殿分解轴测图

Axonometric and exploded drawing of Chenghuang Temple's main hall

0    2    4    8m

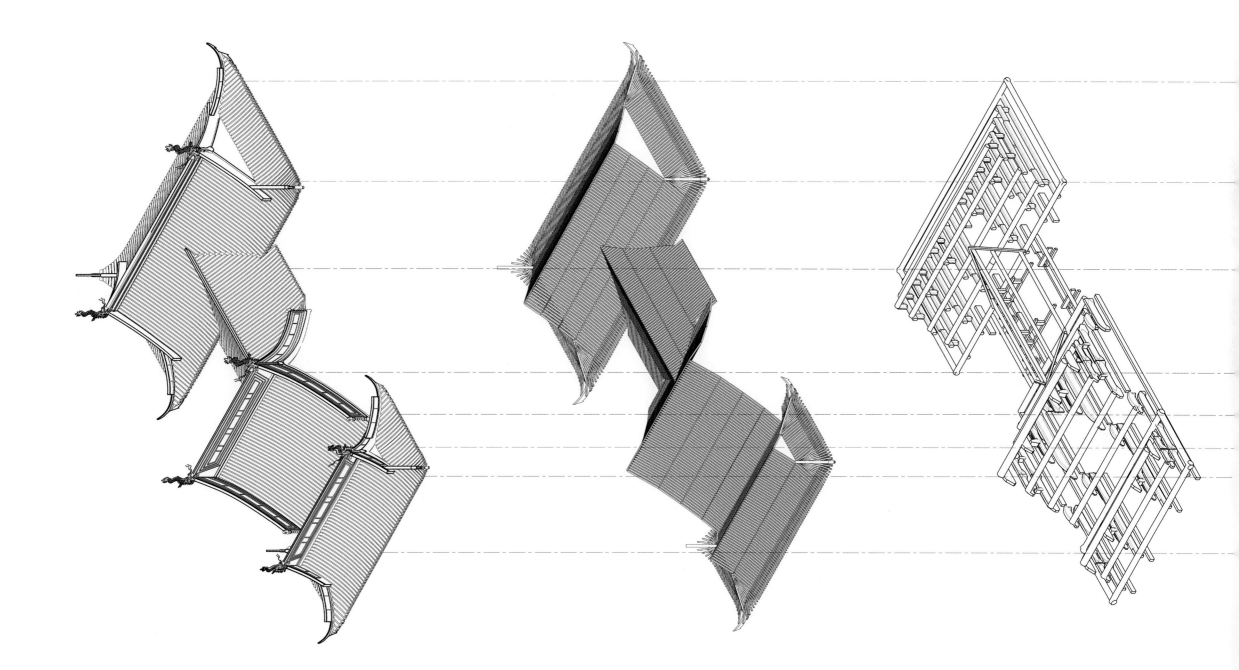

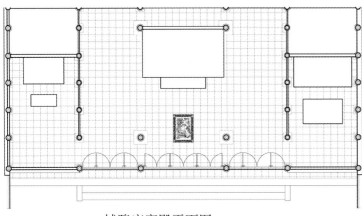

城隍庙寝殿平面图
Plan of Chenghuang Temple's back hall

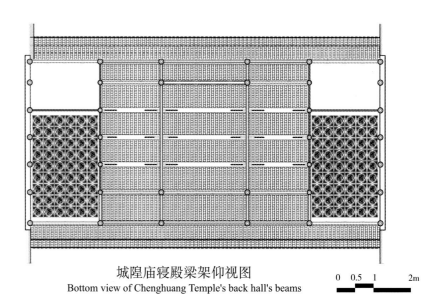

城隍庙寝殿梁架仰视图
Bottom view of Chenghuang Temple's back hall's beams

0  0.5  1    2m

城隍庙寝殿正立面图
Front elevation of Chenghuang Temple's back hall

0     0.5     1        2m

城隍庙"洪武碑亭"复原东立面图
Recovery east elevation of Chenghuang Temple's Hongwu Beiting

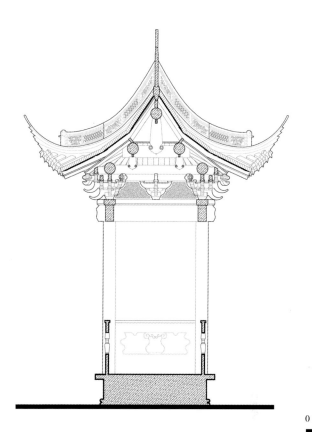

城隍庙"洪武碑亭"复原剖面图
Recovery section of Chenghuang Temple's Hongwu Beiting

注：台基、勾栏疑为明万历三十年（1602年）遗物，木构为清嘉庆十九年（1814年）遗物

城隍庙"洪武碑亭"复原轴测图
Recovery axonometric drawing of Chenghuang Temple's Hongwu Beiting

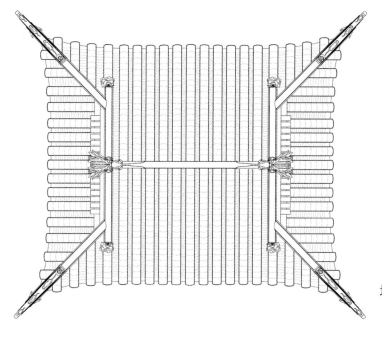

城隍庙"洪武碑亭"复原屋顶平面图
Recovery roof plan of Chenghuang Temple's Hongwu Beiting

Item of survey: Jiading County Chenghuang Temple and Temple Garden (Qiuxiapu
　　　　　　　　Garden and Shenshi Garden)

Address: No314, East Avenue, Jiading Town, Jiading District, Shanghai

Age of construction: Ming Dynasty (Qiuxiapu Garden, also called Gongshi Garden);
　　　　　　　　Southern Song Dynasty (Jiading Chenghuang Temple)

Site area: 25 mu

Competent organization: Management Office of Jiading Qiuxiapu Garden

Survey organization: College of Architecture and Urban Planning, Tongji University

Time of survey: 1998

测绘项目：嘉定县城隍庙及庙园（秋霞圃与沈氏园）

地　　址：嘉定区城乡镇东大街三一四号

始建年代：秋霞圃（龚氏园）始建于明

　　　　　嘉定县城隍庙始建于南宋

占地面积：二十五余亩（其中秋霞圃占地约八亩，沈氏园占地约四亩）

主管单位：嘉定县秋霞圃管理处

测绘单位：同济大学建筑与城规学院

测绘时间：一九九八年

嘉定县城隍庙及庙园（秋霞圃与沈氏园）

# Jiading County Chenghuang Temple and Temple Garden
# (Qiuxiapu Garden and Shenshi Garden)

碧光亭
angting (Pavilion)

明代武康石北山
Wukang Stone North
Rockery of the Ming
Dynasty

即山亭
Jishanting
(Pavilion)

山光潭影馆（碧梧轩）
Shanguangtangyingguan/Biwuxuan (Hall)

山间小径
Mountain path

明代太湖石南山残迹
Taihu Stone South Rockery Remains of the Ming Dynasty

舟而不游轩
Zhouerbuyouxuan (Pavilion)

丛桂轩
Congguixuan (Pavilion)

秋霞圃南北剖面图（自屏山堂望丛桂轩）
South-north section of Qiuxiapu Garden (looking from Pingshantang towards Congguixuan)

花亭
ngting
ilion)

山光潭影馆（碧梧轩）
Shanguangtanyingguan/Biwuxuan
(Hall)

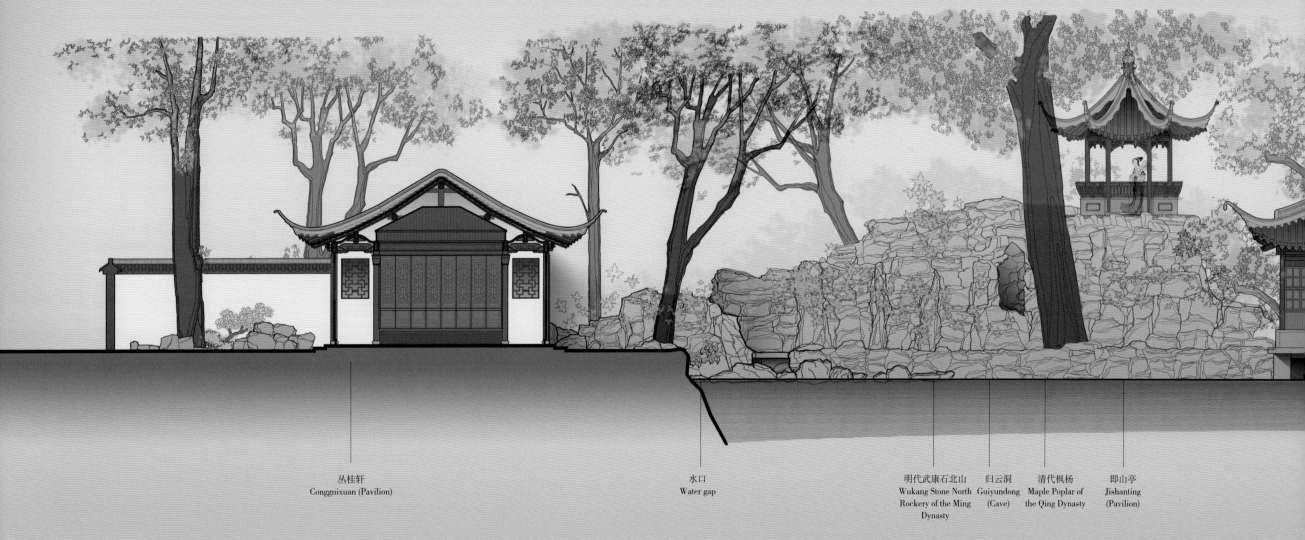

丛桂轩
Congguixuan (Pavilion)

水口
Water gap

明代武康石北山
Wukang Stone North
Rockery of the Ming
Dynasty

归云洞
Guiyundong
(Cave)

清代枫杨
Maple Poplar of
the Qing Dynasty

即山亭
Jishanting
(Pavilion)

嘉定秋霞圃东西剖面（北岸）
East-west section of Qiuxiapu Garden (the north bank)

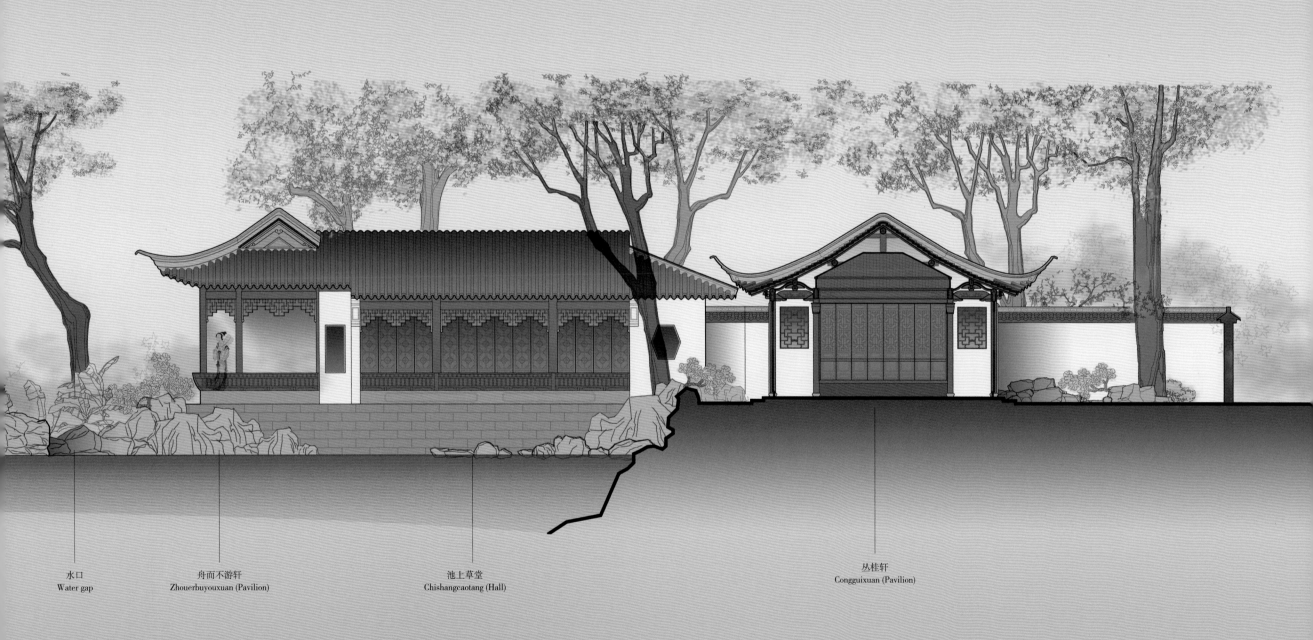

水口
Water gap

舟而不游轩
Zhouerbuyouxuan (Pavilion)

池上草堂
Chishangcaotang (Hall)

丛桂轩
Congguixuan (Pavilion)

明代太湖石南山残迹
Taihu Stone South Rockery Remains of the
Ming Dynasty

明代石梁遗迹
Stone Bridge Remains of the Ming Dynasty

嘉定秋霞圃东西剖面（南岸）
East-west section of Qiuxiapu Garden (the south bank)

# Introduction

The ancient city of Jiading, which was found in 1219, had a water grid as the famous 'crossed double ring'. The main river, north-south Hengli, and east-west Lianchuan, intersected like a cross under Fahua Tower in the city center. The periphery of the city was surrounded by approximately circular inner and outer rivers. This kind of water grid had quite obvious signs of artificial reorganization and seemed to be a kind of constant refinement of nature, which had sed a foundation for certain temperament of the vernacular architecture in Jiading.

Jiading Chenghuang Temple Garden, and its affiliated architecture, located in the north-east of the Jiading old city within the 'crossed double ring'. Facing Dongmen Street and Lianchuan in the south, Jiading Chenghuang Temple Garden echoed Hengli and Fahua Tower in the west, and Jiading Academy and Huilongtan in the south. Not far from the west of the garden, there is a relic tracing back to Song or Yuan Dynasty, namely Taipingyongan Bridge across Lianchuan, which suggests a long history of the area.

There were two gardens attached to the Chenghuang Temple in Jiading County in the mid-Qing Dynasty, including Gongshi Garden of Ming Dynasty, which renamed as Qiuxiapu Garden in the early Qing Dynasty, and Shenshi Garden of Ming Dynasty. And Jinshi Garden of Ming Dynasty located in the north. The overall spatial structure of one temple accompanied by three gardens still exists today, covering an area of more than 45 mu, which is a relatively complete temple and garden complex in Jiangnan area. Among the three gardens, Qiuxiapu Garden covered an area of 8 mu and possessed the most features of literati gardens in the late Ming and early Qing Dynasties, and with a typical and concise 'a river with two banks' layout and a deep artistic conception, it may be comparable with Jichang Garden in Wuxi. Shenshi Garden, also called Shenshi East Garden, was originally a part of Gongshi Garden, covering an area of 4 mu. Its status quo, which symbolizes the 'water' with the flat streets and plain courtyards, and the 'mountain' with the Taihu Stone pond and the peak, surrounds by colonnades and windowed verandas, and has several courtyards. Mr. Chen Congzhou once called it 'dry

南宋嘉定十二年（一二一九年）奠基的嘉定古城，其水网格局为脍炙人口的『十字双环』，即南北干河『横沥』、东西干河『练川』十字相交于城市中心的法华塔下，城市周边则以近似圆形的内、外城河环绕，其人工整饬的痕迹颇为显著，仿佛对自然进行着某种恒久的提炼——这似乎奠定了嘉定风土建筑的某些气质。

嘉定县城隍庙园园建筑群即位于『十字双环』内的老城东北部，南临东门大街与练川，西与横沥、法华塔相望，南与嘉定学宫、汇龙潭建筑群相守。园西不远尚有宋元遗构——太平永安桥飞跨练川，暗示着地脉的悠长。

清中叶臻于极盛后的嘉定县城隍庙曾附有两园，即明龚氏园（清前期改称秋霞圃）、明沈氏园。今一庙与三园的整体空间格局犹存，占地计四十五亩余，为江南现存较完整之庙园建筑群——其中秋霞圃园占地约八亩，留存明末清初文人园面貌较多，其以两山夹一曲潭的『一河两岸』式布局典型而凝练，

garden with water intention'. Which has already been rebuilt again and again in 1920s and 1980s. Jinshi Garden covered an area of 20 mu, which was destroyed by the war of Taiping Army in the late Qing Dynasty, and the current scene was rebuilt in recent years.

Besides, the Shuangjing Pavilion, I-shaped main hall, and the back palace in the main axis of Jiading Chuanghuang Temple still exist today. The I-shaped hall was reconstructed in the late Qing Dynasty, it still has a rare layout of an affiliated house in the front and a corridor in the back, a complete plan and part of the foundation in the early Ming Dynasty. In addition, although the back palace locates in a remote place and has been rebuilt by later generations, its beam still has the Ming style that may be similar to the age of Gongshi Garden, which is a perfect contrast to the back palace of Shanghai Chenghuang Temple one hundred miles away.

Jiading Chenghuang Temple, and Jiading Xuegong which will be talked about later, are both treasures with ancient style and unique charm in the same type of domestic architecture.

意境深邃，足与无锡寄畅园并美于江南。沈氏园，亦称『沈氏东园』，原系龚氏园一部，占地约四亩，

现状以花街平庭象征水意，以湖石花池与立峰象征山意，四周廊轩围绕，院落重重，被陈从周先生称为

『旱园水作』，但已迭经一九二〇年代和一九八〇年代翻建改创。金氏园占地约二十亩，早年毁于晚

清『太平军』战火，今存景象系近年重构。

嘉定县城隍庙今存中轴一线之『双井亭』、工字大殿、寝宫。其中工字大殿虽重建于晚清，仍极

为难得地存留着明早期大型殿宇前厦后廊、一气呵成的平面格局与部分柱础。而貌不惊人、深居僻处的

寝宫虽经后世重修，仍存留着可能与龚氏园年代相近之明风梁架，足与百里外之上海县城隍庙寝宫遥相

辉映。

嘉定县城隍庙与下文将述及的嘉定县学宫均为国内同类型建筑群中高度『存古』而含韵的珍品。

## An Ancient Style Literati Garden Inherited Song Style

The Gong family in the area of Wu, originally coming from Fujian Province, was a literati family who had flourished with scholars and political officials since the Song Dynasty. After SU Shunqin in the Northern Song Dynasty, the Gong family and ZHANG Dun, a powerful minister, separated the Canglangting Garden, a famous garden in Suzhou. The successor of the garden was the famous general HAN Shizhong.

In 1478, the family member Gong Hong (1451-1526) revitalized the family and achieved Jinshi, plunging into the political career. In 1501, as Zhejiangyoucanzheng, GONG Hong, who was famous for his incorruption, returned home to recuperate his disease. Because his parents passed away successively, he was in mourning and unable to back to work, thus living at home for more than ten years. The main part of the Gongshi Garden in Jiading, probably was created behind the residence of GONG Hong in the west of Chenghuang Temple during this melancholic and leisurely period. A few years later, in the city of Suzhou, which was a hundred miles away, WANG Xianchen, who resigned from Jianchayishi, returned to his hometown and created Zhuozheng Garden, which covered a vast area. And decades later, the wealthy Jin family built the Jinshi garden in the north of Gongshi garden. YANG Cheng, as Libushangshu, established Wufeng Garden inside Changmen of Suzhou and the east of the Taibo temple. These were the first cluster of Jiangnan gardens that emerged after the edict of forbidding to build gardens beside residence in the early Ming Dynasty.

Until 1577, as Sichuanyoubuzhengshi, PAN Yunduan resigned and returned to his hometown, and started to build Yu Garden in the west of Shanghai Chenghuang Temple. Sixteen years later, Taipusiqing Xu Taishi returned to Suzhou and built Xushidong Garden (later called Liu Garden) out of Changmen. Later, during 1573-1627, the Xiucai Shen Hongzheng created Sheshi Garden based on the east part of Gongshi Garden. At this time, the entire area of Jiangnan was already immersed in the wave of creating gardens. And Zuibaichi Garden in Songjiang built by GU Dashen, Gongbuzhishi in the early Qing Dynasty and Jin Garden in Changzhou built by YANG Zhaolu, Anchashi of Yanping County, Fujian Province in 1672, were only the ripple of the wave of creating gardens.

# 上承宋风的明中期『古风』文人园

吴地龚氏源出福建，为两宋以来科第鼎盛、进士辈出的文化巨族，北宋时曾在苏舜钦之后，与权相章惇分据苏州名园沧浪亭——继任园主则为名将韩世忠。

明成化十四年（一四七八年），族人龚弘（一四五一至一五二六年）振翮复起，荣登进士巍科，涉足宦海。弘治十四年（一五○一年），廉声颇著的龚弘自浙江右参政任上归乡养疾，因为高堂接连故去，低守孝而不能出仕，连续家居十多年——嘉定龚氏园的主体部分，很可能就在这一忧郁而闲散的时期，调创建于城隍庙西的龚宅后部——大约数年后，一百里外的姑苏城内，从监察御史任上辞官归乡的王献臣创建了面积广袤的拙政园；又或数十年后，豪族金氏在龚氏园北建造了金氏园，吏部尚书杨成则在姑苏阊门内、泰伯庙东创建了山石壁立的五峰园——这是从明初『禁园』阴影里走出的第一批江南园林。

到了明万历五年（一五七七年），上海县潘允端从四川右布政使任上辞归，在县城隍庙西大规模营造豫园；又十六年后，太仆寺卿徐泰时回到苏州阊门外营造徐氏东园（后称『留园』）；再晚一些的万历（一五七三至一六二○年）、天启（一六二一至一六二七年）之间，诸生沈弘正在龚氏园东部景域的基础上创建沈氏园——这时的整片江南，已经沉浸在万历一代的造园浪潮里了。而清初工部主事顾大申营造的松江府醉白池，康熙十一年（一六七二年）福建延平道按察副使杨兆鲁兴建的常州近园，不过是其袅袅余音耳。

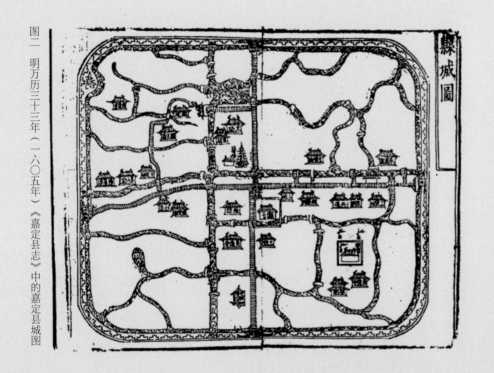

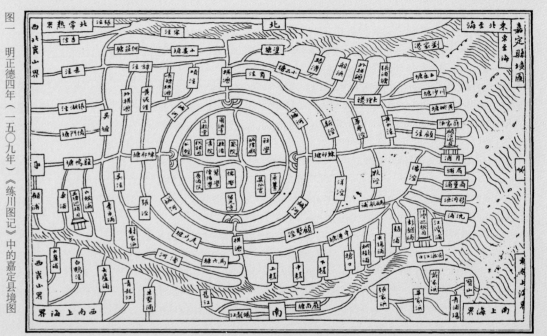

图二 明万历三十三年（一六〇五年）《嘉定县志》中的嘉定县城图

图一 明正德四年（一五〇九年）《练川图记》中的嘉定县境图

Fig.1  The map of Jiading County and its surrounding area in 1509's *Lianchuantuji*

Fig.2  The map of Jiading County in 1605's *Jiadingxianzhi*

143

After resting at home for more than ten years, in 1514, GONG Hong was recruited into the political career at the age of 60s. It was not until 1521 that he resigned under the title of Gongbushangshu. In 1526, he passed away at the age of 70. The second period of staying at home may also be the creation time of Gongshi garden. In the year of GONG Hong's death, in Fujian Province, the birthplace of the GONG family of Wu area, GONG Yongqing,the younger brother of the GONG family, won the first place in imperial examination, and later took the position as Jianjiu in Nanjing Guozijian. And JIN Dayou and JIN Zhaodeng, the son and grandson of JIN Yi, the founder of Jinshi Garden, which was adjacent to Gongshi Garden, passed the imperial exam with high score in 1558 and 1582 respectively. This was really an era that belongs to the literati world.

With a certain inevitable, Jiangnan area in the middle and late Ming Dynasty was also an era of frequent slave rebellion. And the rebellion frequently spread to some famous gardens. For example, the family of LI Yi, the owner of Guyi Garden in Nanxiang Town, Jiading County, was almost destroyed. In 1522-1566, GONG Hong's great-grandson GONG Kexue was also killed in a suspected slave rebellion. The GONG family was once defeated and in 1555, Gongshangshuzhai and its garden was sold to the WANG family, who were the salt merchants in Huizhou. GONG Hong, who just passed away, must be very despaired. Fortunately, GONG Xijue, the great-great-grandson of GONG Hong, who luckily escaped the slave rebellion, won the town examination in 1573 and revitalized the family. As a result, the garden was also returned entirely from the Wang family. The next year, GONG Xijue was blessed to achieve the position of Jinshi, and then became Guangxiyoubuzhengshi, thus building Shigang Garden in the south of Jiading to escape from this bloody wounded place.

This mutable life of Gongshi Garden, also reflected the close connection between the rise and fall of the literati's fame and the life and death of the garden in that era. However, a bigger turning point was yet to come.

家居十余年后的正德九年（一五一四年），龚弘以花甲之年重被征召入仕，到正德十六年（一五二一年）

才以工部尚书荣衔辞归，至嘉靖五年（一五二六年），以古稀之年去世。这第二段家居岁月，也有可能

是龚氏园的兴造年代——龚弘去世之年，吴地龚氏的源出之地、福建龚氏子弟龚用卿又高中状元，后来

官至南京国子监祭酒。而与龚氏园毗邻的金氏园创建人金翊之子金大有则在嘉靖三十七年（一五五八年）

中举，其孙金兆登在万历十年（一五八二年）中举——这实在是一个属于文人世族的时代。

带有某种必然性，中晚明的江南，亦是『奴变』频频波及名园，嘉定南翔镇古漪园

主人李宜之因之几近灭门。嘉靖年间（一五二二至一五六六年），龚弘曾孙龚可学也在疑似『奴变』中

遇害，龚氏一度败落，嘉靖三十四年（一五五五年）将『龚尚书宅』园售给徽州盐商汪氏——尸骨未寒

的龚尚书一定为之怅然，好在『奴变』中侥幸逃出生天的龚弘玄孙龚锡爵在万历元年（一五七三年）乡

试中举，重振门庭，汪氏竟也将宅园璧还原主。第二年，后福不断的龚锡爵联捷高中进士，官至广西右

布政使，又在嘉定城南另建石冈园，回避了这片沾染血痕的伤怀之地。

龚氏园的这段沧桑离合，也足以体现那一时代文人世族的功名起落与园宅兴废的紧密关联。当然，

更大的转折还在后头。

图三 明嘉靖三十年（一五五一年）《拙政园图咏》中的拙政园图

Fig.3  The image of Zhuozheng Garden in 1551's *Zhuozhengyuantuyong*

How different is the Gongshi Garden compared to today's Qiuxiapu Garden? Based on our speculation, the hill located at the south bank of the pond (Songfengling in 10 scenes in the early Qing Dynasty), the mountain trail (Suihanjing) and the waterside road under the mountain (Yingyudi), probably was preserved most of the plain and dense style in the middle Ming Dynasty, which was also coordinate with the status of an incorruptible official who returned home and recuperated his disease. After all, GONG Hong was famous for his incorruption. Besides, the main hall Shanguangtanyinggauan and the waterfront platform in front of the hall may also be relics of that period, and the waterfront platform maybe the grass slope with big stones once upon a time.

Jiangnan gardens in this period may not have formed a relatively condensed, stable and intensive classical layout paradigm in a large number of construction experiences as after the changes in cultural and artistic atmosphere during 1567-1620. However, there were still some traces of the ancient layout of gardens in the Song and Yuan Dynasty. The layout of Suzhou Zhuozheng Garden in ancient literature was another example.

龚氏园的布局面貌较之今日有多大差异？我们只能揣测，园中横亘于池南的土山（或即清初十景中的『松风岭』）以及山间小径（或即『岁寒径』）、山下临水盘道（或即『莺语堤』），可能最多存留了朴茂的明中期风貌，这也符合居丧养疾或致仕乡居的廉吏身份。毕竟，龚弘此前『在兖六年，受代之日，病几不起，诸寮采检厥囊，将备后事，得白金七两，无它物』——几乎廉比海瑞了。而主要厅堂山光潭影馆及其临水平台，也有可能传承自那一时期，只是平台或曾为点石的草坡。

这一时期的江南园林，可能还未能如『隆万风变』之后那样，在大量的营造体验中，形成相对凝练、稳定而集约化的古典布局范式——但也因此多少存留着宋元园林散漫自由的古风布局遗痕，文献中苏州拙政园的布局就是一例。

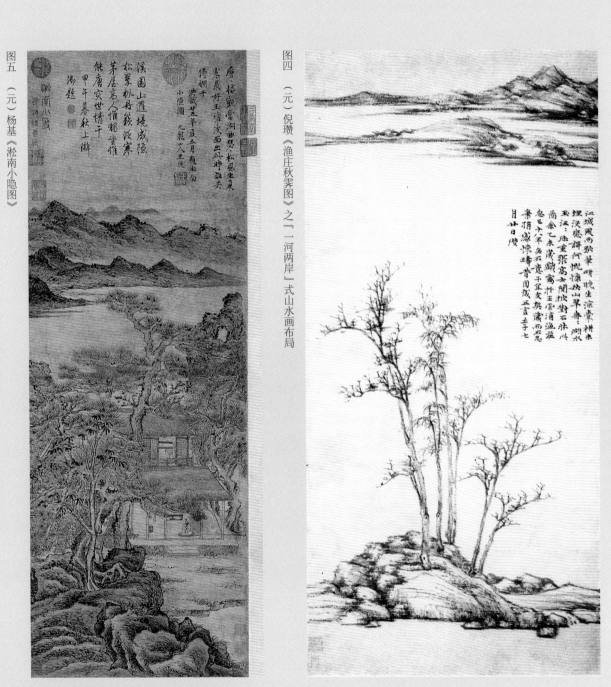

图五 （元）杨基《淞南小隐图》

图四 （元）倪瓒《渔庄秋霁图》之「一河两岸」式山水画布局

Fig.4 'One river and two banks' landscape painting of Ni Zan (Yuan Dynasty)'s *Yuzhuangqiujitu*

Fig.5 Yang Ji (Yuan Dynasty)'s *Songnanxiaoyintu*

## 'One River with Two Banks' and 'One Pool with Two Mountains' Style in the Late Ming Dynasty and Early Qing Dynasty

In 1645, the army of the Qing Dynasty went and fought south, and Jiading had suffered slaughter for three times. The offspring of the Gong family, who had always been stubborn faced with difficulties, struggled to guard the south gate of Jiading City, and the offspring of the JIN family, the owner of Jinshi Garden, guarded the east gate. Both of them died as martyr and fell into the long river of history, which indicated that the complete meaning of the literati garden sometimes needs to be written with life.

At that time, Gongshangshuzhai and its garden were returned to the WANG family, who were the salt merchants in Huizhou. The garden was commonly known as Wangshi Garden.

The WANG family, who received the famous garden for the second time and has a complicated mood, finally made a big effort to repair the old garden. 10 scenes including Songfengling, Yingyudi, Hanxingshi, Baiwutai, Suihanjing, Cengyunshi, Shuyuzhai, Taohuatan, Tiqingdu, Saxuelang. The garden was renamed as Qiuxiapu. Under the operation of the WANG family, Qiuxiapu Garden became even more famous and stepped into its highlight period, becoming the public living room in Jiading City. SONG Wan, a great poet of the Qing dynasty who won the title of Jinshi two years after the GONG family died in the slaughter, praised that in the east of the city, created by the WANG family, the scene of Qiuxiapu Garden was extremely beautiful. The name of Taohuatan derived from the poem that Taohuatan's water was a thousand foot deep, in order to compliment their ancestor WANG Lun. The winter scenes such as Songfeng, Hanxiang, and Suihan may contain the nostalgia for the Gong family. Taohuatan, Yingyudi and Tiqingdu constituted the elegant spring scenery series. Moreover, Shuyuzhai and Cengyunshi seemed summer scene and Baiwutai seemed autumn scene, which made up a remarkable time series.

And the name of Qiuxiapu Garden, may be derived from WANG Bo's sentence, 'Proud eagle flies high in the sky, with rosy clouds floating aside; the river runs far near the horizon, reflecting the same color of blue sky'. making people recall the GONG family who died in the slaughter due to the change of dynasties.

# 明末清初的『一河两岸』与一潭两山

南明弘光元年也即清顺治二年（一六四五年），清兵铁蹄南下，嘉定惨遭『三屠』，一贯临难不苟的龚氏子弟奋守嘉定城南门，金氏园主人金氏子弟则守在东门，两氏均壮烈殉城，就此沦落在历史长河——文人园林的完整意涵有时需用生命来书写。

『龚尚书宅』园此时又归徽商汪氏，俗称『汪氏园』。

二度接收名园，心情复杂的汪氏终于出手对龚氏旧园大加重修，逐渐形成松风岭、莺语堤、寒香室、百五台、岁寒径、层云石、数雨斋、桃花潭、题青渡、洒雪廊十景，并命园名为『秋霞圃』，使之声名鹊起，步入高光时段，成为嘉定城内烜赫的公共客厅。龚氏殉城两年后高中清廷进士的大诗人宋琬即称『秋霞圃在邑之东里，汪氏所辟，木石亭馆极一时之盛』——其中桃花潭之名实出『桃花潭水深千尺』之意，致意先祖汪伦；松风、寒香、岁寒等冬景无形中暗含着对龚氏子弟的追思，桃花潭、莺语堤、题青渡则构成隽逸的春景系列；此外，数雨斋、层云石似夏景，百五台似秋景，景象时序颇为显著。

而秋霞圃之名，或取意于唐人王勃『落霞与孤鹜齐飞，秋水共长天一色』的句子，让人不免追念文明的落霞中依依远去的渺渺孤鹜。

图六 （元）钱选《兰亭观鹅图》

图七 无锡寄畅园全景图（于福辉 摄）

Fig.6  Qian Xuan (Yuan Dynasty)'s *Lantingguanetu*
Fig.7  A panoramic view of Wuxi Jichang Garden (photographed by Yu Fuhui)

During this period, the Jiangnan literati gardens had already accumulated a new style of layout in the past 100 years of intensive practice in building gardens. If it was said that Yu Garden in Shanghai County in the middle and late Ming Dynasty adopted the classical layout of 'main hall facing the mountain across the water'. Then by the end of the Ming Dynasty and the early Qing Dynasty, the sublime 'one river with two banks' layout paradigm gradually formed.

The so-called 'one river with two banks" layout, is centered on a narrow channel-like water, with a small number of buildings on one side of the water and a horizontal mountain on the other side, facing with each other. As for the buildings' side, the small-sized pavilion bursts into the water, as if thrown into the mountains and forests on the other side, forming a focus of 'seeing and being seen' and creating a core dialogue relationship between human and nature. The larger main hall is often placed as far as possible, or even placed in other courtyards, to avoid interference with the condensed pureness of the main scene. This layout style and its variants may have originated from 'one river and two banks' landscape painting of NI Zan in the Yuan Dynasty. From the beginning of the Qing Dynasty to the early period of the mid-Qing Dynasty, it started to influence the literati gardens in Jiangnan. Some famous example includes Wuxi Jichang Garden, the middle part of Suzhou Zhuozheng Garden, Yangzhou Pianshishanfang Garden, even Beihai Jingqingzhai, Songjiang Yi Garden, and also to the later Suzhou Chang Garden and Hu Garden.

The weak point of this layout is that one side of the building may be dull, which should interact with the mountain and forests on the other side of the river so the middle part of Zhuozheng Garden was arranged on both sides of the building with Pipa Garden mountain views and Xiaocanglang water courtyard. Qiuxiapu Garden arranged the water of Taohuatan between the two mountains in the north and the south, with tall tree falling over the water and creating a beautiful scene. The north mountain was slightly vertical and steep (the top of the mountain may be the original site of Baiwutai). The south mountain was slightly horizontal and deep (it may be the original Songfengling). The mountains were stretched, and the water was endless. Once the mountain and water background of 'one pool with two mountains' was completed, and then arranged a small number of buildings on the side of the north mountain. Particularly, Biguanting (also called Pushuiting) burst forward and made the finishing touch, facing the south mountain, leaning against the north mountain, overlooking the water, and glancing right and left, together with Zhiyuzhai of Wuxi Jichang Garden, were the buildings with the best sense and location in the history of Chinese traditional gardens.

这一时期的江南文人园林，已经在百年的密集造园实践中，沉淀出了与此前不同的布局新风。如果说此前明万历年间的上海县豫园，采取的是以主要厅堂『隔水面山』的『古典式』布局，那么到了明末清初，尤其清初，其升华版——『一河两岸』的布局范式可能已悄然登场。

所谓『一河两岸』式布局，即以狭长的河道状水体为中心，水体一侧以少量建筑为主，另一侧以横卧的长山为主，两者隔水相对；建筑一侧并以小体量轩亭突入水面，投向对岸山林，形成对狭窄水上空间的进一步分隔，和『看与被看』的焦点，以及核心的对话关系。体量较大的主要厅堂则尽可能退后放置，甚至置于别院，以避免干扰主体景象空间的凝练纯粹——这一布局样式及其变体，可能始自元人倪瓒的『一河两岸』式山水画布局，至清初至清中前期，始染及江南文人园林，一时触手成春，名作迭出，如无锡寄畅园、苏州拙政园中部、扬州片石山房，甚至北海镜清斋，并流脉至更晚近的苏州畅园、壶园。

这一布局的薄弱点，在于建筑一侧易显平淡，须与对岸山林适度互渗，所以拙政园中部即在建筑两旁布置枇杷园山景和『小沧浪』水院。而秋霞圃则素性以南北两山夹桃花潭水而峙，高林俯水，参差交盖。其中北山稍稍纵向而峻拔（山巅疑即『百五台』旧址），南山稍稍横长而深厂（疑即『松风岭』），山则绵长迤逦，水则悠远不尽——『一潭两山』的景象背景既成，再于北山一侧，添加一堂一亭，加以统驭，其中以『碧光』一亭（亦称『扑水亭』）突前，挑临于桃花潭上，前横南岭，后倚北冈，俯鉴清泓，左右顾盼，与无锡寄畅园知鱼槛同为中国传统园林中位置感最佳的点睛之笔。

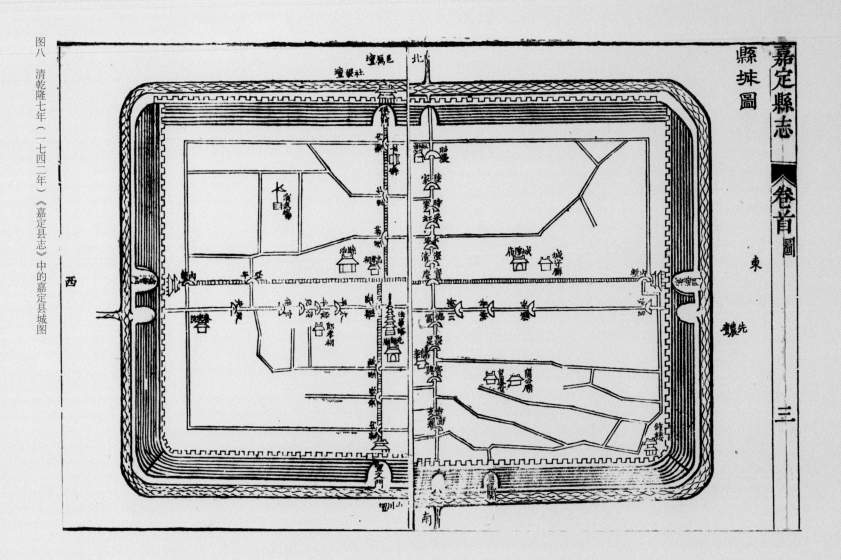

图八　清乾隆七年（一七四二年）《嘉定县志》中的嘉定县城图

Fig.8  The map of Jiading County in 1742's *Jiadingxianzhi*

At the west bank of Taohuatan, it ended with Congguixuan, which was suspected the original site of Hanxiangshi. Zhouerbuyouxuan (which seemed like a barge with Chishangcaotang), came forward slightly, forming the south-west port's and endless levels and becoming the second center. Certainly, Congguixuan and Zhouerbuyouxuan were not separated from Biguangting, forming an excellent spatial enclosure relationship. Several hundred-year-old maple poplars were decorated among the water and the mountains, forming a 'crab claw'-like outline effect in paintings of the Song Dynasty. The branches and leaves were sweeping, pure and powerful, which was one of the best examples of planting in the existing famous gardens. Moreover, it can never be replaced by other magnificent or colorful trees.

From the perspective of the old pictures in the 1950s, the east bank of Taohuatan, apart from the port of the north and south direction, the straight back eaves of Pingshantang in Shenshi Garden was hidden behind the trees as a finishing touch. The south of Pingshantang stretched out the straight white wall with the atmosphere of the mountains, and the straight corridor roof was slightly exposed outside the wall indicating a sense of deep and subtle courtyard, which was closer to the appearance of the early Qing Dynasty. Unfortunately, today the back eaves of Pingshantang are added with a gable and hip roof, and welcome Taohuatan with a pretty face, which may be overdone and suspected of grabbing the spotlight.

'One pool with two mountains' style of Qiuxiapu Garden may be established in the Ming Dynasty and finally took shape in the early Qing Dynasty. The Wukang Stone mountain in the north of Taohuatan and the stones extending to the middle of the pool were more likely to be relics of this period. The core buildings in the garden today, especially Biguangting, Congguixuan at the west bank of the pool and Zhouerbuyouxuan (Chishangcaotang), which give people a sense of irreversibility, are also very likely to follow the architecture from this period, as Hangxiangshi and etc. As for the location of Shanguangtanyingguan (Biwuxuan), the main hall of the whole garden, has been retreated from the north bank of Taohuatan as far as possible and concealed itself, which met the routines in the Ming and early Qing Dynasties, but its Qing Stone drum shape column bases with entwined vines decoration proved to be at the original site. As for its large size, high 'raise and depress', and the location pushed forwards to two mountains, there may be still helpless functional requirements and remains of the rebuilding in the Republic of China.

The current pattern of Jiading Qiuxiapu Garden and Wuxi Jichang Garden were both laid in the early Qing Dynasty, and their quite similar plain and simple mountain and water view, convincing that it should be influenced by and develop the work of ZHANG Nanyuan's faction. Despite the builder of the garden lost for a long time in the history of Chinese traditional garden, the 'one pool with two mountains' style survived, and Biguangting stood slightly, which has become a permanent monument.

桃花潭西尽端，则以『丛桂轩』（疑即『寒香室』旧址）隐约收尾；而『舟而不游轩』（与『池上

草堂』共成写意型画舫）稍稍突前，形成西南港汊和不尽层次，成为第二构图中心；两者并与『碧光亭』

疏落层递，形成极佳的空间围合关系。数株百年枫杨点缀于山间水际，形成宋画中『蟹爪』般的线条勾

勒效果，枝叶横扫，精纯有力，亦为现存名园中配植景象最佳的范例之一，非其他徒具绚烂的堂皇名木

或一味缤纷的蓊郁杂树所能替代。

从一九五〇年代旧影看，桃花潭东尽端除向南北分出港汊外，系借沈氏园屏山堂平直的后檐隐于乔

木之后，作为收尾，堂南复延出山野气息的平直粉墙，墙外微露平直的廊顶，有隔院楼台，深沉隐约之

感，这一影像或更接近清初旧貌——惜今日屏山堂后檐添建歇山顶抱厦，以俏面直迎桃花潭，反觉殷勤

过度，有抢镜之嫌。堂南亦改为起伏云墙，亦觉妖娆，反似不及桃花潭之幽邃存真。

秋霞圃的『一潭两山』格局可能奠基于明而最终成形于清前期，其中桃花潭北的武康石掇山及其延

至潭中的散石矶，较有可能是这一时期的遗物，而今日园中核心建筑，尤其碧光亭及池西丛桂轩，舟而

不游轩（池上草堂）三者位置，给人以不可改易之感，亦极可能沿袭自这一时期的建筑，如『寒香室』

等。至于全园主要厅堂山光潭影馆（碧梧轩）的位置，已尽可能自桃花潭北岸退后而消隐，既符合明至

清初的主要厅堂常规，也可由其现存青石缠枝纹鼓形柱础证明为明至清初原址。至于其体量偏大，举折

高峻，陵迫两山，则或是无可奈何的功能需求及民国改建的遗痕。

嘉定秋霞圃的今日格局与无锡寄畅园同奠定于清初，且山浑水朴，千里如一，令人笃信其应受到张

南垣一脉影响，而自出新意。虽然造园史久湮作者之名，但『一潭两山』洒然长存，碧光之亭无言静峙，

已足成其永久丰碑。

图十 1950 年代（2020 年代）秋霞圃山光潭影馆，其屋脊简素流畅，回廊四面舒展，台前竹篱围合，临水石栏飞跨有力，均与现状有一定差别而更存文人园林古风。

图九 1950 年代（2020 年代）秋霞圃桃花潭东景象，碧光亭作为「舫意」暖亭似更合适，远处建筑、墙体之平直低调尤为得体，似存古风。

Fig.9 In the 1950s (2020s), in the east scene of Taohuatan of Qiuxiapu Garden, Biguangting seems to be more suitable as a "boat-like" warm pavilion. The buildings and walls in the distance are straight and low, which seem to preserve the ancient style.

Fig.10 In the 1950s (2020s), Shanguangtanyingguan in Qiuxiapu Garden had simple and smooth ridges, with corridors stretching on all sides, the platform enclosed by bamboo fences, and the stone fence near the water leaping powerfully, all of which were different from the status quo and preserved the ancient style of a literati garden.

## The Fragile Sense of Belonging to Chenghuang Temple Garden since the Early Qing Dynasty

In the early Qing Dynasty, Jiangnan once developed rapidly in the economic wave due to the opening of the coast.

In 1709, the first-benefited gentry and merchant in Shanghai County had built Lingyuan in the east of Chenghuang Temple, as a place for the Chenghuang gods to visit, so as to better protect its citizens. Seventeen years later, in 1726, the WANG family who may have been unable to maintain the famous garden, also donated Qiuxiapu Garden to the Chenghuang Temple in Jiading County as a Lingyuan. And thirty-three years later, in 1579, Shenshi Garden in the east also followed the pace and was donated to the Chenghuang gods of the city. Yu Garden owned by the PAN family was also acquired as the West Garden of Chenghuang Temple in Shanghai County in the next year. Eventually, in the rolling wave of the times, literati gardens had changed owners one after another, and the pastoral era of the bureaucrats was gone forever.

However, different from the fate of Yu Garden, which was classified as a large-scale barren garden donated to Shanghai Chenghuang Temple, and almost completely rebuilt into a 'collection style' merchant garden, Qiuxiapu Garden, with a precise and intensive scale and a complete present situation, had a very high level of artistic standard attracting people all the time. Moreover, the cultural background of Jiading was quite different from that of Shanghai County after all, so after its identity change, the scene may not change too much. The two mountains were still dense and flourishing. The pool was deep. The trees were tall and covering the space, and Biguangting alone faced the water. From the current situation, Shenshi Garden has a high density in architecture, while WANG Garden (GONG Garden) has scattered mountain and water landscape. The two gardens, which were a whole originally, merged and donated to the same temple, which may be the most successful combination in the history. Only the vicissitudes of the world and the fragile sense of belonging, make it inevitable that GONG Shangshu and the WANG and SHEN family must be depressed after death.

# 清前期以来『一身如寄』的城隍庙园

清初的江南，在一度开海的经济浪潮中，曾经发展迅猛。

康熙四十八年（一七〇九年），首蒙其惠的上海县绅商已在县城隍庙东兴筑灵苑，作为城市之神游目骋怀之所，以利其更好地恩庇市民。十七年后的雍正四年（一七二六年），可能已无力维持名园的汪氏也把秋霞圃捐入嘉定县城隍庙，以作灵苑。又三十三年后的乾隆二十四年（一七五九年），东邻沈氏园也追随其步伐，投入城市之神麾下，再一年后，潘氏豫园亦被收购为上海县城隍庙西园——时代的滚滚浪潮中，文人园林纷纷易主，士大夫的田园时代一去不复返了。

只是与豫园以大面积荒园状态归入上海县城隍庙，而几乎被彻底重建为『集锦式』商贾园林的命运不同，秋霞圃规模精约，存留或亦完整，艺术水准尤其深入人心，且嘉定一城的人文底色毕竟与上海县有异，故其身份转变后，仍是两山郁郁，潭影深长，高树如盖，孤轩枕流，景象或许变化不大。且从现状看，沈园建筑绵密，汪园（龚园）山水散淡，二者原为一体，又同归一庙，珠联璧合，或是史上最成功的归并之一。只是世事沧桑，一身如寄，不免令九泉下的龚尚书和汪、沈二氏先人们伤怀不已。

图十一　一八六七年的嘉定西门城墙，（英）亨利·坎米奇摄

Fig.11  The city wall of the west gate of Jiading County in 1867, photographed by Henry Cammidge

The war of Taiping Army in 1860 was a common test for the famous gardens in Jiangnan area. LIU Garden in Suzhou was named after the survival of the war. The buildings in Qiuxiapu Garden and Shenshi Gardens in Jiading may have been destroyed in the war. Fortunately, the mountain and water was unharmed and the structure of scene still remained. Without the power as fearless as the commercial guilds in Shanghai County to dismember the temple garden for profit, the two gardens therefore avoided the fate of Yu Garden, and were gradually restored during the revival in 1862-1908 after the war. Chishangcaotang was rebuilt in 1876, covered with a round ridge and flush gable roof, facing the water with side scuttle and Meirenkao. It seemed like a humble barge and had an obvious Ming style. Congguixuan was rebuilt in 1886, covered with single eave round ridge and gable and hip roof. It had an exquisite and concise style and an accurate grasp of scale, which constituted a picture with the surrounding, while the inner ceiling was simple and concise, full of simple and wild feelings. Shanguangtanyingguan was rebuilt in 1922, covered with single eave gable and hip roof and having high 'raise and depress', which was full of the atmosphere of a temple, while recent restoration has enhanced the expression of the ridges and accentuated its conflict of scale with the overall landscape. Biguangting (Liuyiting) and Jishanting were both rebuilt in 1981. Biguangting was a little bit blunt, while Jishanting seemed to retain the Gualuo in the Republic of China.

Time flies while the garden still kept the original look. Today, when wandering along the bank of Lianchuan, turning from the mighty Chenghuang Temple into courtyards of SHEN Garden and then turning from Pingshantang of SHEN Garden into Qiuxiapu Garden, we can still feel that with the blue sky and the strong passion, the atmosphere of the mountain and the water coming forward, and the spirit of the garden has been kept and continued for centuries.

咸丰十年（一八六〇年）的『太平军』战事对江南名园是普遍的考验。苏州留园即得名于战火后的硕果仅存。而嘉定秋霞圃、沈氏园中建筑可能均于这次战火中损毁殆尽。好在山池无恙，景象格局犹存，嘉定城中亦无势力如上海县诸公所那般无所顾忌，以肢解庙园牟利，故两园避免了豫园的命运，于战火后的『同光中兴』（指同治与光绪两朝，一八六二至一九〇八年——本书责编注）之际逐渐修复。

其中池上草堂重建于光绪二年（一八七六年），卷棚硬山，侧面临水处作舷窗、美人靠，寥寥数笔，即具写意型船舫意匠，平淡谦抑，尤近明风。丛桂轩重建于光绪十二年（一八八六年），单檐卷棚歇山顶，精雅凝练，四顾成图，体量把握精确，内天花则素洁简练，富于山野趣味。『山光潭影馆』重建于民国十一年（一九二二年），单檐歇山，举折偏大，庙堂气偏重，近年的修缮强化了屋脊表达，又凸显了其与全园景物的尺度冲突。碧光（宜六）、即山两亭均重建于一九八一年，前者稍觉生硬，后者似沿用了民国年间的挂落。

岁月更迭，名园幸而如故。时至今日，由淙淙练川河畔，嘉定县城隍庙的巍巍殿宇转入沈氏园的重重庭院，复由沈氏园屏山堂后转入秋霞圃时，仍觉一片水色山光扑面而来，空翠如洗，浓情欲滴，数百年文人志士的耿耿精魂，至今不灭。

图十三　嘉定秋霞圃山光潭影馆（碧梧轩）内青石缠枝纹鼓形柱础

图十二　嘉定秋霞圃碧光亭

Fig.12  Biguangting in Jiading Qiuxiapu Garden

Fig.13  The Qing Stone drum shape column base with entwined vines decoration in Shanguangtanyingguan (Biwuxuan) of Jiading Qiuxiapu Garden

# Jiading County Chenghuang Temple: The Remaining Layout of an Affiliated House in the Front and a Corridor in the Back

Jiading County Chenghuang Temple was first built during 1208-1224 when Jiading County was first established in the Southern Song Dynasty, which was similar to the construction time of Jiading Academy. The original site was at Fu, anfang on Nanmen Street, which is now the area around Lijialong. It was relocated on Dongmen Street in 1370 when the belief of Chenghuang gods was emphasized and Chenghuang Temples were constructed throughout the country. Nowadays, three existing buildings on the central axis are a pair of Jingting, the main hall and the back palace. The first gate and the second gate (Yimen, which had Xilou in the back side) between Jingting and the main hall have been destroyed. The buildings behind the back palace, the affiliated halls and corridors on the two sides of the axis and the east and west line of buildings have been destroyed too.

The pair of Jinting, like Chuiguting, located at the south of the axis confront on the two sides of the original first gate, which is the legacy of Quelou on both sides of the palace gate. Two pavilions both cover with the single eave pyramidal roof and adopt stone column and wooden tie beam. Dingzipaike were used between the tie beams and the pattern of Sifengpenglong was used in the ceiling inside, which is gorgeous and delicate. The Huanggang Stone square pillars are large and rugged, and have obvious Cejiao, which were suspected the relic of the Ming Dynasty, while the wooden beams are steep and have the characteristics of the late Qing Dynasty, which are suspected the relic of the reconstruction in 1882.

The main hall is five-room wide and covers with single eave gable and hip roof. The central part protrudes forward with a single eave gable and hip roof affiliated house, which is parallel to the main hall and at the same height, but the width narrows to three rooms. The affiliated house then protrudes forward with a single eave round ridge roof windowed veranda, which is parallel to the affiliated house with the same width but the height cuts in half and steps on the wide platform in front of the hall. The diverse roof forms which combined together with each other, is full of a mature design sense, and the space in the hall is virtually divided into three parts, the front, middle and back, which strengthens the spatial hierarchy. The front windowed veranda creating a 'gray space', is used for indoor and outdoor transition and buffering. The affiliated house in the middle

# 嘉定县城隍庙：前厦后廊、腾蛟起凤的布局遗音

嘉定县城隍庙创建于嘉定立县之初的南宋嘉定年间（一二〇八至一二二四年），与县学宫建造时间相去不远，当时位置在南门大街富安坊（今李家弄一带）。明洪武三年（一三七〇年）整顿城隍信仰，大修天下城隍庙时，将其移建至东门大街今址——现存中轴线上双井亭、大殿与寝宫三座建筑，而井亭与大殿间的头门、二门（即仪门，背面应为戏楼）两建筑已不存，寝宫后原似有之建筑，以及中轴两侧配殿、廊庑及东西路建筑亦早付阙如。

最南的双井亭如吹鼓亭般对峙于原头门外两侧，仿佛双阙遗制。两亭均为单檐四方攒尖顶，石柱木枋，桁枋间施丁字牌科出两跳，内天花剔地起突『四凤捧龙』图案，华美而工细——亭下部的花岗石抹角方柱则较为粗犷，侧脚明显，疑为明初原物，而其上的木构屋架陡峻飞扬，具有晚清特征，或为光绪八年（一八八二年）重建时的遗物。

其北的大殿面阔五间，单檐歇山，其中部向前凸出单檐歇山厦屋，与正身平行同高，而缩减面阔至三间；厦屋复向前凸出单檐卷棚歇山前轩，与厦屋平行同面阔，而降低高度至其半，并跨上殿前的宽阔月台——其缤纷的屋顶组合饱含着成熟的设计感，无形中将殿内空间分为前、中、后三部，强化了空间层次。前部副阶『灰空间』用于室内外过渡与缓冲，中部抱厦为主要祭拜空间，后部正屋为神

图十四 嘉定城隍庙寝宫明风梁架，由徐瑞彤绘制

Fig.14 The Ming style beam of Jiading County Chenghuang Temple's back palace, drawn by XU Ruitong is the main worship space, while the main hall in the back furnishes the sculpture for the gods. The back of the main hall is connected with a single eave flush gable roof and three-room wide back palace with a wide I-shaped corridor, which is combined into an ancient style of 'I-shaped hall'.

Walking through the first gate and Yimen in the past, the whole group of 'I-shaped hall' was majestic due to the straight perspective and the protruding affiliated house increased the volume and level of the building. When looked back from the temple garden the ups and downs of the skyline would greatly weaken its sense of volume, adding a little charm to the solemnity. Before the main hall there were two affiliated house protruded forward and then stretched forward again with a platform and behind the I-shaped corridor there was the back palace, which were all in one go and beyond the simple rainproof function, making people think of the similar appearance of original Jiading County Chenghuang Temple belonging to Suzhou Province in the early Ming Dynasty, and the relics of many 'one-piece' building layouts in the older era, and even the exotic Hoodo Pavilion of The Byodo-in in Kyoto with Tang style, and eventually making people recall the similar layouts and special forms full of charm and spiritual significance in Chinese history.

According to the documents, the main hall, including the affiliated house in the front and the I-shaped corridor in the back were all rebuilt in 1882. However, according to the current situation, it is more likely that the affiliated house and windowed veranda before the main hall have retained more components of the reconstruction after the fire in 1788. Some wood and stone components can even be traced back to the early Ming Dynasty. The beam work of the back palace, which is simple, gentle and beautiful, although it is recorded the relic of the reconstruction after the fire in 1750, it may still be earlier.

The Chenghuang God enshrined in Jiading County Chenghuang Temple was born quite late. LU Longqi (1630-1692), a famous scholar of the Qing dynasty and a native of Pinghu, Zhejiang (anciently called Danghu), became a Jinshi in 1670 and took the position of Jiading Zhixian in 1675. He Advocated that the good governance was not as good as the good education. With good temptations and high reputation, he was known as Danghuxiansheng at that time. Not far from the temple, there is still an architecture called Danghushuyuan in Jiading Academy. This is obviously different from the customs of respecting the old and inclusion in Shanghai Chenghuang Temple that worshipped Huo Guang, the god who defends the sea in the front hall, and QIN Yubo, the Confucian official in the back hall.

159

像陈设空间——大殿后部复以宽阔的工字连廊连接单檐硬山、面阔三间的寝宫，组合为古风犹存的『工字殿』样式。

自昔日头门、仪门步入时，整组『工字殿』因视角端正而巍峨雄壮，前凸的层叠抱厦更加大了建筑的体量与层次。自庙园回望工字殿侧后时，起伏跌宕的天际线则会大大削弱其体量感，于庄严外平添几分妩媚——而大殿正身之前出两重抱厦再展为月台，后曳工字长廊而展为寝宫，仿佛腾蛟起凤，一气呵成，超越了简单防雨功能，令人想起嘉定县旧属苏州府城隍庙之明初工字殿的类似面貌，和更古老时代的诸多『连体式』建筑布局遗风，甚至异域而存唐风的京都平等院凤凰堂，更令人想起中华历史上，类似布局传统的特殊形态魅力和精神意义。

考诸文献，大殿、前廊及殿后工字廊均重建于光绪八年（一八八二年）。但从现状看，大殿之前厦、前轩较可能留存了较多乾隆五十三年（一七八八年）年大火后重建的构件，局部木、石构件甚至可追溯至明初。而寝宫之扁作月梁简劲舒缓，而又流丽动人，虽按史载系乾隆十五年（一七五〇年）火灾后重建之物，却可能存留着更早期梁架。

嘉定城隍庙所奉城隍之神出世颇晚，为清代理学名家、浙江平湖（古称当湖）人陆陇其（一六三〇至一六九二年），其人为清康熙九年（一六七〇年）进士，十四年（一六七五年）出任嘉定知县，主张『善政不如善教』，循循善诱，垂誉久长，世称『当湖先生』。庙南不远处的嘉定县学宫尚有以其为名的当湖书院——这与东邻上海县城隍庙于前殿祀捍海之神霍光，后殿祀儒臣秦裕伯的崇古而兼容不同。

嘉定县城隍庙及庙园（秋霞园、沈氏园）、金氏园总轴测图
Axonometric drawing of Jiading County Chenghuang Temple and Temple Garden
(Qiuxiapu Garden, Shenshi Garden) and Jinshi Garden

0　5　10　　20m

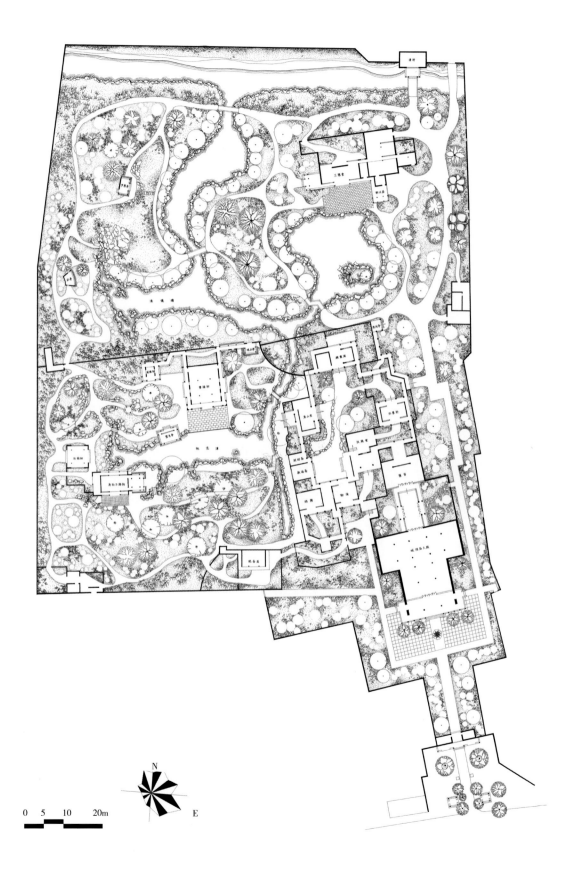

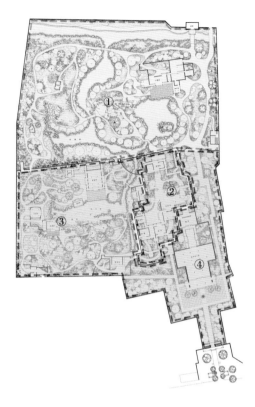

①金氏园 Jinshi Garden
②沈氏园 Shenshi Garden
③秋霞圃（明龚氏园） Qiuxiapu Garden (Gongshi Garden of the Ming Dynasty)
④嘉定县城隍庙 Jiading County Chenghuang Temple

N

E

0　5　10　　20m

嘉定县城隍庙及庙园（秋霞园、沈氏园）、金氏园总平面图
Master plan of Jiading County Chenghuang Temple and Temple Garden
(Qiuxiapu Garden, Shenshi Garden) and Jinshi Garden

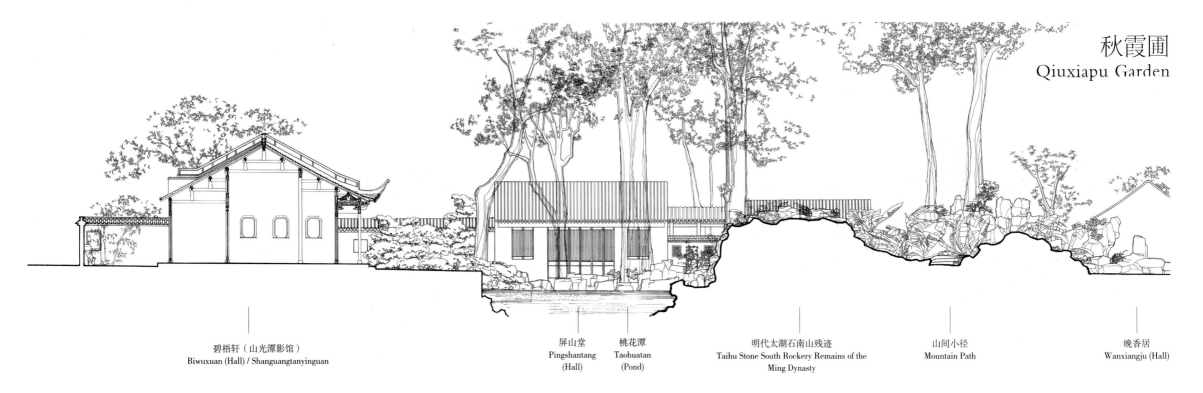

碧梧轩（山光潭影馆）
Biwuxuan (Hall) / Shanguangtanyinguan

屏山堂
Pingshantang
(Hall)

桃花潭
Taohuatan
(Pond)

明代太湖石南山残迹
Taihu Stone South Rockery Remains of the
Ming Dynasty

山间小径
Mountain Path

晚香居
Wanxiangju (Hall)

秋霞圃南北剖面复原图（自丛桂轩望屏山堂）
Restored East-west section of Qiuxiapu Garden (watching Pingshantang from Congguixuan)

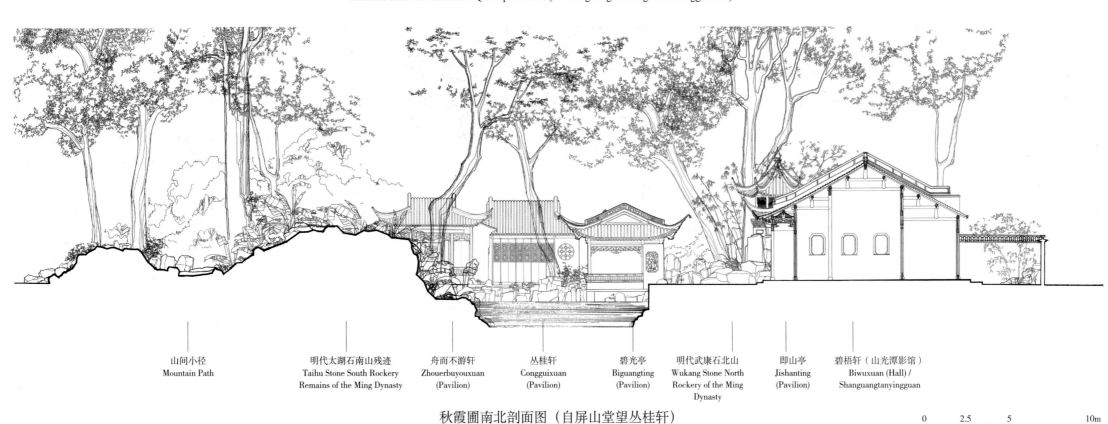

山间小径
Mountain Path

明代太湖石南山残迹
Taihu Stone South Rockery
Remains of the Ming Dynasty

舟而不游轩
Zhouerbuyouxuan
(Pavilion)

丛桂轩
Congguixuan
(Pavilion)

碧光亭
Biguangting
(Pavilion)

明代武康石北山
Wukang Stone North
Rockery of the Ming
Dynasty

即山亭
Jishanting
(Pavilion)

碧梧轩（山光潭影馆）
Biwuxuan (Hall) /
Shanguangtanyingguan

秋霞圃南北剖面图（自屏山堂望丛桂轩）
East-west section of Qiuxiapu Garden (watching Congguixuan from Pingshantang)

0  2.5  5  10m

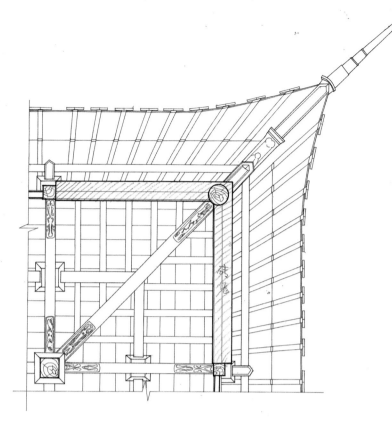

0    0.3    0.6      1.2m

丛桂轩屋架转角仰视图
Bottom view of Congguixuan's roof corner beams

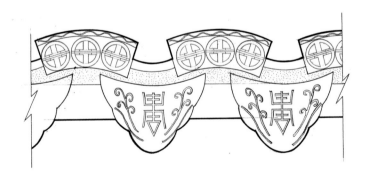

0   0.04   0.08     0.16m

丛桂轩花边滴水大样图
Huabian Dishui of Congguixuan

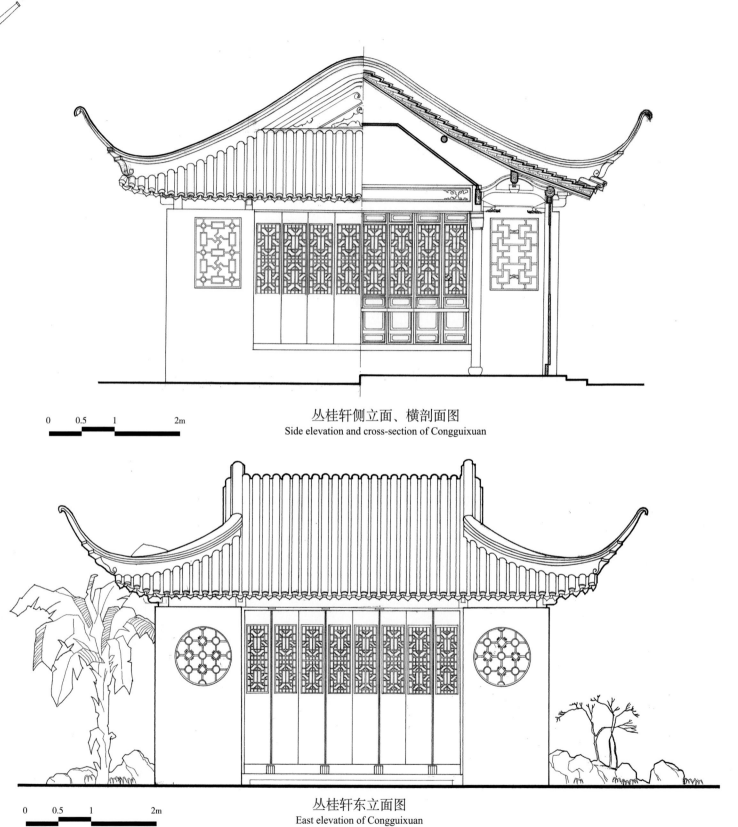

0    0.5    1      2m

丛桂轩侧立面、横剖面图
Side elevation and cross-section of Congguixuan

0    0.5    1      2m

丛桂轩东立面图
East elevation of Congguixuan

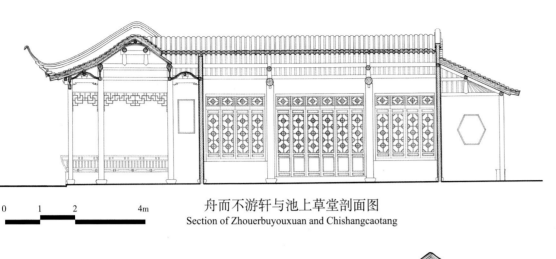

舟而不游轩与池上草堂剖面图
Section of Zhouerbuyouxuan and Chishangcaotang

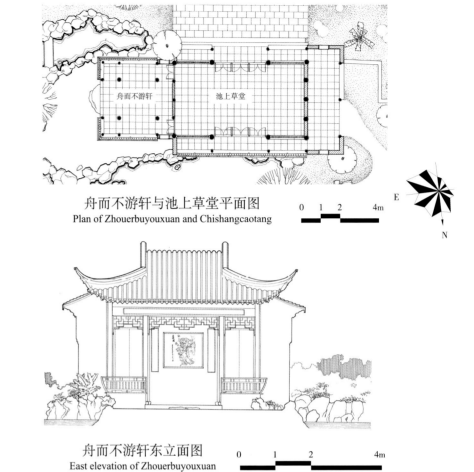

舟而不游轩与池上草堂平面图
Plan of Zhouerbuyouxuan and Chishangcaotang

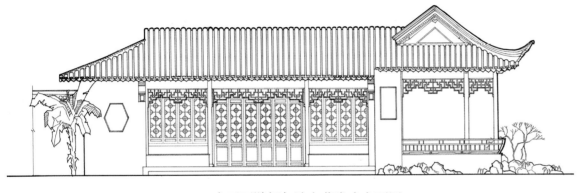

舟而不游轩与池上草堂南立面图
South elevation of Zhouerbuyouxuan and Chishangcaotang

舟而不游轩东立面图
East elevation of Zhouerbuyouxuan

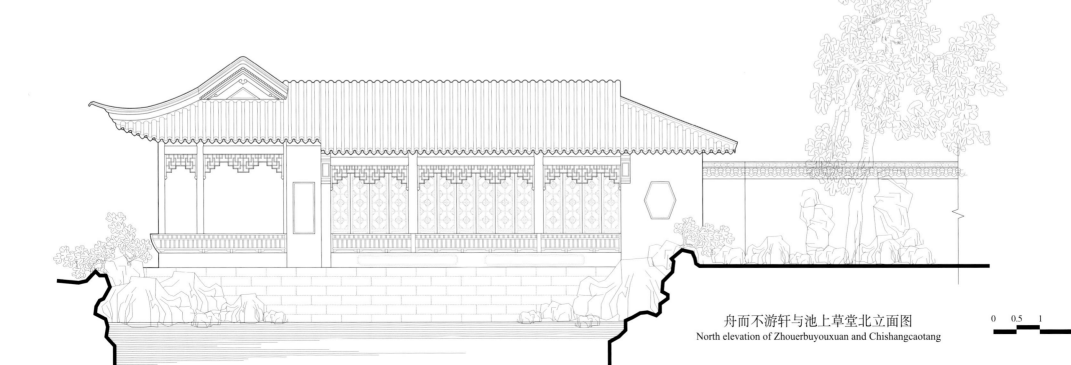

舟而不游轩与池上草堂北立面图
North elevation of Zhouerbuyouxuan and Chishangcaotang

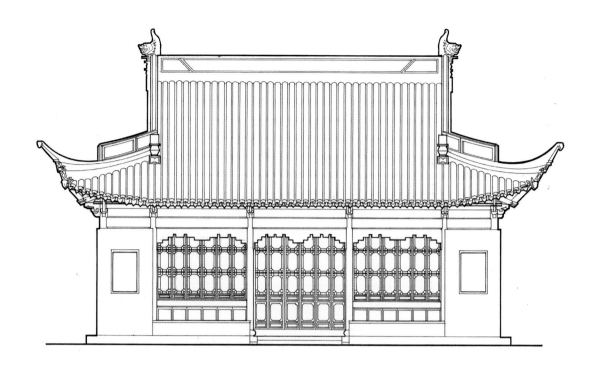

碧梧轩（山光潭影馆）南立面图

South elevation of Biwuxuan (Shanguangtanyingguan)

碧梧轩（山光潭影馆）西立面图

West elevation of Biwuxuan (Shanguangtanyingguan)

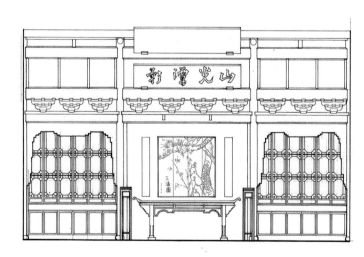

碧梧轩（山光潭影馆）内屏门、飞罩立面图

Elevation of Biwuxuan (Shanguangtanyingguan)'s Pingmen and Feizhao

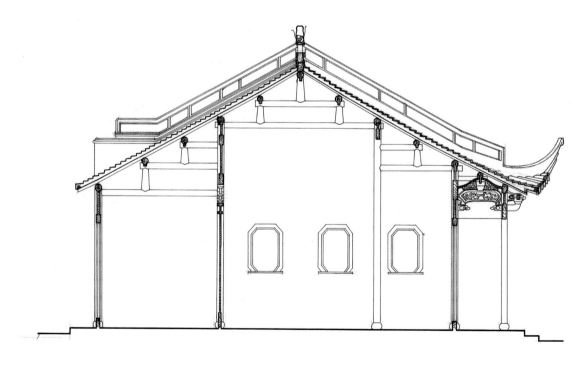

0  1  2  4m

碧梧轩（山光潭影馆）横剖面图

Cross- section of Biwuxuan (Shanguangtanyingguan)

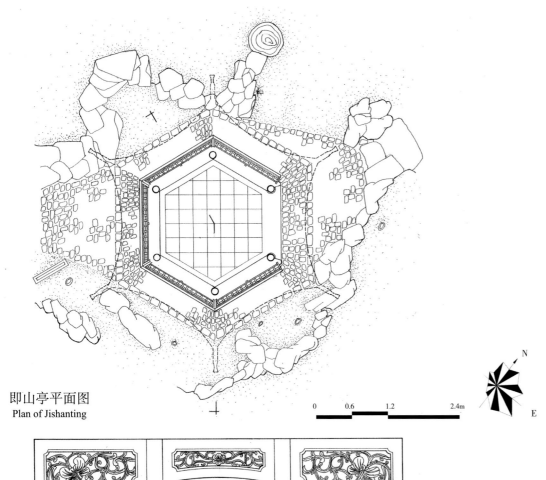

即山亭平面图
Plan of Jishanting

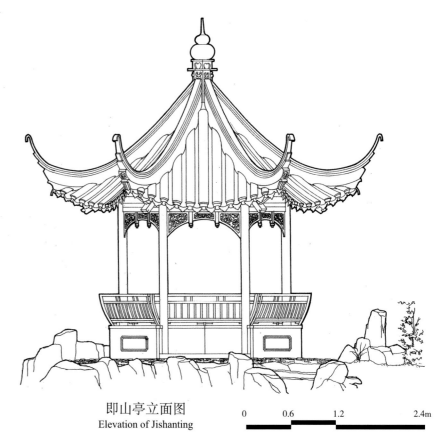

即山亭立面图
Elevation of Jishanting

即山亭挂落大样图
Gualuo of Jishanting

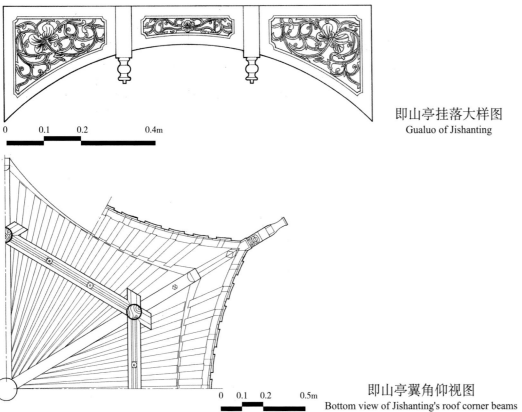

即山亭翼角仰视图
Bottom view of Jishanting's roof corner beams

即山亭剖面图
Section of Jishanting

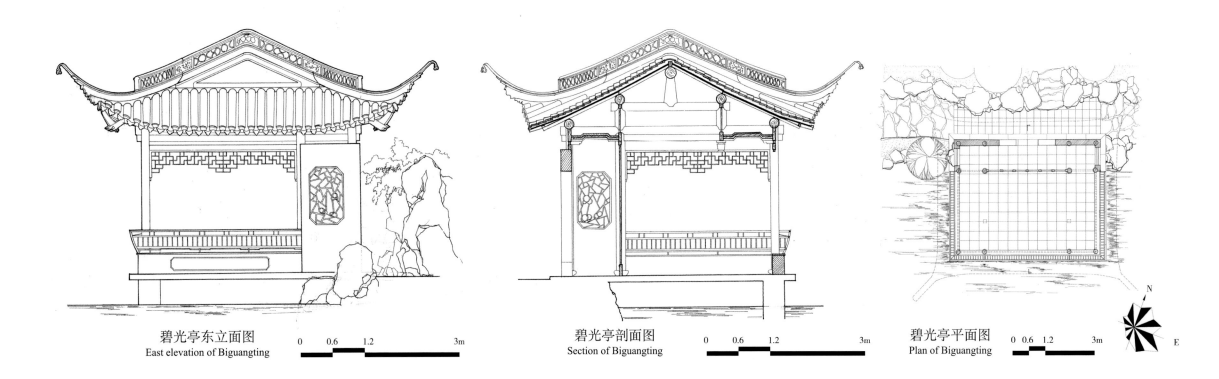

碧光亭东立面图
East elevation of Biguangting

0 0.6 1.2 3m

碧光亭剖面图
Section of Biguangting

0 0.6 1.2 3m

碧光亭平面图
Plan of Biguangting

0 0.6 1.2 3m

N
E

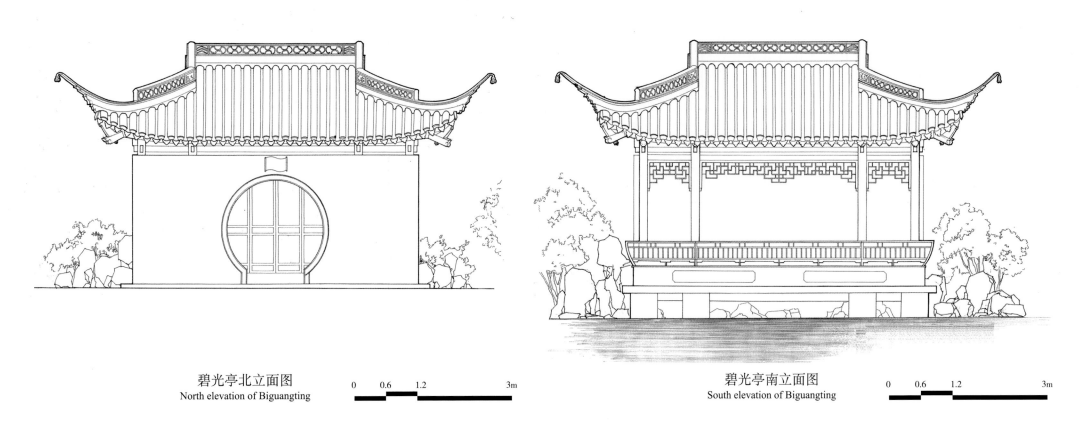

碧光亭北立面图
North elevation of Biguangting

0 0.6 1.2 3m

碧光亭南立面图
South elevation of Biguangting

0 0.6 1.2 3m

沈氏园
Shenshi Garden

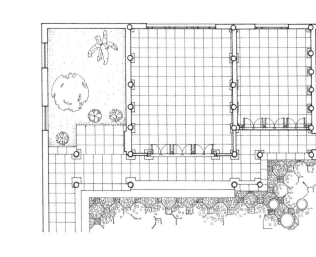

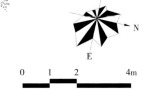

闲研斋剖面图
Section of Xianyanzhai

0 0.5 1 2m

数雨斋（左）闲研斋（右）平面图
Plan of Shuyuzhai (left) and Xianyanzhai (right)

0 1 2 4m

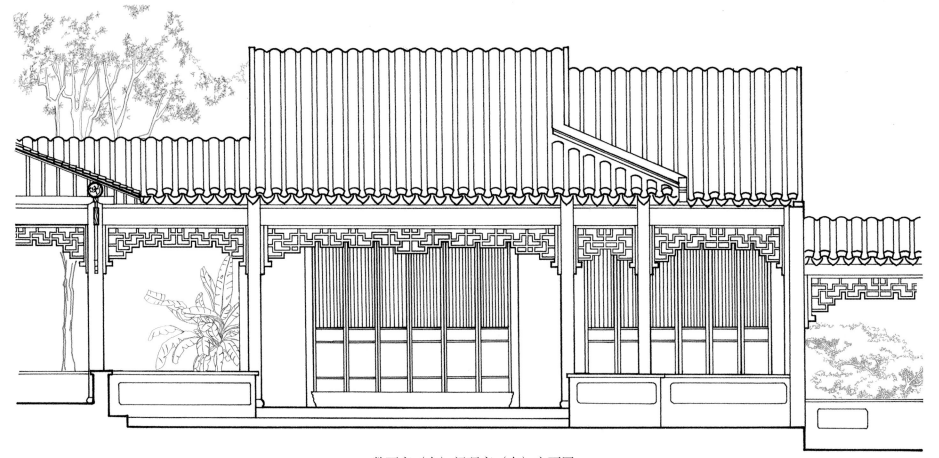

数雨斋（左）闲研斋（右）立面图
Elevation of Shuyuzhai (left) and Xianyanzhai (right)

0 0.5 1 2m

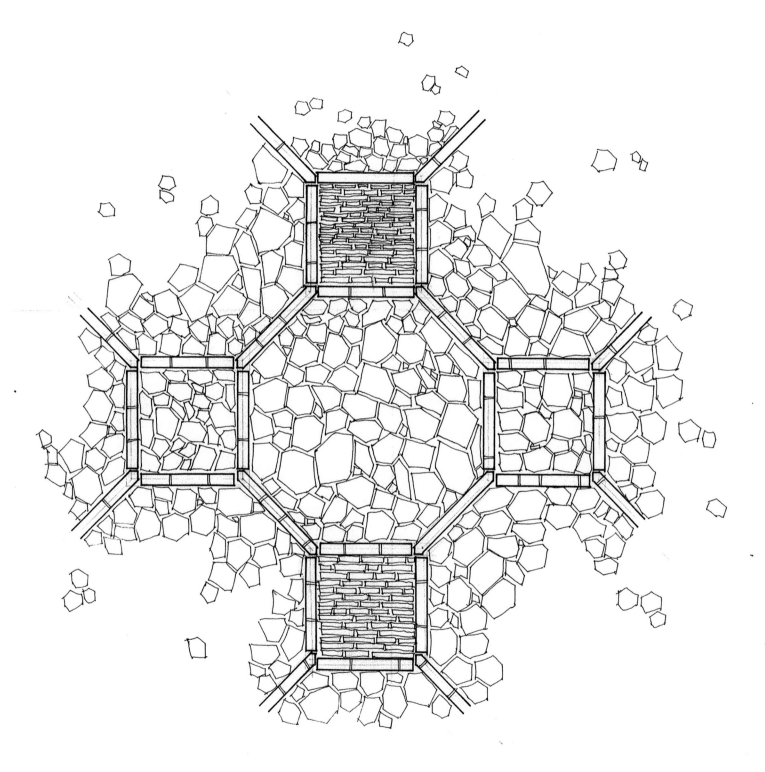

沈氏园花街铺地大样图
Huajie Pudi of Shenshi Garden

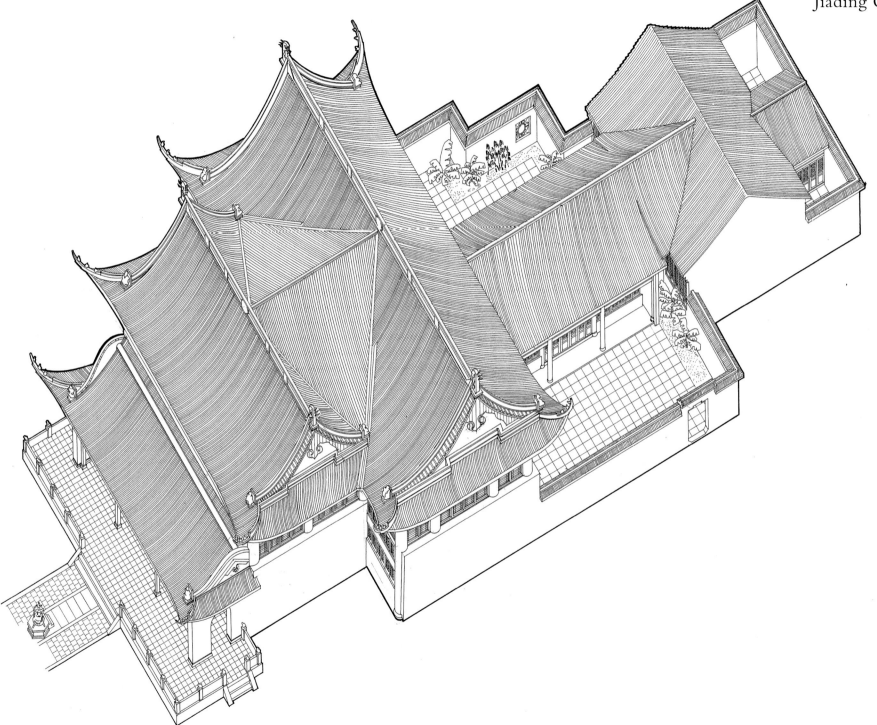

嘉定县城隍庙大殿、工字廊及寝宫轴测图
Axonometric drawing of Jiading County Chenghuang Temple's main hall, I-shaped hall and back palace

0  2  4  10m

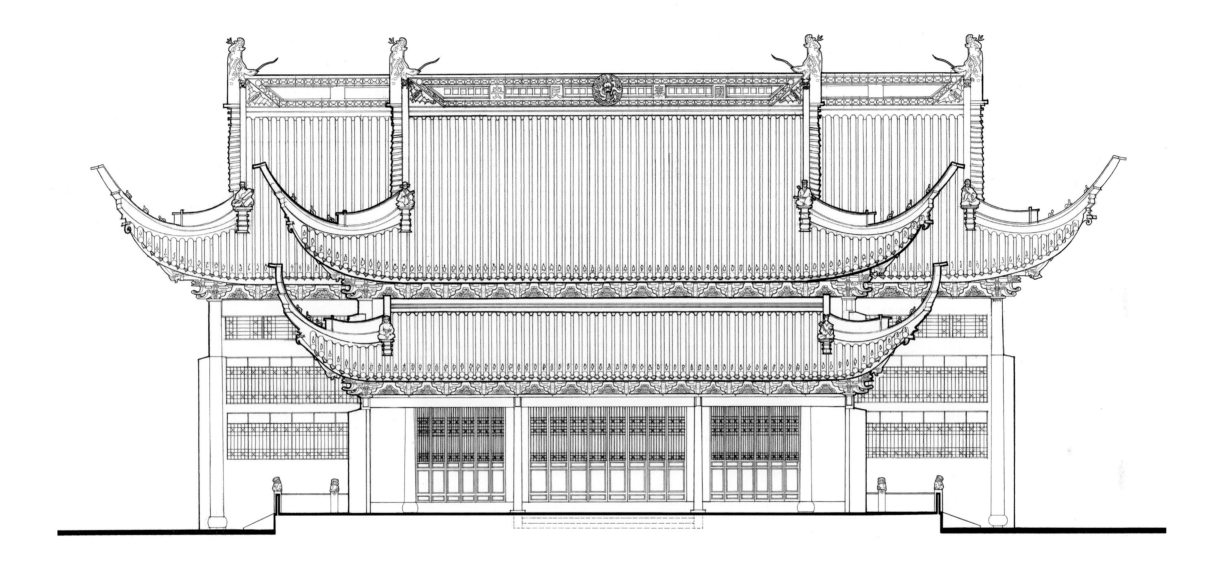

0　　1　　2　　　　4m

嘉定县城隍庙大殿正立面图
Front elevation of Jiading County Chenghuang Temple's main hall

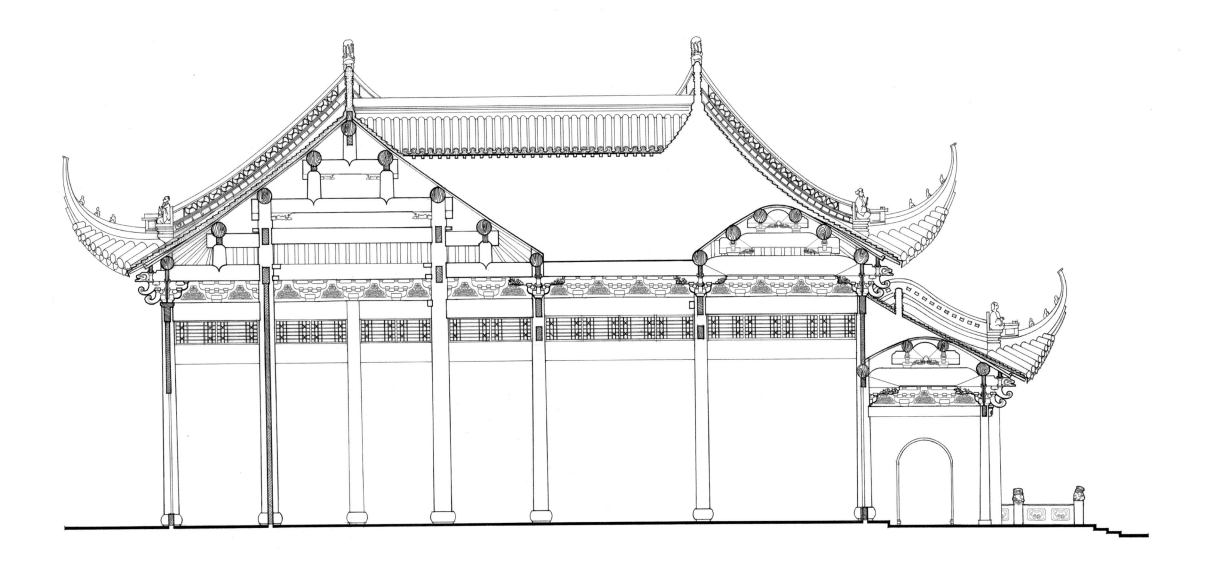

嘉定县城隍庙大殿横剖面图
Cross-section of Jiading County Chenghuang Temple's main hall

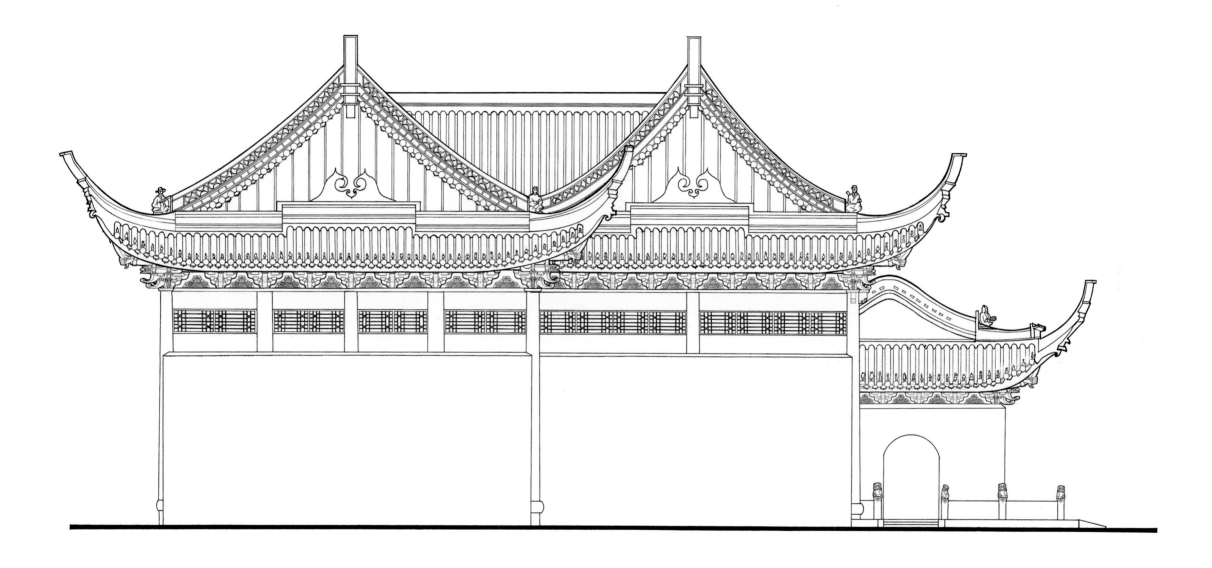

嘉定县城隍庙大殿侧立面图
Side elevation of Jiading County Chenghuang Temple's main hall

0  1  2  4m

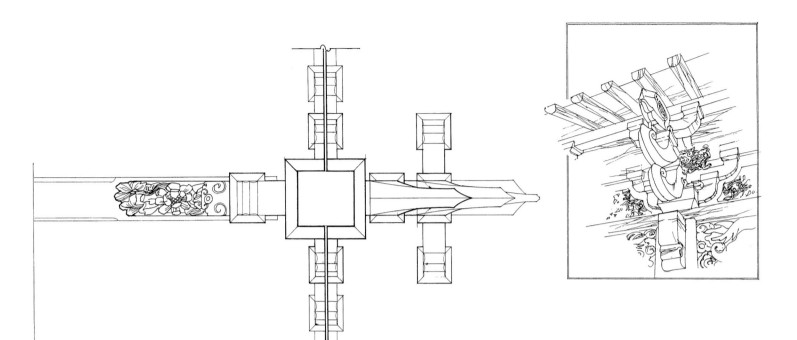

大殿轩廊柱头铺作透视图
Perspective view of Zhutou Puzuo of main hall's Xuanlang

大殿轩廊柱头铺作平面仰视图
Bottom view of Zhutou Puzuo of main hall's Xuanlang

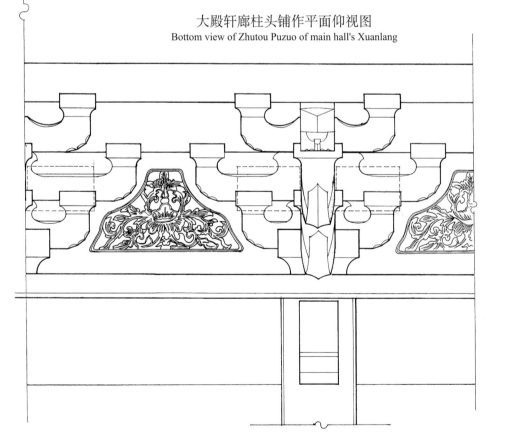

大殿轩廊柱头铺作正面立面图
Front elevation of Zhutou Puzuo of main hall's Xuanlang

大殿轩廊柱头铺作剖面图
Section of Zhutou Puzuo of main hall's Xuanlang

0  0.06  0.12      0.3m

大殿轩廊补间铺作透视图
Perspective view of Bujian Puzuo of main hall's Xuanlang

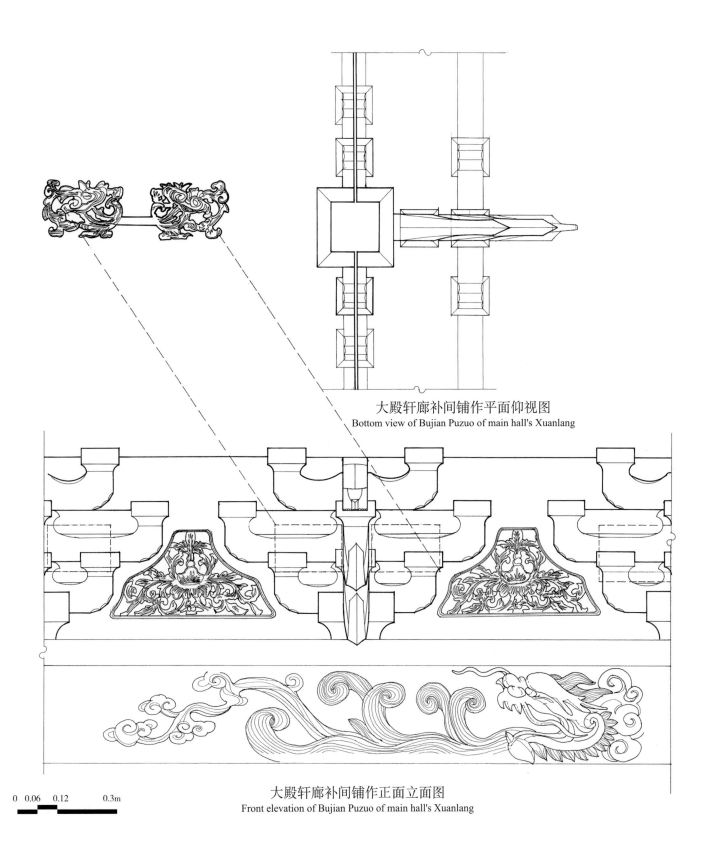

大殿轩廊补间铺作平面仰视图
Bottom view of Bujian Puzuo of main hall's Xuanlang

0    0.06    0.12        0.3m

大殿轩廊补间铺作正面立面图
Front elevation of Bujian Puzuo of main hall's Xuanlang

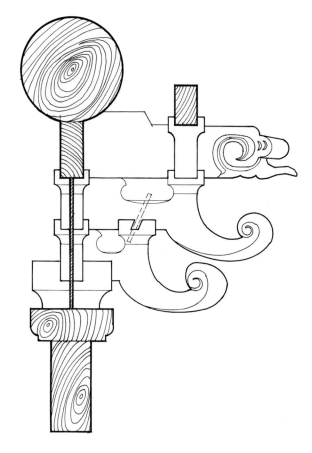

大殿轩廊补间铺作剖面图
Section of Bujian Puzuo of main hall's Xuanlang

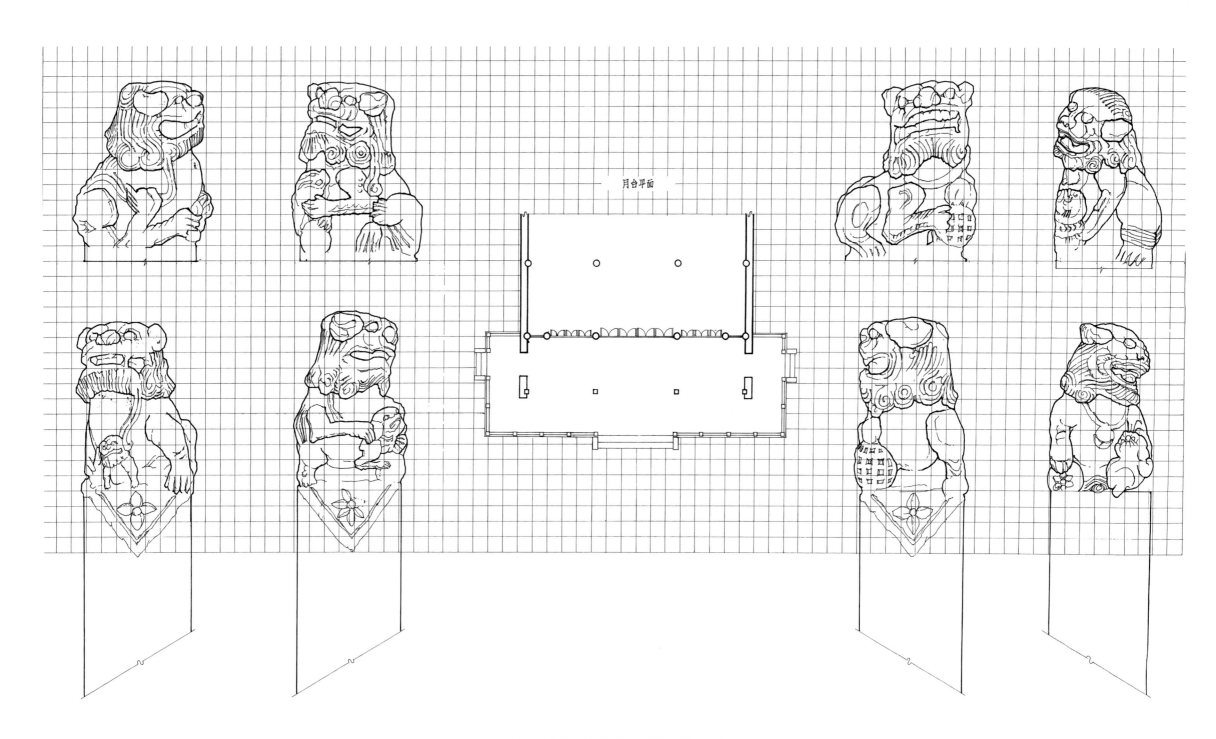

月台平面

大殿月台清代花岗石望柱石狮大样图
Qing Dynasty Huagang Stone Yuetai and Wangzhu lions of main hall

0  0.03  0.06      0.15m

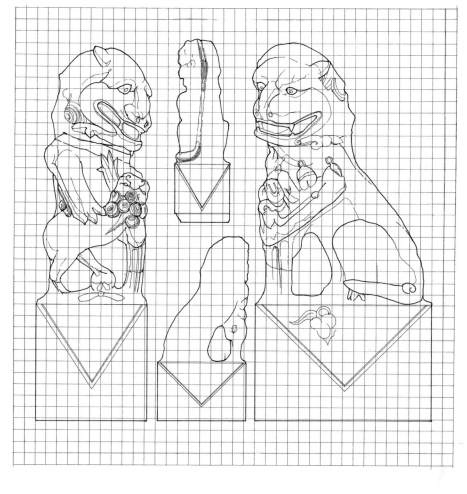

0    0.15    0.3        0.6m

大殿轩廊月梁雕花大样图
Yueliang Diaohua of main hall's Xuanlang

庙门前明风青石雌狮立面图
Elevation of Ming style Qing Stone female lion

寝宫梁架透视图
Perspective view of back palace's beams

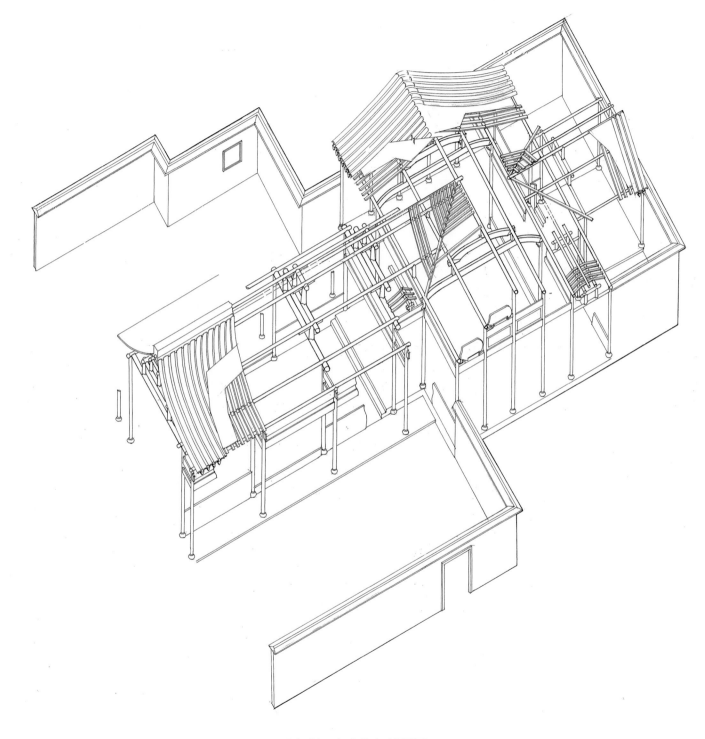

工字廊、寝宫构架轴测图
Axonometric drawing of I-shaped hall and back palace's structure

0　2　4　　　　　10m

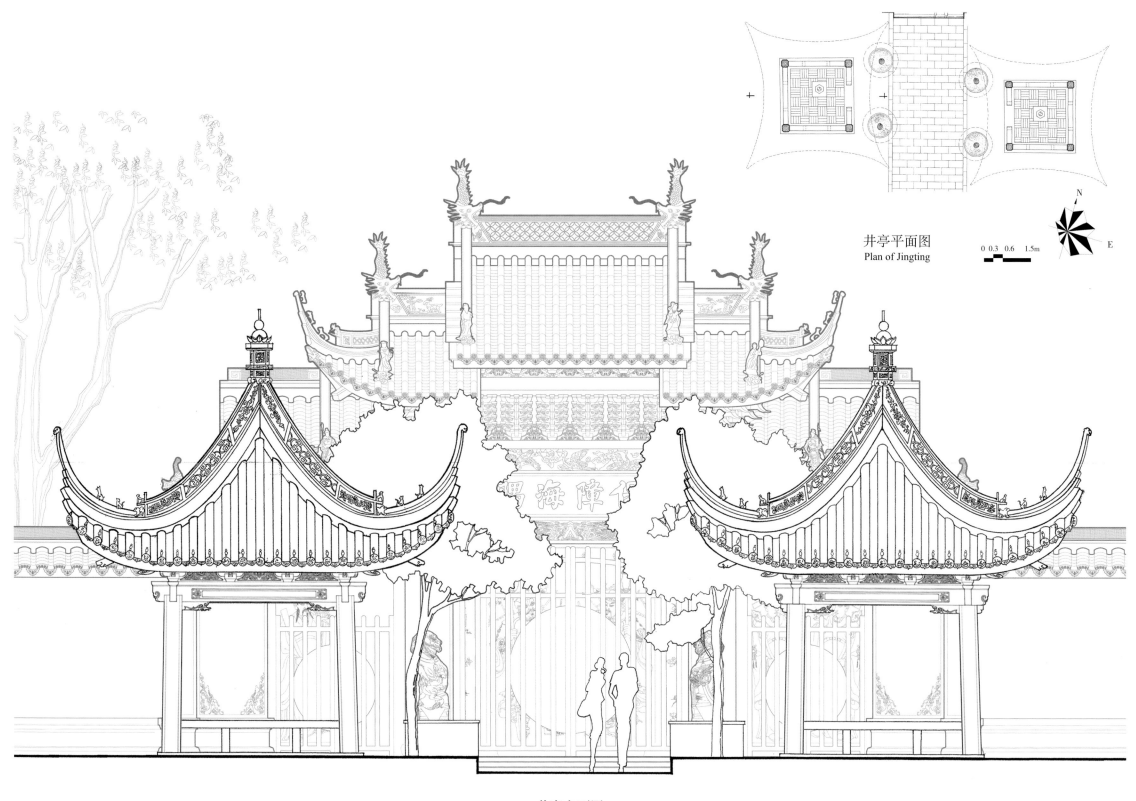

井亭平面图
Plan of Jingting

0 0.3 0.6 1.5m

注：庙门已毁，据上海县城隍庙示意

井亭立面图
Elevation of Jingting

0 0.3 0.6 1.5m

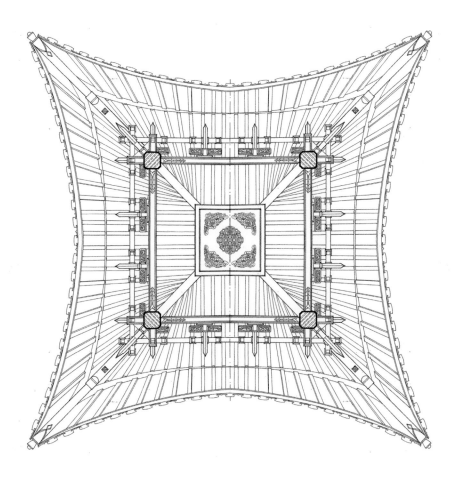

井亭屋架、藻井仰视图
Bottom view of Jingting's beams and Zaojing

西侧井亭剖面图
Section of the west Jingting

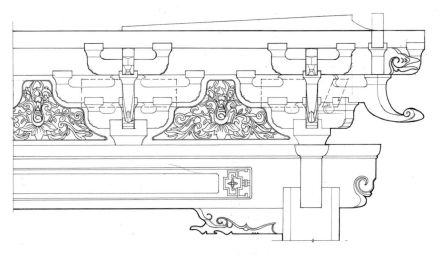

0  0.3  0.6     1.5m

井亭补间及转角铺作立面图
Elevation of Jingting's Bujian and Zhuanjiao Puzuo

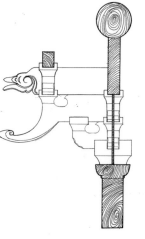

井亭补间铺作侧面图
Side view of Jingting's Bujian Puzuo

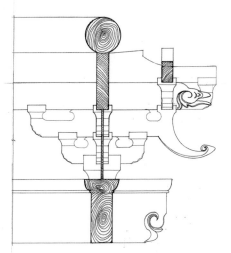

井亭转角铺作侧面图
Side view of Jingting's Zhuanjiao Puzuo

**Item of survey:** Qingpu County Chenghuang Temple and Temple Garden （Qushui Garden）

**Address:** No612, Gongyuan Road, Qingpu District, Shanghai

**Age of construction:** 1745 (year 10 of Qianlong Emperor of Qing Dynasty)

**Site area:** More than 30 mu

**Competent organization:** Management Office of Qingpu Qushui Garden

**Survey organization:** College of Architecture and Urban Planning, Tongji University

**Time of survey:** 1999

测绘项目：青浦县城隍庙及庙园（曲水园）

地　　址：青浦城厢镇公园路六五〇号

始建年代：清乾隆十年（一七四五年）

占地面积：三十余亩

主管单位：青浦具曲水园管理处

测绘单位：同济大学建筑与城规学院

测绘时间：一九九九年

Qingpu County Chenghuang Temple
and Temple Garden (Qushui Garden)

青浦县城隍庙及庙园（曲水园）

# Introduction

Qingpu Chenghuang Temple Garden was located in the original north-east corner of the turtle-shaped Qingpu County, covering an area of nearly 30 mu. The overall layout was the temple in the west and the garden in the east (the garden was called Qushui Garden after 1796-1820), standing across the river called Yinjiadou. On the east and north sides, it leant against the city wall, and beyond the wall was the vast and magnificent moat, which connected Dayingpu in the north and directly flowed into Wusongjiang, with ships sailing and bird flying.

Qingpu Chenghuang Temple was built in 1578, covering an area of nearly 4 mu. The temple was used to worship SHEN En, who took the position of Sichuanzuobuzhengshi in the Ming Dynasty. Today, only the second gate on the middle axis (Yimen, which had Xilou in the back side), which was rebuilt in 1884, and Niangniangdian on the east axis remained. The first gate, the main hall and the back palace were all destroyed, while the magnificent and colorful temple seen today were mostly rebuilt in recent years.

Qushui Garden in the east was established in 1745, specializing in amusing the Chenghuang Gods. It was full of city temperament since it was first built. And after the rapid expansion, it covered an area of nearly 27 mu. With the abundant and colorful scene in the garden, it gradually became a Shanghai style garden. Today, the whole garden still retain the landscape of the middle Qing Dynasty, while the architecture was all reconstructed in the late Qing Dynasty. And more new buildings were added in recent years, which started a new trend. However, on the west side of the city river, more than sixty meters of Qing Stone barge under Deyuexuan, is still the relic of 1745 when Qushui Garden was first built. Youjuetang in the west corner of the garden also seems to retain Qing Stone drum-shaped foundation of the same period. Ninghetang in the middle part of the garden, together with Dianchuntang in Yu Garden, can be regarded as a magnificent building in the garden halls of the late Qing Dynasty.

青浦县城隍庙园建筑群位于原龟背形的青浦县城东北隅，占地约三十亩，西庙东园（清嘉庆后称『曲水园』），跨市河『殷家兜』而峙，其东、北两面倚靠城垣，巍巍一墙之外，即是宽阔浩荡的城濠，北接大盈浦，风帆白鸟，直入吴淞江。

其中城隍庙创建于明万历六年（一五七八），占地近四亩，祀主为明四川左布政使沈恩。今仅存清末光绪十年（一八八四年）重建的中路二门（即仪门，背面即戏楼）与东路娘娘殿，其余头门、大殿、寝宫等建筑均废，今日所见堂皇斑斓之庙貌多系近年重建。

东部曲水园系清乾隆十年（一七四五年）专为城隍之神游目骋怀而建，自立身之日起，即饱含市井色彩，叠经扩充后，占地近二十七亩，景象缤纷罗列，渐具『海派园林』之风——今全园犹存清中叶山水格局，建筑则纯系晚清重构，近年更不断添筑，别开新风。但市河西侧、得月轩下六十余米青石驳岸，至迟还是清乾隆十年（一七四五年）建园之初的遗物；园西『有觉堂』亦似存留同时期青石鼓形柱础；园中部凝和堂则堪与上海豫园点春堂并称晚清园林厅堂中的磅礴巨制。

图一　青浦县曲水园核心景域总平面图（朱宇晖 1997 年步测测绘制，部分系推测复原）

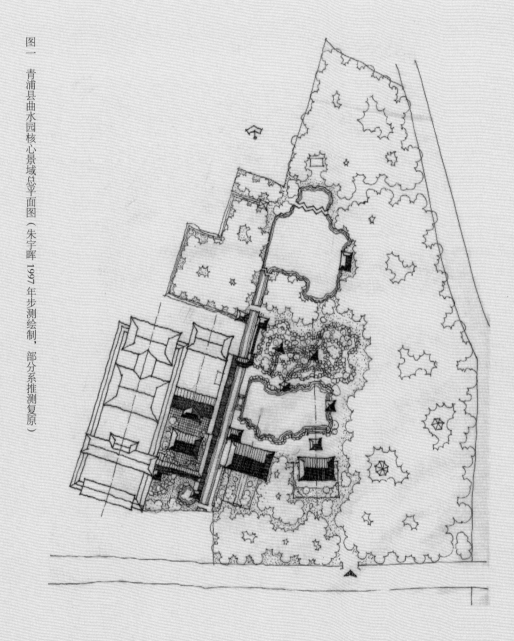

Fig.1  Master plan of Qingpu County Qushui Garden's core area, (drawn by ZHU Yuhui in 1997, measured by steps, partly restored).

## Past Life in the Tang and Song Dynasties and Present Life of a Marketplace

In 1542 in the middle and late Ming Dynasty, the rising Qingpu County was established in the tide of economic recovery in Jiangnan area. The government of Qingpu County was initially established in Qinglong Town, which was a famous port and a giant town in the south-east area in the Tang and Song dynasties. Following the trend, the Longping Temple, a famous temple in the north of the county, was changed into Chenghuang Temple of Qingpu County. However, with Wusong River silted up and the coast line moved east, the famous town that once owned 'three pavilions, seven towers and thirty-six workshops', and governed by the great calligrapher MI Fu, which can be called the former life of Shanghai Port, had finally ceased to be prosperous, and even abandoned the county for a while. Till 1573, when Qingpu County was rebuilt, it had to be moved to the current site where five rivers met. In 1577, with a circumference of only 8 Li, Qingpu City Wall, which is more exquisite and pocket-sized than Shanghai County, rose from the ground. The next year, the new Chenghuang Temple was completed. It was said that the official plan to relocate the old Chenghuang Temple of Qinglong Town to the new site was failed due to opposition from the people in the town. But the new temple leaning against a majestic wall to the north and facing the river and overlooking the sea, continued the heritage of the famous port to some extent.

In 1633 in the late Ming Dynasty, the thoughtful emperor Chongzhen honored newly passed away Sichuanzuobuzheng SHEN En (1472-1533) as Qingpu Chenghuang God, thus continuing the delicate relationship between Qingpu City and Shanghai Yu Garden. Shen En achieved *Jinshi* in 1496 and he was from Shanghai County, the former owner of the residence of PAN Yunduan, who was the owner of Yu Garden.

# 唐宋前世与市井今生

明中后期的嘉靖二十一年（一五四二年），在江南经济复苏的春潮中，后起的青浦县应运而生，县治起初设立于唐宋名港、东南巨镇青龙镇，并因势将镇北名刹隆平寺改为县城隍庙。但随着吴淞淤塞，海岸东移，曾拥有『三亭七塔三十六坊』、以大书画家米芾为『镇监』、堪称上海港前世的皇皇名镇终究不复旧日繁华，竟至一度废县弃治。待到万历元年（一五七三年）重新恢复青浦县建置时，才不得不将县城迁至五浦交汇、曲水萦绕的今址。万历五年（一五七七年），周长仅八里，比上海县更加玲珑袖珍的青浦城垣拔地而起，翌年，新城隍庙落成——据说官方原拟将青龙镇上的城隍故庙迁来新城，因镇民反对而未遂。但新庙北倚巍垣，临江眺海，也算接续起了一代名港的数百年传承。

明末的崇祯六年（一六三三年），心事重重的崇祯帝册封刚刚故去的四川左布政沈恩（一四七二至一五三三年）为青浦城隍，又接续起青浦一城与上海豫园的微妙关系——此公系弘治九年（一四九六年）进士，上海县人，豫园主人潘允端宅第的前任主人。

图二　上海青浦县城隍庙曲水园『天光云影』水岸全景

图四　一度为青浦县城隍庙东侧界河，最终被跨越而成为庙园内河的『殷家兜』

图三　上海青浦县城隍庙曲水园新建大门及围廊

Fig.2　A panoramic view of Shanghai Qingpu County Qushui Garden's 'Tianguangyunying' lake and it bank

Fig.3　The newly-built gate and its corridor of Shanghai Qingpu County Qushui Garden

Fig.4　'Yinjiadou', the original east boundary river of Shanghai Qingpu County Chenghuang Temple, which was eventually crossed to become the inland river of Qushui Garden

In 1745, under the influence of opening the sea during 1662-1722, Qingpu County also established a temple garden in the east of its Chenghuang Temple (the west part of Qushui Garden nowadays). At that time, it had been 36 years since Shanghai County Chenghuang Temple's East Garden was built, and 19 years since Jiading County Chenghuang Temple received the Qiuxiapu Garden. In the middle and late Qianlong period, with the conservatism of the society, the accumulation of wealth, the rise of civic awareness, and the increasing demand for urban public space, the garden gradually developed eastward, crossing Yinjiadou, and reaching the city wall. Therefore, from the deep courtyard only for pilgrims to rest, the garden gradually developed and formed the landscape of mountains and rivers, enough for the county people to wander. At the same time, Shanghai County Chenghuang Temple accommodated the east Yu Garden owned by the PAN family to form a "super temple garden", and Jiading County Chenghuang Temple also accommodated the north Shenshi Garden.

The expansion and maintenance costs of the garden come from the 'Yi Wen Juan', every person of the county donated a little money. When it was rebuilt in the late Qing Dynasty, the government also collected money from the people's land tax to raise funds. Therefore, the garden was also known as Yiwen Garden, which reminds people of the Yiwen Hall built by Suzhou citizens at the end of the Ming Dynasty a hundred years ago for the wise officials who fought against eunuchs. This made this garden different from pure scholar-official gardens such as Jiading Qiuxiapu Garden from the beginning of its establishment. It was full of public genes and sentiments from the market, and embodied the aesthetic taste of the merchants and citizens. It has become a dazzling representative of the early Shanghai-style gardens.

清乾隆十年（一七四五年），『康熙开海』的经济涟漪所及，青浦县亦于城隍庙东创建庙园（今曲水园河西部分），此时前距上海县城隍庙筑东园已经三十六年、距嘉定县城隍庙收纳秋霞圃亦已十九年。至乾隆中后期，社会因循，财富累积，市民意识崛起，城市公共空间需求增大，此园亦渐渐向东发展，跨越殷家兜，直至城垣脚下。由庭院深邃，仅容香客小憩，而渐至山水横陈，足供县民徜徉——同一时刻，东邻上海县城隍庙收纳了潘氏豫园，形成『超级庙园』，北邻嘉定县城隍庙则收纳了沈氏园。

此园扩建与维护费用来自县民每人一文的『一文捐』，而清末重修时，亦曾『于地丁项下每钱加纳制钱一文』以筹经费，故此园又有『一文园』之称，让人油然忆起百年前明末苏州市民为抗击权阉的贤臣建造的『一文厅』。这令此园从立身之初，即与纯粹的士大夫园林如嘉定秋霞圃有别，而饱含着公众基因与市井情怀，较多体现着商贾市民阶层的审美趣味，成为早期『海派园林』的夺目代表。

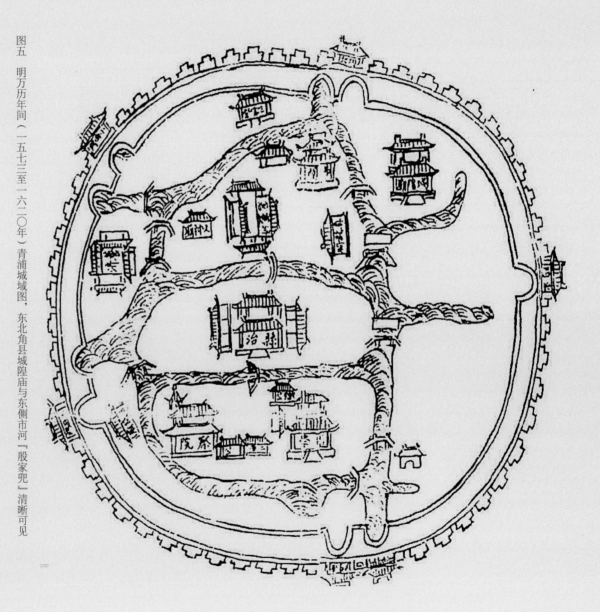

图六　青浦县城隍庙曾改建白青龙镇古刹隆平寺，此即隆平寺北宋柱础

图五　明万历年间（一五七三至一六二〇年）青浦城域图，东北角县城隍庙与东侧市河「殷家兜」清晰可见

Fig.5　The map of Qingpu County in 1573－1620, in which Qingpu County Chenghuang Temple in the north－east corner and the city river "Yingjiadou" in the east are very clear

Fig.6　Qingpu County Chenghuang Temple was once rebuild on the original site of Longping Temple in Qinglong Town. This is the column base of Northern Song Dynasty of Longping Temple.

## Deep Courtyard with Curved Water, Leaning against River and Facing to Mountains, the Age of 'God' in the Early Qing Dynasty

The overall formation of Qushui Garden can be roughly divided into two stages.

The first stage is the age of 'god', which characterized by the deep courtyard with curved water, leaning against river and facing to mountains in the west part of the garden. The first generation of Qushui Garden (there was no name at the time) in the early years of Qianlong period should be rebuilt from the original buildings on the east axis of Qingpu County Chenghuang Temple, which was only slightly interspersed, twisted and softened parallel to the strict middle axis.

At that time, the buildings included Youjuetang, Gexunlou, Yinghuige and so on. Judging from the current appearance, Youjuetang should be at the present site, and it was the main hall of the whole garden. There was Yushulou in the north, Gexunlou maybe in the south (suspected that it should be a theatre building facing Youjuetang in the north, later renamed or rebuilt as Yingxiange), and Yinghuige on the east side (suspected that it was rebuilt into Xiyanghongbanlou afterwards). The south of the building was the main scene of the whole garden. It was only next to Yinghuige 'chiseling a small pool, stacking stones and building bridges on the clear water', which formed a small 'facing the mountain across the water' layout.

The east wing of the whole garden was a line of waterside pavilions close to the city river Yinjiadou, namely Deyuexuan Building, which imitated the city building in water lane of Jiangnan area. The river was as straight as an arrow, and the building was also straight along the bank, together along the river. With its own necessary simplicity and monotony, the landscape on both sides was set off particularly bright and moving. Stepping up the building, you can look around the whole scene, leading the limited garden space to the open woodland in the east of the river and the lonely moon at the top of the city.

# 清前期深院曲水、倚河面山的『附神』时代

曲水园之全局形成可大致分为两个阶段。

第一阶段为河西深院曲水、倚河面山的『附神』时代——乾隆初年的第一代曲水园（当时未有其名）应改建自县城隍庙东路的原有附属建筑群，只在与中路平行的森严轴线上，稍作穿插、扭转与柔化。

当时建筑包括有觉堂、歌薰楼、迎晖阁等前后数进，结合今貌判断，有觉堂应即在今址，为全园主要厅堂，其北有御书楼，南面或即歌薰楼（原应为面向北侧之有觉堂的戏楼，后改称或改建为迎仙阁），并连接面东的迎晖小阁（疑后改建为兼顾东西的『夕阳红半楼』）。楼南即为全园主要景象空间，仅于迎晖阁旁『凿小池，叠石池中，桥架碧水』，形成小型的『隔水面山』之局。

全园东厢，则为紧临市河『殷家兜』的一线『近水楼台』，即仿江南水巷市楼的『得月轩』楼。河道笔直如矢，楼身亦傍河直行，共河而远，以自身必要的质朴单调，将两侧景象映衬得格外明艳动人。登楼则可推蓬眺远，临流顾盼，将有限的园内空间引向河东的疏林旷地与城头的孤月。

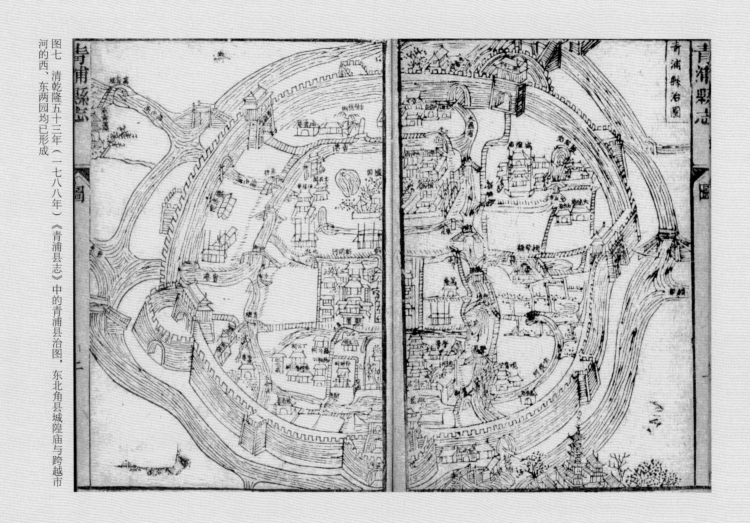

Fig.7  The map of Qingpu County in 1788's *Qingpuxianzhi*, in which Qingpu County Chenghuang Temple and West Garden and East Garden cross the city river all took shape

20 years later in 1767, the residents of the county who have been enthusiastic about the construction of public space follow the trend and gradually built (or maybe reconstructed) the Xiyanghongbanlou in the main area of the south of Youjuetang, and become the visual focus of the garden. In addition, a freehand boat was built at the south end of Deyuexuan Building, connecting the building as its tail cabin, giving the original building the double meaning of 'playing the moon' and 'rafting on the river', making it a dazzling stroke of straight rivers, traversing the entire garden, wading through mountains and rivers, and dividing flowers and willows, concise and colorful, full of the flavor of the times.

Due to the overlapping of the halls and buildings, leaving no gaps in the west area, the construction of this period finally crossed the city river, and Ninghetang (the volume at that time should not be as huge as today) was built in the idle land near the city wall in the east of the river, laying the groundwork for the overall eastward expansion in the future. After the large-scale eastward expansion of Qushui Garden across the river, this area was called West Garden.

二十余年后的乾隆三十二年（一七六七年）间，公共空间建设热情未衰的县民又踵事增华，逐步在有觉堂南的主体景象空间内兴建（改建）『夕阳红半楼』，或成园中视觉焦点；并于『得月轩』楼南首增建写意型船舫，连接轩楼作为其尾舱，赋予原有轩楼以『玩月』与『泛舟』的双重意涵，使其成为襟带直河、纵贯全园、跋山涉水、分花拂柳的华彩一笔，斩截利落而又缤纷多姿，富含时代气息。

由于河西堂楼重叠，再无隙地，这一期建设最终跨越市河，在河东贴近城垣的闲散之地兴建了凝和堂（当时体量应不若今日之巨），为未来的整体东扩埋下了伏笔。在曲水园大规模跨河东扩后，这一区被称为『西园』。

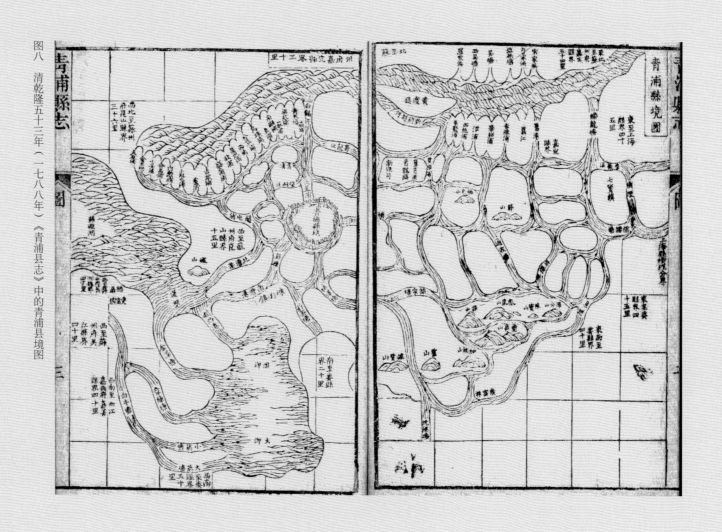

图八　清乾隆五十三年（一七八八年）《青浦县志》中的青浦县境图

Fig.8  The map of Qingpu County and its surrounding area in 1788's *Qingpuxianzhi*

## Crossing the River into the Market, Facing the Mountain across the Water, the Age of 'City' in the Prosperous Qing Dynasty

The second stage is the age of 'city', which characterized by the layout of crossing the river into the market and facing the mountain across the water. In 1784, when Shanghai County residents completed the reconstruction of the large-scale Chenghuang Temple West Garden, and raised two thousand taels of silver to build the radiant city landmark Huxin Pavilion which still remains today, the residents of Qingpu County was also unwilling to be left behind. They enthusiastically sculpted a large-scale pond and mountain in the east of the city river, so that the temple garden's landscape and space were completely free from the divine shackles of the courtyard and the axis, blooming colorfully near to the bright world of the city.

The layout of this area was centered on the large earth-rock mountain Xiaofeilaifeng. The north and the south gather water to form a pool, connected the city river in the west, and surrounded the mountains from three sides, giving the mountain the landmark meaning just like Huxin Pavilion. The mountain top borrowed the scene from Jiufeng Yanlan and Yingpu Fengfan outside the city.

The north of the mountain, which was relatively minor was treated as the deep water scene, featuring big mountains and swamp. The south of the mountain, as the main scene, still inherited the classical layout of 'main hall facing the mountain across the water'. Ninghetang and Huashenci (which was abandoned after the Republic of China, called Huashentang) stood side by side on two axes, facing the street and in front of the water. They were grave and solemn, feeling more like temples and lack interaction with the landscape. Actually, the main hall was the 'Huangduifeilai' (commonly known as Hehuating, which was rebuilt as a pavilion during the Anti-Japanese War) in the north of Huashenci, which penetrated into the main water surface in a high profile. The location of Yongcuiting at the south end of the long bank was quite similar to Biguangting in Jading Qiuxiapu Garden, but it deliberately retreated away from the water and hided in the shade, making the north bank lean against the mountain and the pond, and the beautiful Xiaohaoliang (Pavilion) a well-deserved visual focus in the deep green plants around the pond. Although the overall scene management was quite complicated, there were still distinctions between the primary and the secondary, corresponding the classic style.

# 清盛期跨河入市、隔水面山的『附城』时代

曲水园的第二个布局阶段为跨河入市、隔水面山的『附城』时代——乾隆四十九年（一七八四年），在上海县民完成规模浩大的城隍庙西园重建，并集资白银二千两，兴建起光芒四射、留存至今的城市地标『湖心亭』之际，青浦县也不甘人后，于市河以东大规模凿池掇山，令庙园景象空间全面挣脱了深院与轴线的神性桎梏，缤纷绽放在更贴近城市的明朗天地。

此区布局系以大型土石掇山『小飞来峰』为核心，南北各自汇水成池，而以西侧市河勾连，自三面环拥山体——令此山也仿佛具有『湖心亭』般的地标意味。山巅则借景于城外九峰烟岚、盈浦风帆。

其中相对次要的山北被处理为深山大泽的渊潭景象，山南则为主体景象空间，仍承继着古典时代主要厅堂『隔水面山』的布局。其中凝和堂与花神祠（民国后废祠，称『花神堂』）双轴并立，面街背水，庄敬端肃，庙堂气重而烟云气少，实际并未与山水发生互动。真正意义上的主要厅堂系花神祠北、高调突入主体水面的『恍对飞来』（俗称『荷花厅』，二十世纪四十年代被重建为亭）。而长堤南首的涌翠亭位置颇似嘉定秋霞圃之碧光亭，却刻意退离水面，隐入浓荫，使北岸倚山临池、盈盈俏立的『小濠梁』（亭）成为池周一片浓绿中当之无愧的视觉焦点——其景象经营虽觉繁密，但尚有主次之分，不失古典遗范。

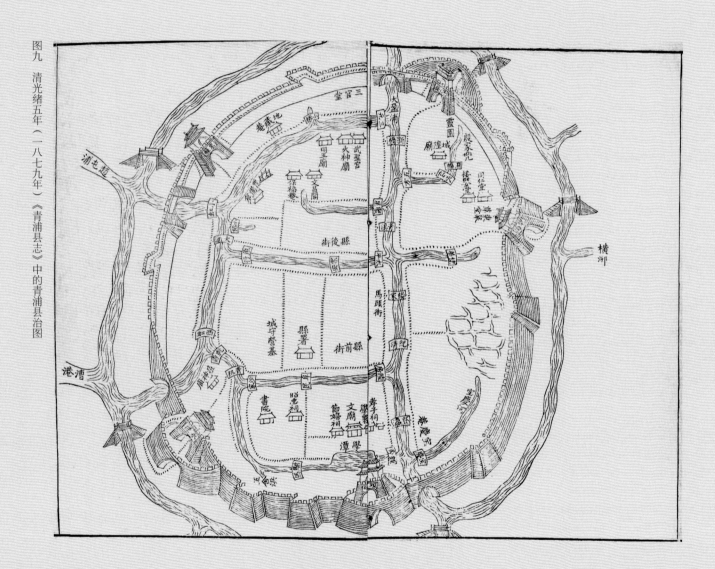

Fig.9  The map of Qingpu County in *Qingpuxianzhi*, 1879

It was particularly eye-catching that on the west side of the water (named Tianguangyunying) in this area, there was a long bank running from the north to the south, which separates it from Yinjiadou. The river outside the bank was straight, but the river was still a river, while the river inside the bank was full of sight, just like a big lake. There was a high arch bridge on the bank, which could pass the boat. In this way, the scene of the river and lake separated by the bank and bridges blended the different hints of painted bridges, willow bank and narrow bridges, and takes into account the convenience of land travel and boat travel, showing the wonderful interpenetration between the temple garden and the city, thus occupying a unique position in the city gardens in Jiangnan area.

In 1798, when the scenic garden was first completed, the magistrate of Qingpu County once hosted LIU Quanzhi, who was a Xuezheng, in the garden. This future Tirengedaxueshi borrowed the allusion of Lanting-Yaji and Qushui-Liushang, and named the garden 'Qushui Garden'. This so-called 'using the new title to destroy the evil' allowed the garden to be incorporated into the mainstream literati garden's discourse system in addition to 'god' and 'city' spirit.

尤为引人注目的是，此区水面（名『天光云影』）西侧，有长堤南北纵亘，使其与『殷家兜』相隔——堤外一水修直，依然河道，堤内满目弥漫，宛然湖泊。堤上尚有拱桥高耸，可通舟楫。这样以堤桥分隔河湖的景象，交融了画桥柳堤与纤桥纤道的不同暗示，兼顾了陆游、舟游之便，展现出庙园与尘世间的美妙互渗，在江南城市地园林中别具一格。

佳景初成的清嘉庆三年（一七九八年），青浦县令在园中款待学政刘权之，这位未来的体仁阁大学士借晋人兰亭雅集、曲水流觞的韵事，命园名为『曲水园』。所谓『顿教题额新，并将恶札毁』，让此园在『附神』和『附城』之外，又被不动声色地纳入了主流文人园林的话语体系。

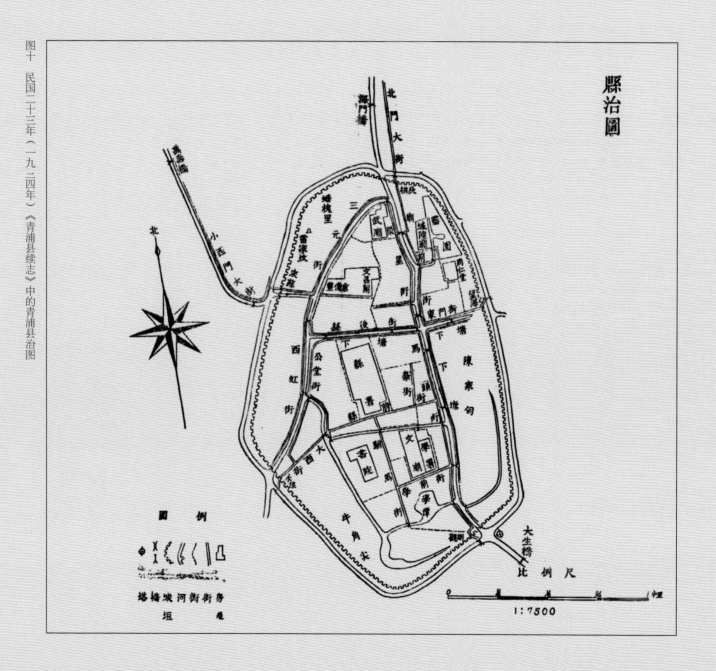

縣治圖

Fig.10 The map of Qingpu County in *Jiadingxianxuzhi*, 1934

## A Pool with Four Pavilions, the Age of 'Shanghai-Style' since the Late Qing Dynasty

With the long-lasting expansion and decoration lead by the residents in Qingpu County, Qushui Garden gradually formed twenty-four scene in high density. The number of buildings and scene density were both quite amazing. For example, Huanbilou was added to the top of Xiaofeilaifeng and Ying (Yin) xiting was built on the east bank of Tianguangyunying. In fact, from the description of 1784, 'There was Yongcuiting, Zhuojin Stone and Xiyu Bridge on the bank. From the foothills to the top, there were rocks, peaks and lush trees. Along the way there were Canxiating, Xieyuelou, Yinxiting and Yulexuan', we can see that at that time, the top of the mountain seemed to have no buildings, and Yinxiting seemed to be at the north of the mountain. But by this time, Yinxiting was already moved to the east bank of Tianguangyunying.

The war of Taiping Army in 1860 destroyed all buildings in the garden. Since 1883, the residents in the county had raised funds for the continuous reconstruction. After fifteen years, the building was finally renewed and the plants and trees bloomed again, laying the foundation for today's layout.

In 1911, the whole garden officially separated from Chenghuang Temple and became a purer citizen park. In 1927, it was renamed as Zhongshan Park. The gentry ZHANG Jingzhou renovated Xiaofeilaifeng, and his son later built a pavilion at the summit, borrowing the scene to be more lofty, and continuing to write the history of the glorious garden that the whole people care for and endure.

# 清后期以来一池四亭的『海派』时代

在长期持续的全民增饰中，曲水园逐步形成繁密的二十四景，建筑数量与景象密度都是颇为惊人的。

如小飞来峰巅增建了环碧楼，『天光云影』东岸建造了迎（寅）曦亭——其实从清乾隆四十九年（一七八四年）的描述看：『堤有亭曰「涌翠」，矶曰「濯锦」，石梁曰「喜雨」。由此而麓而巅，重峦复嶂，树木蓊郁，有亭曰「餐霞」，后障以楼曰「偕乐」、亭曰「寅曦」，轩曰「鱼乐国」从九曲廊而上，凭高纵目……』，则当时山巅似无楼阁，而『寅曦』亭似应在山北，而非今日山南池东之『孤亭翼然跂』了。

咸丰十年（一八六〇年）的『太平军』战事照例摧毁了园中所有建筑。晚清光绪九年（一八八三年）起，县民集资次第重建，历十五载，终于楼台鼎新，花木重光，奠定今日规制。

宣统三年（一九一一年），全园正式脱离邑庙，成为更纯粹的市民公园，民国十六年（一九二七年）更名为『中山公园』，邑绅张景周修复小飞来峰，其子后于峰巅建阁，借景更加高远，续写着全民呵护、耿耿长存的光荣园史。

图十一　民国二十三年（一九三四年）《青浦县续志》中的曲水园图，与现状最为接近

Fig.11  The image of Qushi Garden in *Jiadingxianxuzhi*, 1934, which is the closest one to the status quo

Huangduifeilai was destroyed in the anti-Japanese war. Japanese puppet government focused on partial changes and rebuild it into a pavilion. Therefore at that time, there were four pavilions, Xiaolianghao, Yingxi, Huangduifeilai and Yongcui, staying around the pool, making this area further present the appearance of a Shanghai-style garden with multiple competitions and sparse piles, completely saying goodbye to the classical era of clear rules and restraint. And for Qushui Garden, this may be a destiny that has been destined since the day of its birth.

If it is said that Songjiang-Zuibaichi Garden, which will be discussed later, seemed to be the lingering sound of the colorful movements of the Jiangnan literati gardens in the middle and late Ming Dynasty, then the Qingpu Qushui Garden showed the strong desire and aesthetic appeals for public recreational space by the rising citizens in the middle Qing Dynasty, indicating that the modern public gardens were in full swing before their debut.

抗日战争炮火中『恍对飞来』遭毁，日伪时着眼于局部的变化，将其改建为亭——盈盈一池畔，最终竟有小濠梁、迎曦、恍对飞来、涌翠四亭争胜，这令此区进一步呈现多元竞秀而又疏散堆砌的『海派园林』面貌，彻底挥别了章法分明而又精约克制的古典时代——这对曲水园来说，或许是从出生之日起就已注定的宿命吧。

如果说，后文将述及的松江醉白池仿佛是明中晚期江南文人园华彩乐章的袅袅余音，那么青浦曲水园则昭示着清中叶日益崛起的市民阶层对公众游憩空间的强烈渴望与审美诉求，宛然是现代公共园林粉墨登场前的紧锣密鼓了。

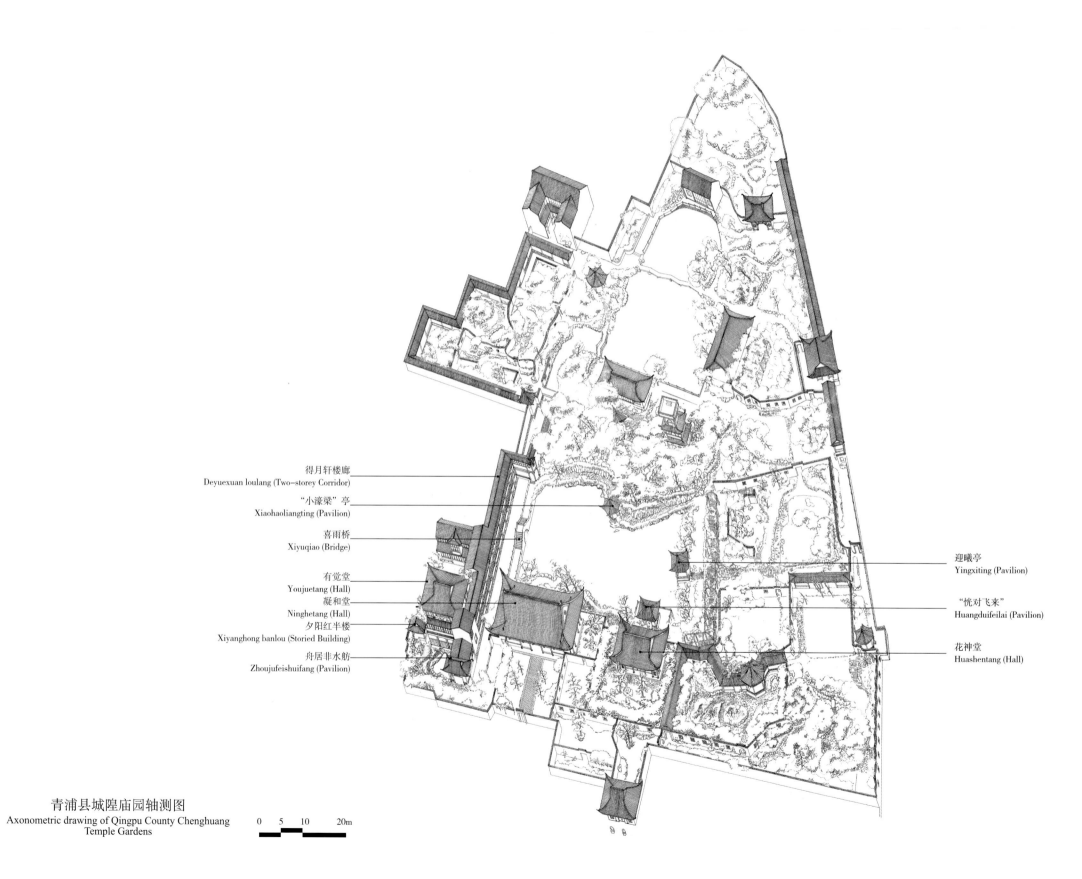

得月轩楼廊
Deyuexuan loulang (Two-storey Corridor)

"小濠梁" 亭
Xiaohaoliangting (Pavilion)

喜雨桥
Xiyuqiao (Bridge)

有觉堂
Youjuetang (Hall)

凝和堂
Ninghetang (Hall)

夕阳红半楼
Xiyanghong banlou (Storied Building)

舟居非水舫
Zhoujufeishuifang (Pavilion)

迎曦亭
Yingxiting (Pavilion)

"恍对飞来"
Huangduifeilai (Pavilion)

花神堂
Huashentang (Hall)

青浦县城隍庙园轴测图
Axonometric drawing of Qingpu County Chenghuang
Temple Gardens

0    5    10        20m

青浦县城隍庙园一层平面图
Ground floor plan of Qingpu County Chenghuang Temple Garden

0  5  10    20m

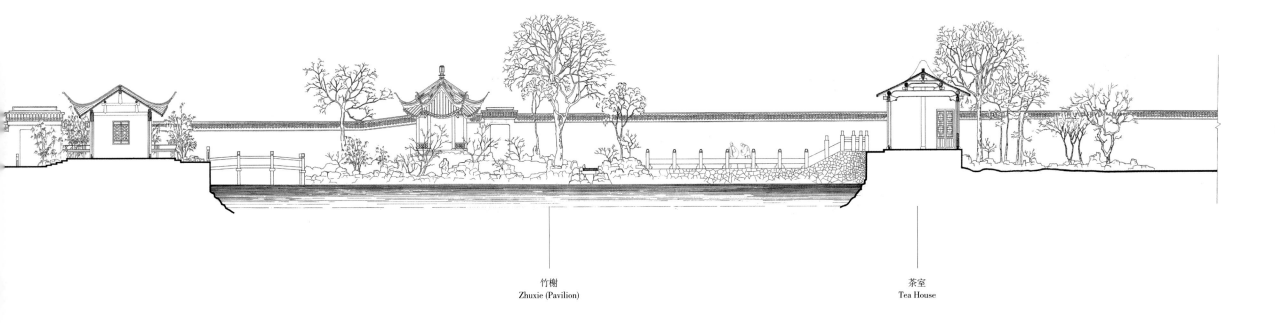

竹榭
Zhuxie (Pavilion)

茶室
Tea House

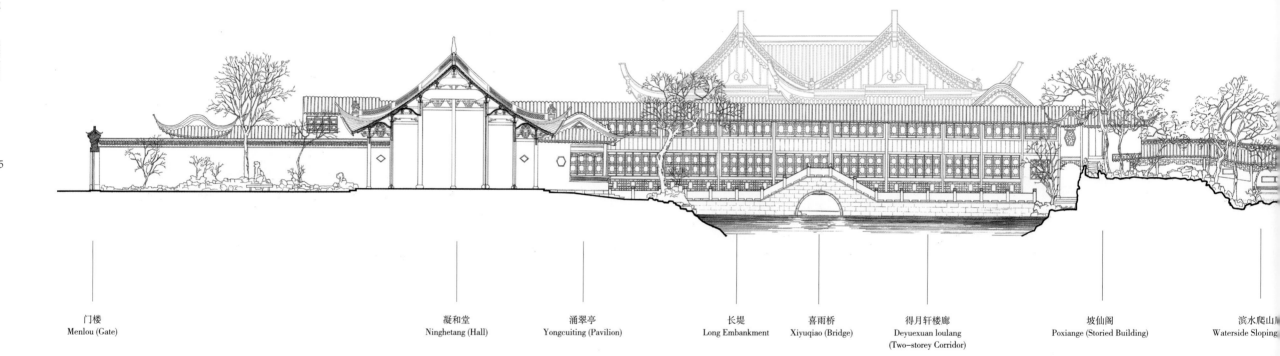

门楼
Menlou (Gate)

滨水爬山廊

凝和堂
Ninghetang (Hall)

涌翠亭
Yongcuiting (Pavilion)

长堤
Long Embankment

喜雨桥
Xiyuqiao (Bridge)

得月轩楼廊
Deyuexuan loulang
(Two-storey Corridor)

坡仙阁
Poxiange (Storied Building)

滨水爬山廊
Waterside Sloping

注：青浦县城隍庙大殿已毁，据嘉定县城隍庙示意

曲水园南北剖面图
South-north section of Qushui Garden

0    1    2         5m

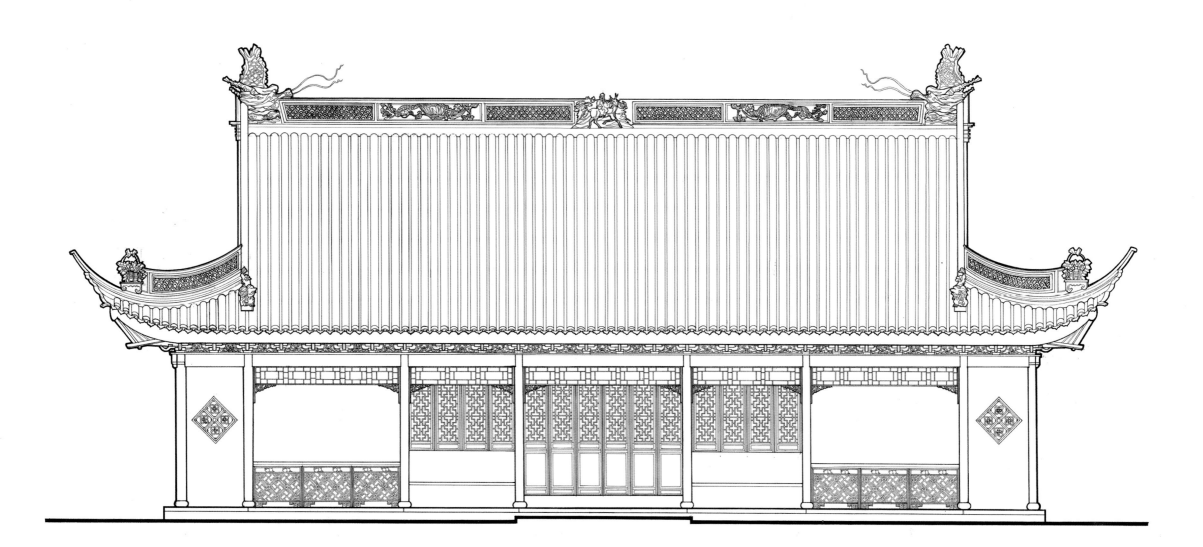

凝和堂南立面图
South elevation of Ninghetang

0  1  2  5m

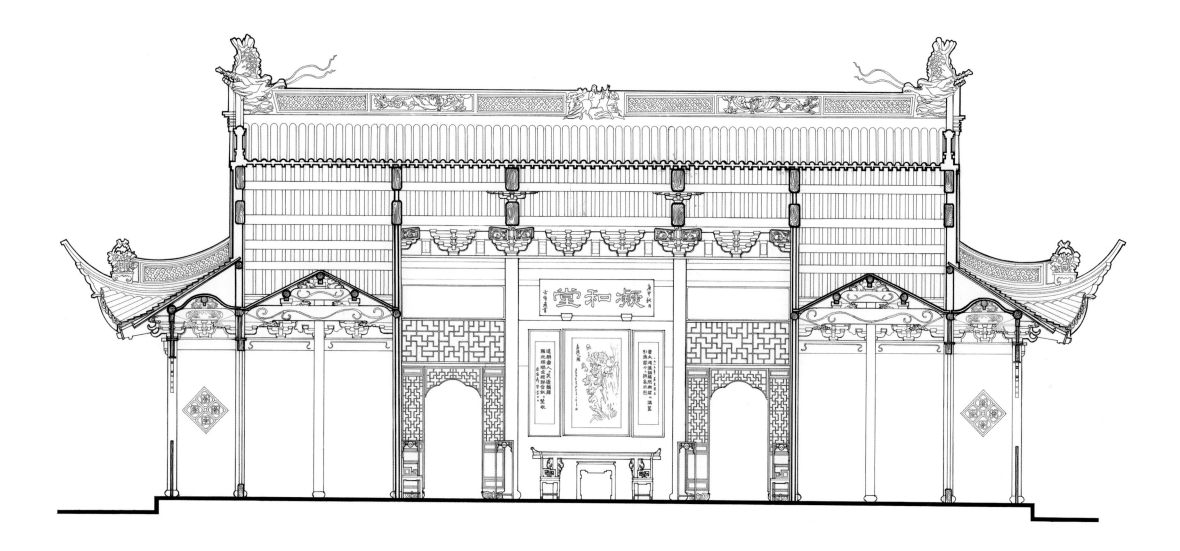

凝和堂纵剖面图
Longitudinal section of Ninghetang

0 1 2 5m

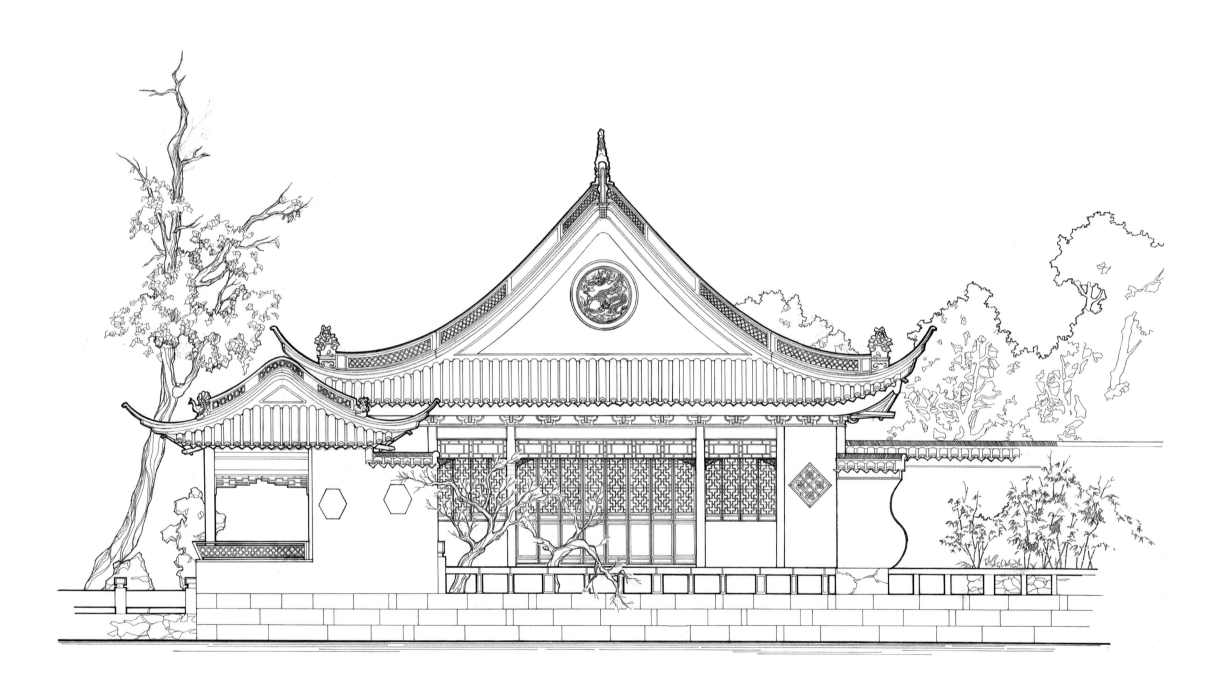

凝和堂与涌翠亭西立面图
West elevation of Ninghetang and Yongcuiting

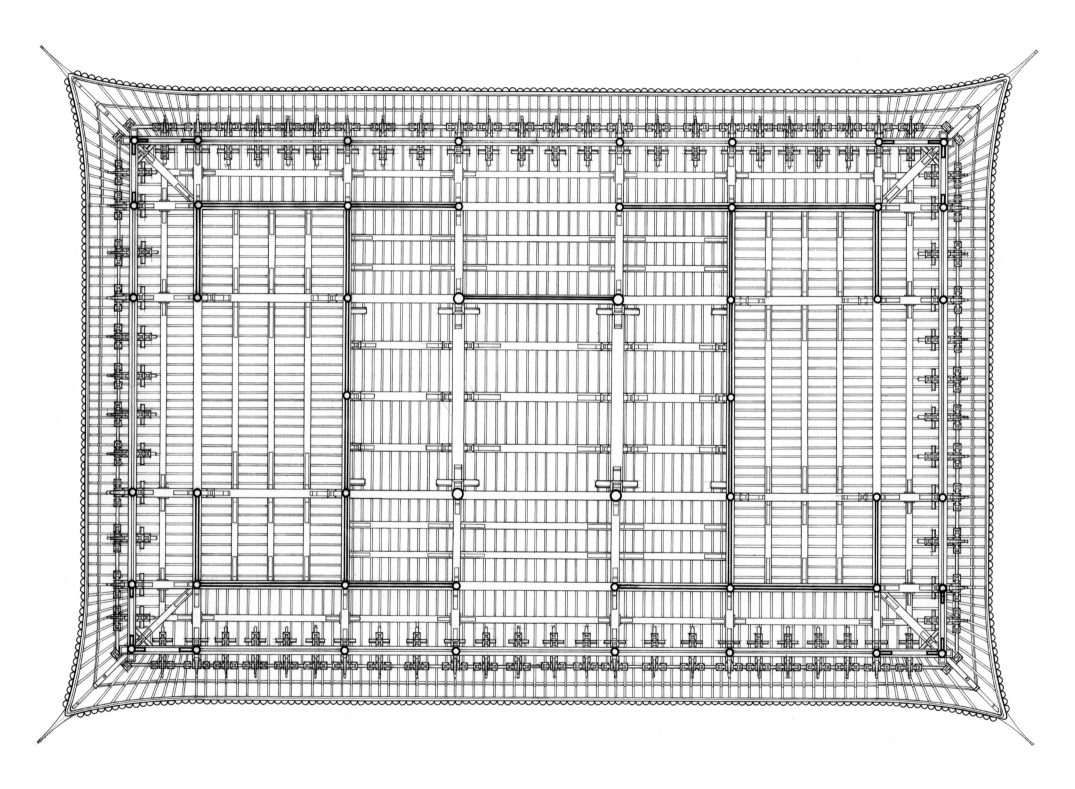

0 1 2 5m

凝和堂梁架仰视图
Bottom view of Ninghetang's beams

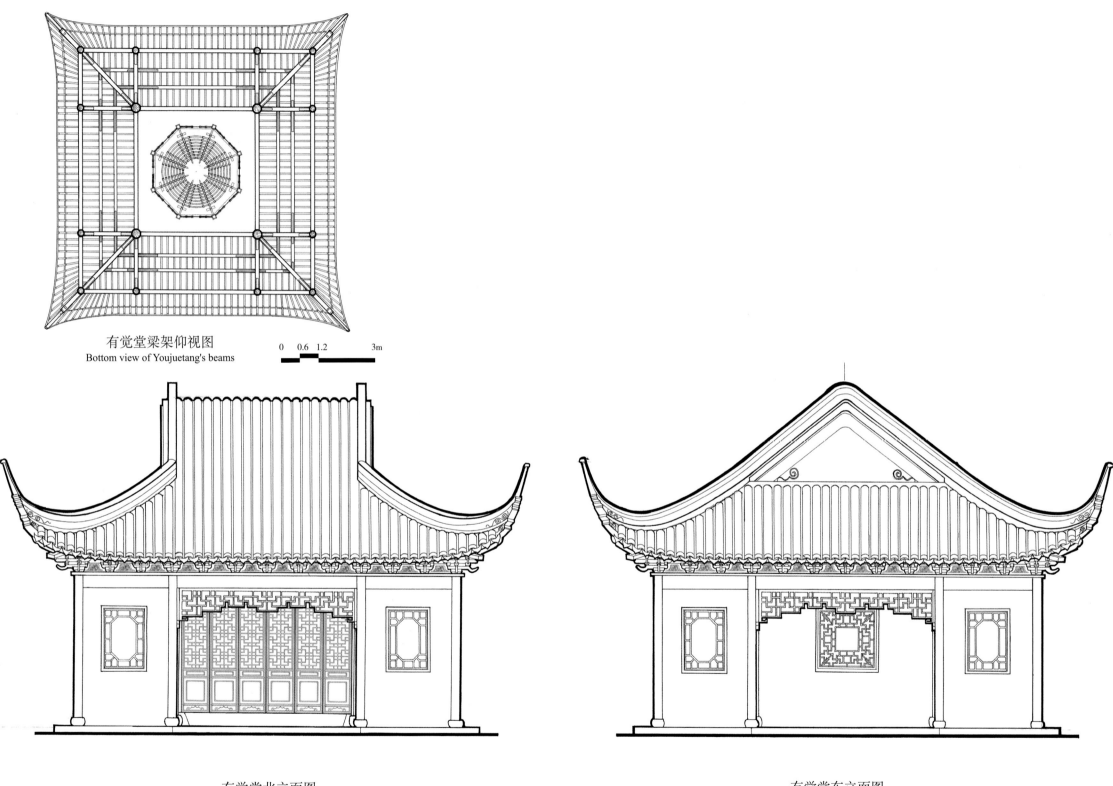

有觉堂梁架仰视图
Bottom view of Youjuetang's beams

0　0.6　1.2　　　3m

有觉堂北立面图
North elevation of Youjuetang

有觉堂东立面图
East elevation of Youjuetang

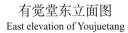

0　　0.6　　1.2　　　　　3m

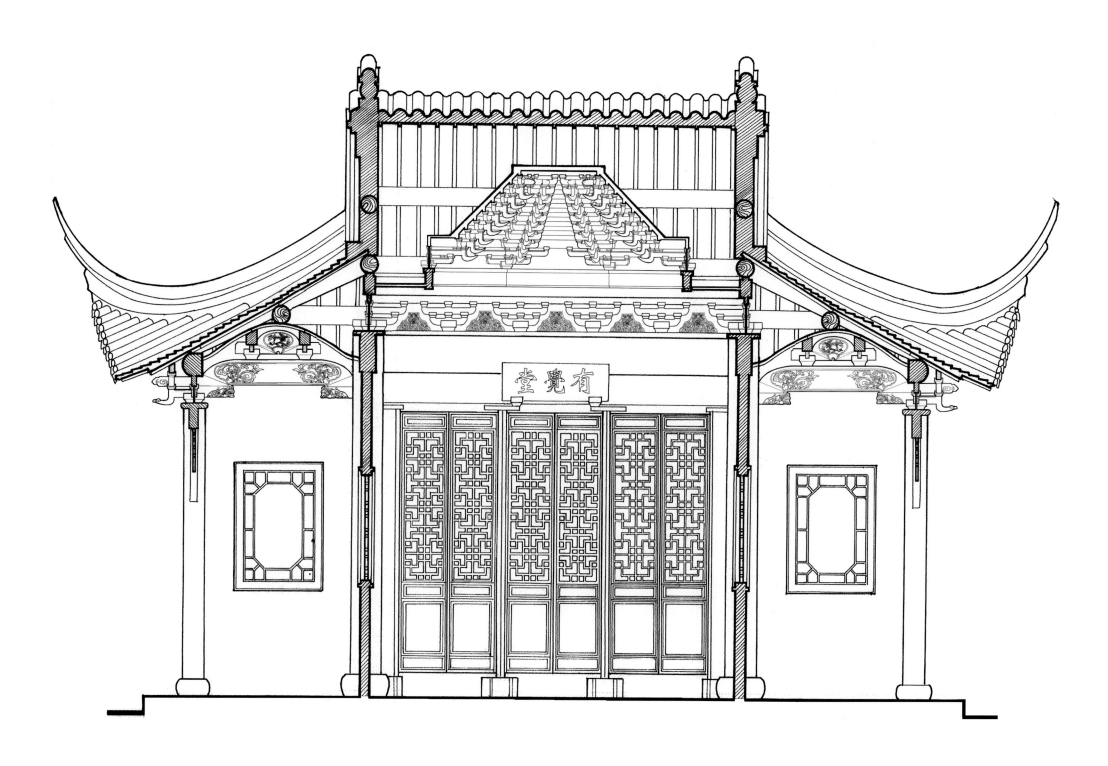

有觉堂

0    0.6    1.2                    3m

有觉堂纵剖面图
Longitudinal section of Youjuetang

有觉堂横剖面图
Longitudinal section of Youjuetang

0    0.6    1.2    3m

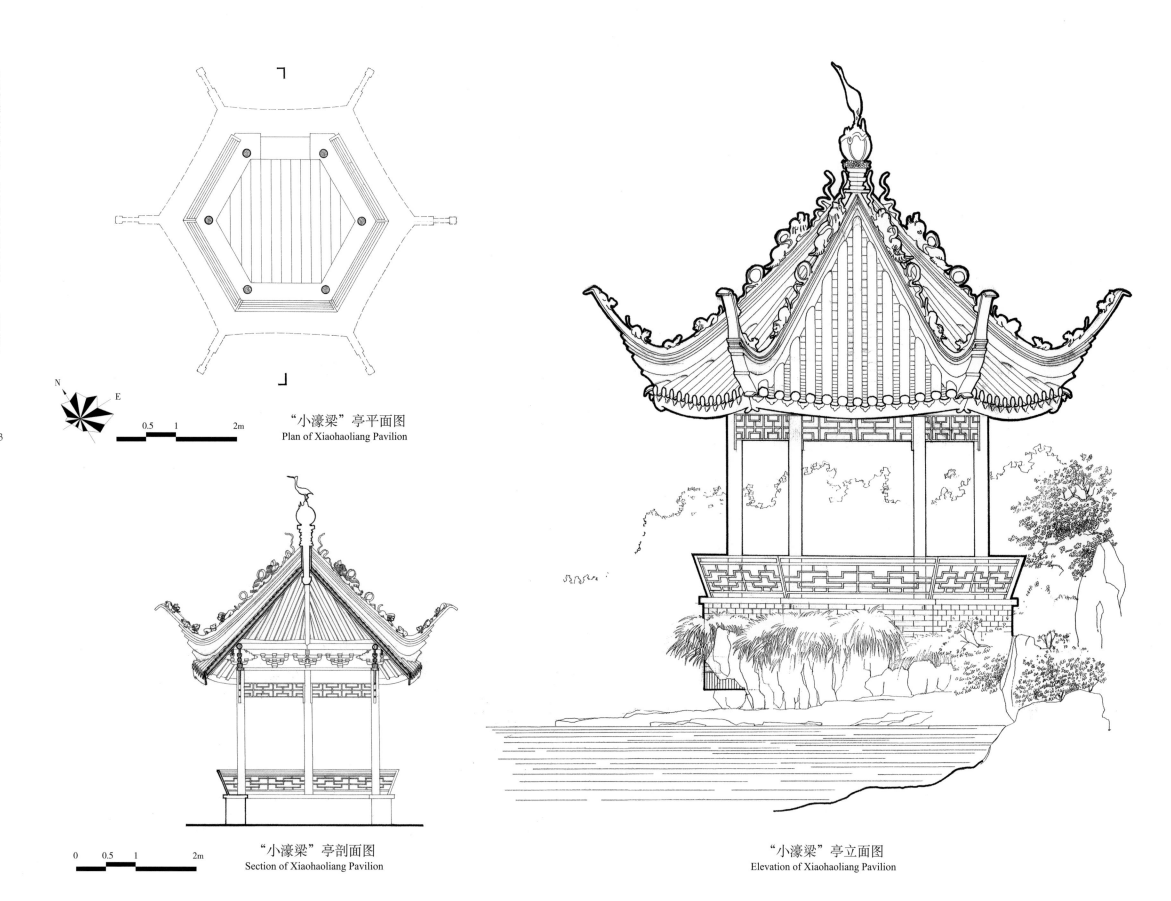

N E

"小濠梁"亭平面图
Plan of Xiaohaoliang Pavilion

0.5　1　2m

"小濠梁"亭剖面图
Section of Xiaohaoliang Pavilion

0　0.5　1　2m

"小濠梁"亭立面图
Elevation of Xiaohaoliang Pavilion

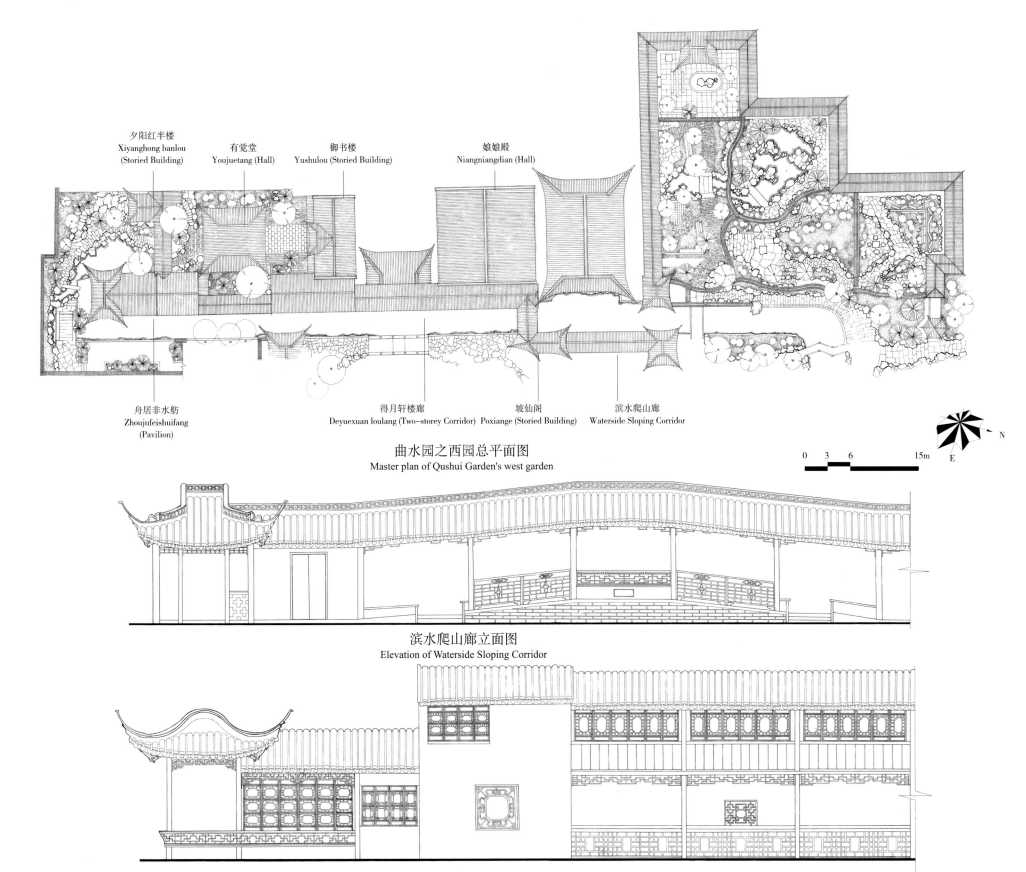

夕阳红半楼
Xiyanghong banlou
(Storied Building)

有觉堂
Youjuetang (Hall)

御书楼
Yushulou (Storied Building)

娘娘殿
Niangniangdian (Hall)

舟居非水舫
Zhoujufeishuifang
(Pavilion)

得月轩楼廊
Deyuexuan loulang (Two–storey Corridor)

坡仙阁
Poxiange (Storied Building)

滨水爬山廊
Waterside Sloping Corridor

**曲水园之西园总平面图**
Master plan of Qushui Garden's west garden

0  3  6       15m

N

E

**滨水爬山廊立面图**
Elevation of Waterside Sloping Corridor

**舟居非水舫及得月轩楼廊侧立面图**
Side elevation of Zhoujufeishuifang and Deyuexuanloulang

0  0.6  1.2      3m

215

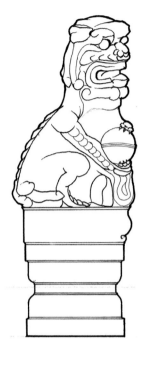

雄狮西立面图
West elevation of male lion

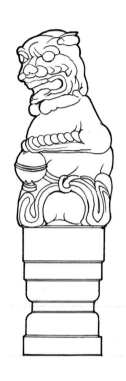

雄狮南立面图
South elevation of male lion

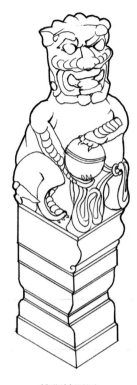

雄狮轴测图
Axonometric drawing of male lion

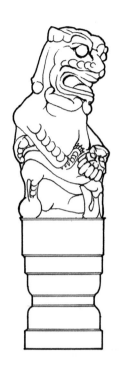

雌狮南立面图
South elevation of female lion

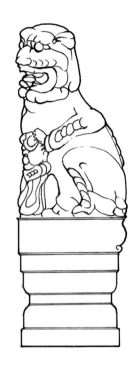

雌狮东立面图
East elevation of female lion

雌狮轴测图
Axonometric drawing of female lion

清代花岗石狮大样图
Huagang Stone lions of the Qing Dynasty

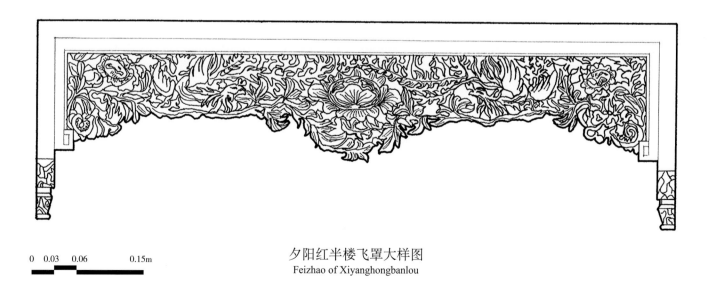

0  0.03  0.06      0.15m

夕阳红半楼飞罩大样图
Feizhao of Xiyanghongbanlou

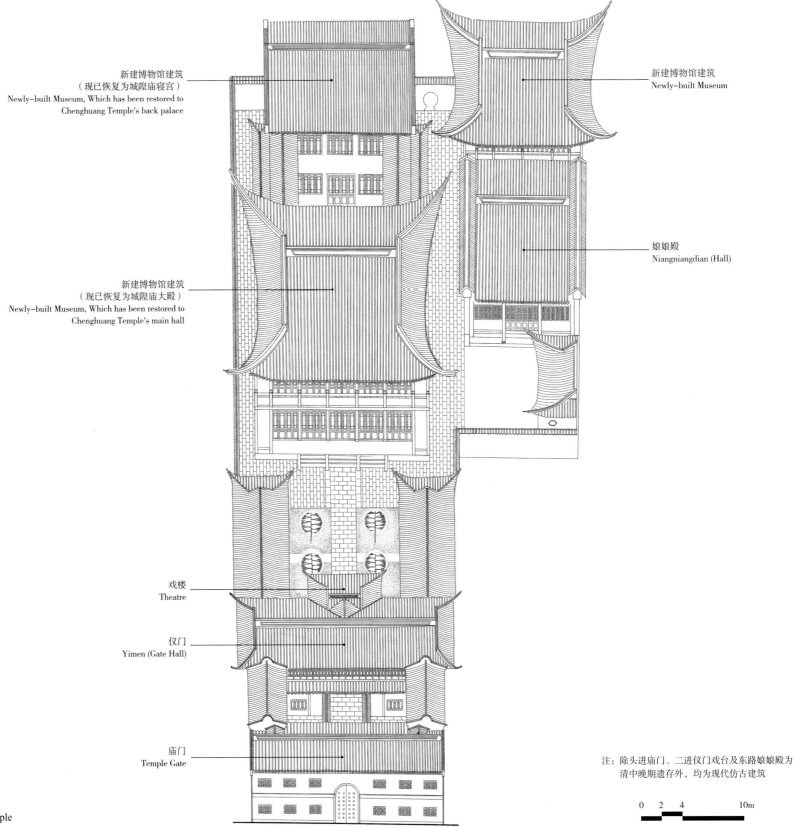

新建博物馆建筑
（现已恢复为城隍庙寝宫）
Newly-built Museum, Which has been restored to
Chenghuang Temple's back palace

新建博物馆建筑
Newly-built Museum

娘娘殿
Niangniangdian (Hall)

新建博物馆建筑
（现已恢复为城隍庙大殿）
Newly-built Museum, Which has been restored to
Chenghuang Temple's main hall

戏楼
Theatre

仪门
Yimen (Gate Hall)

庙门
Temple Gate

注：除头进庙门、二进仪门戏台及东路娘娘殿为
清中晚期遗存外，均为现代仿古建筑

0  2  4        10m

青浦县城隍庙总体轴测图
Axonometric drawing of Qingpu County Chenghuang Temple

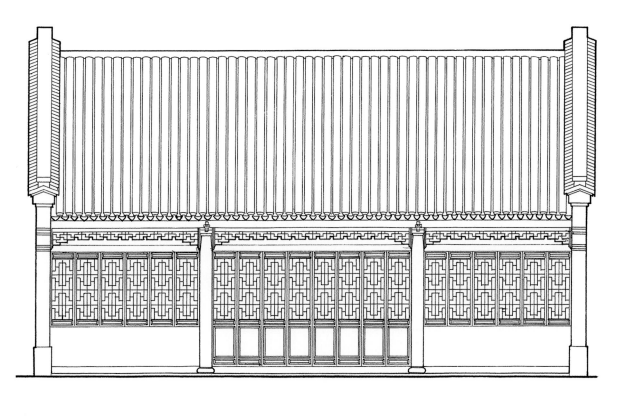

娘娘殿立面图

Elevation of Niangniangdian

0　0.5　1　　　2.5m

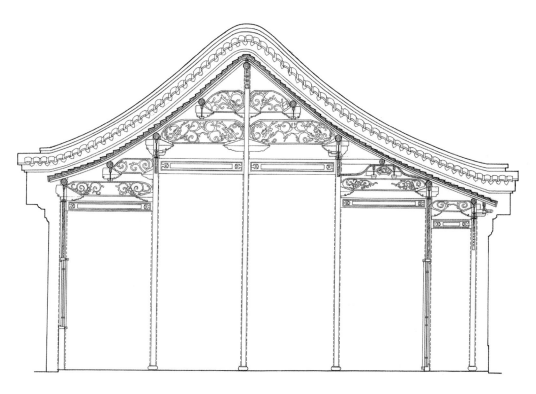

娘娘殿横剖面图

Cross-section of Niangniangdian

0　0.5　1　　　2.5m

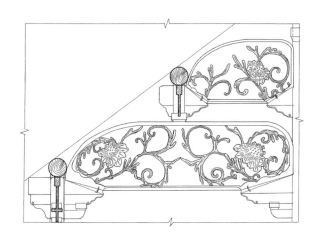
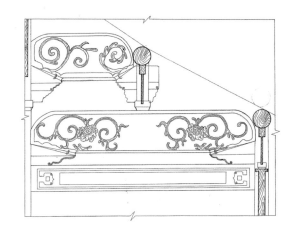
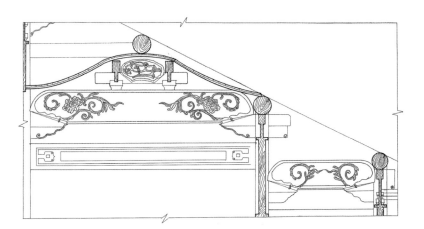

娘娘殿月梁大样图

Yueliang of Niangniangdian

0　0.25　0.5　　1m

**Item of survey:** Jiading County Xuegong and Xuegong Garden (Yingkui Mountain and Huilong Pond)

**Address:** No183, South Avenue, Jiading Town, Jiading District, Shanghai

**Age of construction:** 1219 (year 12 of Jiading Emperor of southern Song Dynasty)

**Site area:** 43.3 mu

**Competent organization:** Jiading Museum

**Survey organization:** College of Architecture and Urban Planning, Tongji University

**Time of survey:** 1997

测绘项目：嘉定县学宫及学宫苑区（应奎山与汇龙潭）

地　　址：嘉定区南大街一八三号

始建年代：宋嘉定十二年（一二一九年）

占地面积：四十三·三亩

主管单位：嘉定县博物馆

测绘单位：同济大学建筑与城规学院

测绘时间：一九九七年

# Jiading County Xuegong and Xuegong Garden
# (Yingkui Mountain and Huilong Pond)

嘉定县学宫及学宫苑区（应奎山与汇龙潭）

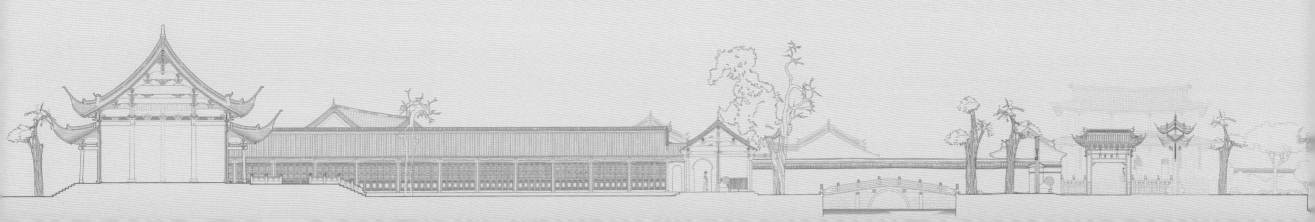

# Introduction

导　言

Jiading Xuegong is located on the original east side of Nanmen Street and the west bank of the north-south main river 'Hengli' in Jiading County. In the north, it faces 'Lianchuan' (the east-west main river) and Fahua Tower. In the north-east, it meets Jiading Chenghuang Temple of the early Ming Dynasty and Qiuxiapu Garden (Gongshi Garden in the Ming Dynasty) of the early Qing Dynasty. In the south, it faces Liuguang Temple of the Song Dynasty (later changed to Lianchuan Shuyuan, which does not exist today) across the mountain. Not far in the south is the old site of Nanmen (which called Chengjiangmen in the Yuan Dynasty, and renamed as Xuanwenmen during 1522-1566) and the city moat.

The whole building complex are composed of four parts, Wenmiao on the middle axis facing the south, Xianxue on the east axis, Danghushuyuan outside the east axis and its garden, which consists of Huilongtan and Yingkuishan, in the south. It originally covered an area of 43.3 mu. With multiple gates horizontally spreading, axes solemnly displaying, corridors and courtyards being pure, mountains and ponds lushly growing, Jiading Xuegong is one of the most typical example whose regulation and style are relatively complete and preserving the ancient feature, and whose scene is extremely beautiful and vivid among the similar building complex in Jiangnan area.

嘉定学宫建筑群坐落于原嘉定县城内南门大街东侧、南北干河『横沥』西岸，北与盈盈『练川』（东西干河）清波、亭亭法华塔塔影相望，东北与明初县城隍庙、清初秋霞圃（明龚氏园）燕雀往还，南与宋留光寺（后改为练川书院，今不存）隔山相对，南行不远即是南门（元代称『澄江门』，明嘉靖后称『宣文门』）旧址与汤汤城濠。

该建筑群由面南的中路文庙、东路县学与外东路当湖书院三路建筑，及其南部的汇龙潭、应奎山苑区共四部组成，原占地四十三·三亩，其重门横舒，纵轴秀挺，廊院纯净，山潭盘郁，为江南诸多府、县级学宫建筑群中，规制、样式较完备而『存古』，景象极丰美而生动的一例。

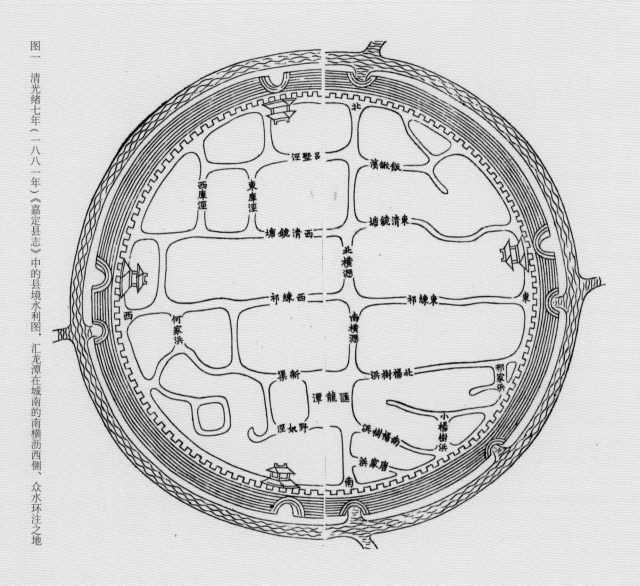

图一 清光绪七年（一八八一年）《嘉定县志》中的县境水利图，汇龙潭在城南的南横沥西侧、众水环注之地

Fig.1 The map of Jiading County's water lines in *Jiadingxianzhi,* 1881, in which Huilongtan is located at the west side of South Hengli River, surrounded by several rivers

## Overall Expression of 'Ming Style' Xuegong and Dynamic Landscape

Jianding Wenmiao and Xianxue were first built in the spring of 1219, when Jiading County was first established. Xuegong was in the left while Wenmiao was in the right. Together with Jading County Chenghuang Temple, which was also inside the South Gate and moved to the current site of Dongmen Street in the early Ming Dynasty, Jiading Xuegong opened the prelude to the narrative of a city's 'divine space'. After that, it gradually covered an area of 26.5 mu. In Jiading Wenmiao, Yanggaofang (originally called Yingkuifang, Yangzhifang), Lingxingmen (including Panchi and Panqiao), Dachengmen, Dachengdian (once called Xianshimiao during 1522-1566) still exists today while the original Yanjudian (which was changed to Jingyiting in 1535) in the back had been destroyed. In Jiading Xianxue, Limen and Mingluntang still exists today while Ruxuemen, and Zunjingge, Chongshengci and etc. had been destroyed. Although Jiading Xuegong was rebuilt on a large scale after the war of Taiping Army in the late Qing Dynasty, there are still many remains of 1460 when it was first established, and Xianxue still maintain the old layout of 1358 which can be regarded as a rare 'Ming style' Xuegong building complex. The stone structure part seems to have even more old relics of the Song and Yuan Dynasties, especially the Yunlong pattern imperial stone of Panqiao and the stone-imitating-wood fence of Dachengdian's platform are the most exquisite and valuable.

Danghushuyuan, in the south-east of Xianxue, was the original Wenchangci which was first built in 1460. In 1540, it was changed to Qishengci. In 1552, it was changed to Xundaoshu. In 1723, it was changed to Wenxingshuyuan. In 1755, it was changed to Yingkuishuyuan. And in 1765, it was eventually changed to its current name, in order to honor LU Longqi, who was Chenghuang God of Jiading County and was called Danghuxiansheng, existing three buildings on the same axis. The first gate's beams may be the relic of the Ming Dynasty, and there are also a pair of stone boot-shaped pillars similar to Dachengdian and a pair of Ming Dynasty Qing Stone gate lions. In front of to the third lecture hall, there is still a brick-carved gatehouse imitating the wood style from 1765, which has been rebuilt, but still preserves the Ming style.

『明式』学宫与动态山水的『互文』

嘉定县文庙、县学始建于嘉定立县之初的南宋嘉定十二年（一二一九年）春，左学而右庙，与同在南门内的县城隍庙（明初迁至东门大街今址）共同开启了一城『神性空间』叙事的序幕，后渐占地至二十六·五亩。其文庙部分今存仰高坊（旧称『应奎坊』『仰止坊』）、棂星门（及泮池泮桥）、大成门、大成殿（嘉靖时一度称『先师庙』）一路四进，后部原燕居殿（嘉靖十四年改为敬一亭）等建筑今不存。县学部分今存礼门、明伦堂一路两进，首进儒学门，后部尊经阁、崇圣祠等建筑均不存——庙、学二部虽经清太晚清太平军战事后的大规模重修重建，仍存留明天顺四年（一四六〇年）学宫『定制』时之旧式旧貌极多，县学更多存元末至正十八年（一三五八年）旧制，故仍可视为难得的『明式』学宫建筑群。其石构部分更似多存宋元旧物，尤以泮桥云龙纹御路石、大成殿月台石仿木勾栏最为精湛可贵。

县学东南隅的当湖书院，原系明天顺四年（一四六〇年）肇建之文昌祠，嘉靖十九年（一五四〇年）改启圣祠，三十一年（一五五二年）改训导署，清雍正元年（一七二三年）改兴文书院，乾隆二十年（一七五五年）改应奎书院，三十年（一七六五年）为纪念嘉定县城隍、当湖先生陆珑其而改称今名——现存一路三进。其首进大门存与大成殿相似石质的靴形柱礩及明代青石门狮一对，梁架亦为明代遗物，其第三进讲堂前，尚残存乾隆三十年（一七六五年）款砖雕仿木门楼，虽经重修，仍存部分明风。

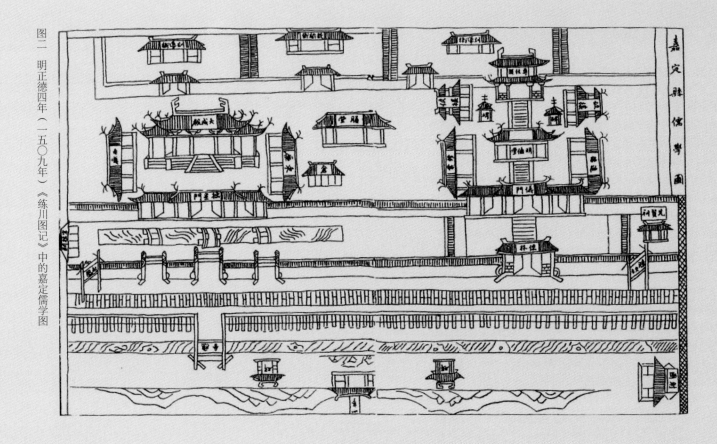

图二　明正德四年（一五〇九年）《练川图记》中的嘉定儒学图

Fig.2  The map of Jiading County Xuegong in *Lianchuantuji*, 1509

Wenmiao, Xianxue and Danghushuyuan stand in silence. The towering halls are like mountains, the horizontal gates are like water, and the corridors are like ripples. They resonate with the scenic beauty of more than 16.6 mu in the south (the east of Hengli was not included), namely Huilongtan and Yingkuishan.

Huilongtan was excavated in 1588. The lush Yingkuishan, which is quietly standing in the middle of the lake, was built by increasing the original soil slope, in order to shade the original temple (Liuguang Temple) in the south in 1460. The landscapes have gone through hundred years of changes, and they have been modified repeatedly, and eventually formed the depressed and secluded scene, without traces, as if the context of the nature were deeply embedded in the southern corner of the city. The mountain condensed the image of stars, and the lake gathered the five streams, were known as the 'five dragons embracing the pearl' pattern. It composed the generosity of Jiading County at the turn of the Ming and Qing Dynasties and the myth that numbers of scholars came into being as brilliant as the stars in the Qing Dynasty.

庙、学与书院肃然静峙，巍殿如山，横门如水，廊庑如漪，共振着南部占地十六·六余亩的山水胜景（不计入横沥以东部分），即汇龙潭与应奎山。

汇龙潭凿于明万历十六年（一五八八年），潭中静峙如案的郁郁应奎山，则系明天顺四年（一四六〇年）为屏蔽庙南曾经的佛刹（留光寺），增培原有土坡而成。山水均历经百年演替，反复改易，终至沉郁幽邃，朴茂无痕，仿佛八方地脉深情潜注于城南一隅，山凝星象，潭汇五溪，蔚为『五龙抱珠』之局，谱写出嘉定一城明清之交的慷慨节烈和有清一代状元、巨儒迭出，『科名』如星汉灿烂的神话。

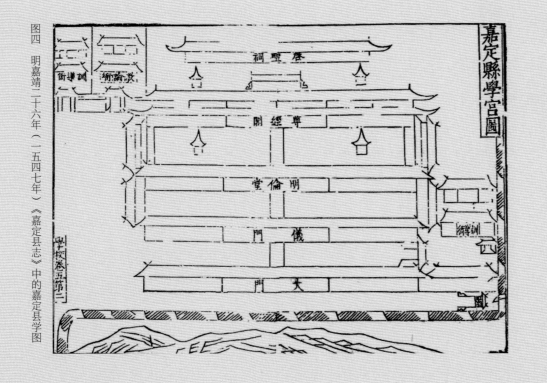

图四 明嘉靖二十六年（一五四七年）《嘉定县志》中的嘉定县学图

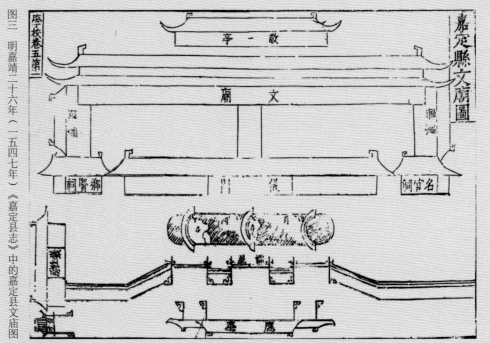

图三 明嘉靖二十六年（一五四七年）《嘉定县志》中的嘉定县文庙图

Fig.3 The map of Jiading County Wenmiao in *Jiadingxianzhi*, 1547
Fig.4 The map of Jiading County Xianxue in *Jiadingxianzhi*, 1547

## Between the Holy Land and the Landscape: The Space Permeation from the Divine Path to 'Chazi' and Fence

Yanggaofang (once called Yingkuifang or Yangzhifang) at the very south of the main axis of Wenmiao, Xingxianfang (once called Bingxingfang) facing the east, Yucaifang (once called Rulinfang) facing the south, and the the waterfront stone fence standing highly on the stone-imitating-wood 'Chazi', constituted the spatial guidance of the holy land of Xuegong and its wonderful transition with the southern landscape. Particularly, two stone pillars of Yanggaofang's sub-room are still old relics of the Ming Dynasty in 1506.

The stone fence was originally a closed wall, forming a pure path with the walls on both sides of Lingxingmen in the north, and the path, which was common before the Ming Dynasty, is clear and leads to the sanctuary. In 1506 when Yingkuishan was added, it was changed into wooden fence, which was suspected to be 'Chazi' of the style in the Song Dynasty, in order to moderately welcome the landscapes. And at the southern end of the middle axis of Wenmiao and outside Lingxingmen, the steep Yingkuifang was built, which became the focus and frame of the dialogue between architecture and landscape. In 1516, the wooden fence was changed to stronger and more lasting stone fence. In order to achieve sufficient height and sense of ritual, the lower part was still the dense vertical stone-imitating-wood mullion, which moderately maintained the strict feelings, while the upper part was the horizontal stone-imitating-wood fence, taking into account the imagination of leaning on the fence and overlooking the landscapes. The wonderful combination of the vertical and the horizontal, seemed to seek a certain balance between divinity and poetry.

Nowadays, Some Qing Stone fence and 'Chazi' on the east side of Yanggaofang are mostly old relics of the period from this time to the early Qing Dynasty. Counting from the west, the first and the seventh pillars are probably the original ones in 1516, while the Huagang Stone fence on the west side of Yanggaofang seems to be the result of restoration in 1879 and recent years.

The high 'Chazi' and fence leaning against the palace wall and tall trees, lined up in a row, with good space limitation and penetration, while outside the fence are Huilongtan, a clear lake, and Yingkui Mountain, which is shadowed by the lake.

# 圣境与山水之间：从神性夹道到『叉子』勾栏的空间渗透

文庙中轴南端的仰高坊（曾称『应奎坊』『仰止坊』）、其两侧面东的兴贤坊（曾称『宾兴坊』）与面西的育才坊（旧称『儒林坊』），以及三坊间高矗在石仿木『叉子』上的临水石质勾栏，构成了学宫圣境的空间引导及其与南部山水间的美妙过渡。其中仰高坊次间石柱犹为明正德元年（一五〇六年）始建时的旧物。

该石栏原为封闭式垣墙，与北面棂星门两侧墙体形成明以前常见的纯净夹道，一径清肃，导向圣域。正德元年增筑应奎山时，为适度迎纳山光水色，将其改为木栏，疑为近宋风的『叉子』样式，并于其中文庙中轴线南端、棂星门外构筑峭拔的应奎坊，成为建筑与山水间对话的焦点与画框。正德十一年（一五一六年），又将木栏改为更坚固恒久的石栏。为求足够的高度与仪式感，其下部仍为密密的纵向石仿木直棂，适度保持了圣境的森严郁闭，上部则为横向的石仿木勾栏，兼顾了凭栏眺览的想象——奇妙的纵横组合，在神性与诗情间寻求着某种平衡。

至今仰高坊东侧西端部分青石勾栏、叉子多为这一时期至清初的旧物，其西首第一、七根望柱更似为正德十一年创始原物。其余花岗石栏及叉子则似为清光绪五年（一八七九年）和近年修复后的结果。

数十丈叉子勾栏背倚宫墙与高林，森然如阵而列，具有良好的空间限定与渗透力。栏外则是一水涵碧的汇龙潭和影入清潭的应奎山。

图七　仰高仿东侧勾栏西七望柱石狮，亦似仍为明正德十一年（一五一六年）原物

图六　仰高坊东侧勾栏西一望柱石狮，似仍为明正德十一年（一五一六年）原物

图五　上海嘉定学宫仰高坊东侧勾栏，远端似多为明代原物

Fig.5　The fence in the east of Yanggaofang of Shanghai Jiading County Xuegong, and the distant ones were mainly the relics of the Ming Dynasty

Fig.6　The first Wangzhu stone lion counting from the west of the fence in the east of Yanggaofang, which was the relic of 1516 (the 11st year of Zhengde Emperor in Ming Dynasty).

Fig.7　The seventh Wangzhu stone lion counting from the west of the fence in the east of Yanggaofang, which was the relic of 1516 (the 11st year of Zhengde Emperor in Ming Dynasty).

## Song Style and Yuan Scale: the Horizontal Control of the 'Super' Gate

Jiading has become a city (Lianqi City) because of the water, and it has become a county because of the city. When it was first built in the Southern Song Dynasty, Chenghuang Temple and Xuegong of the county were already built. In 1296, due to the economic recovery in Jiangnan area, the population proliferated and the county was promoted to Zhongzhou. Till 1308, the prefectural government was expanded and the prefectural Xuegong was expanded since 1324. The grand Qing Stone Panchi in front of Dachengmen today was the legacy after this renovation and expansion, but it may not be completed until the end of Yuan Dynasty, as it was said, 'There were six layers of stacked stones on the four banks of the Panchi, with a total perimeter of forty-four zhang and four chi, and two thousand five hundred and three pine stakes.' Despite the war in the late Yuan Dynasty and the austerity in the early Ming Dynasty, Jiading Prefecture was reluctantly reduced to a county in 1369. But the former golden age had been condensed in this 'super' Panchi with a width of more than 40 meters and full of Yuan style, which set the prefecture-level scale and layout of vertical and horizontal rows for the future 'Super' Yimen (Dachengmen), Jialang, and the entire Jiading Confucian Temple on the north side.

In the next thirty years, Jiading Xuegong was continued to be expanded, which was recorded in *Chongjian mingluntang ji* (following called Yang's text) written by YANG Weizhen, a great writer in the late Yuan Dynasty. 'In the summer of 1352, Zhizhou GUO Liangbi arrived the position and first visited Xianshengmiao. Feeling that the two Langwus were crude and not worth watching, he decided to repair them…The next year (1353)… Dachengdian was rebuilt and Jiadian and qianying was added… The original two Langwus were demolished. The base of the architecture was heightened. The two Langwus were rebuilt into twelve rooms and Yimen and Jialang was also rebuilt… Panqiao was built by bricks, with Shizhou and Danchi of several Chi long. Three Lingxingmen were added… Till 1358, Mingluntang with five rooms and three rooms of Qianxuan was established. The original Yimen with five rooms, five rooms on each sides were added… With Wenmiao in the right and Xianxue in the left, the scale of the building complex was extremely grand.' This article can be contrasted with Yang's another article *Jiadingzhou chongjian ruxueji.*

# 宋魄与元度：超级门廊的空间横扼

嘉定因水成市（练祁市），因市成县。南宋立县之初，便有县城隍庙、县学宫的建设。到了元代元贞二年（一二九六年），又因江南经济恢复，人口繁衍，由县升为中州，再到至大元年（一三〇八年）扩建州衙，自泰定元年（一三二四年）扩建学宫。今日大成门前所存的超大型青石泮池，就是这次改扩建后的遗制，但可能直至元末才最终完成——『泮池四岸叠石六层，通计四十四丈四尺，松桩计二千五百六十三根』。虽然历经元末战乱和明初紧缩，嘉定州于洪武二年（一三六九年）无奈地复降为县。但曾经的黄金时代，已经凝聚在这座面阔四十余米、元风盎然的『超级』泮池里，为其北侧未来的『超级』仪门（大成门）、挟廊，以及整座嘉定孔庙定下了州府级的尺度和纵横排宫的布局方式。

此后三十余年后的学宫增拓，历见于元末大文学家杨维桢的《重建明伦堂记》（以下称『杨文』）：

『至正壬辰（一三五二年）夏，知州郭良弼到任，首谒先圣庙，为觉廊庑卑陋，殊失观瞻，留心修葺……次年（一三五三年）……翻瓴大成殿、增创挟殿、前楹……撤旧两庑，高筑基址，改造廊庑各十二间及仪门挟廊……砖砌泮桥，石甃丹墀若干丈，又建棂星门三座……至正戊戌（一三五八年）立明伦堂五间，轩三间，前建仪门五间，左右斋舍各五间……文庙居右，儒林居左，规模宏大』。此文可与杨氏《嘉定州重建儒学记》一文对照印证。

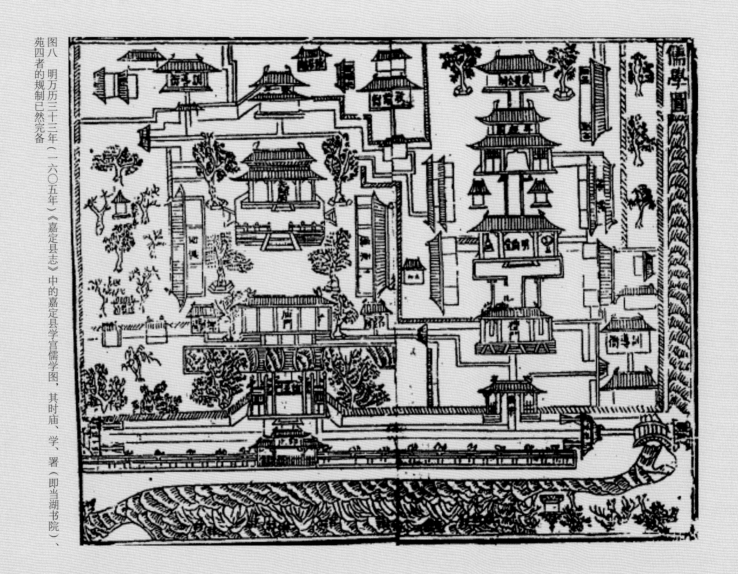

图八　明万历三十三年（一六〇五年）《嘉定其志》中的嘉定县学宫儒学图，其时庙、学、署（即当湖书院）、苑四者的规制已然完备

Fig.8  The map of Jiading County Xuegong Ruxue in *Jiadingxianzhi*, 1605, in which the regulations of the temple, school, academy (namely Danghu Shuyuan) and garden were already complete.

The four main transformations in the article were all to expand the width of the entire Wenmiao complex horizontally to reach the prefecture-level scale, and to be commensurate with the previously expanded Panchi. Firstly, Jiawu and Qianxuan were added to Dachengdian (this could also verify from the side that the style of Jiading County Chenghuang Temple is quite 'preserving the ancient'). Secondly, the two langwus, with twelve rooms each, was demolished and rebuilt. Thirdly, Yimen and Jialang was rebuilt. Fourthly, three parallel Lingxingmen was built. The scale and the style of Wenmiao were then laid down. The two *Langwus* of Dachengdian, which had tewlve rooms each, and even the five rooms Mingluntang and the three rooms Qianxuan in the east axis have magically survived to today. Only a hundred years later, in 1460, Dachengdian would be integrated into a giant hall, and the composing of this vigorous movement was finally completed.

It is worth mentioning that in the later period of this expansion, when the five room Yimen (also called Limen) of Zhouxue on the east axis were added, there was lack of space for adding and to avoid the ramifications and trivialities, it was very likely that it would be the same as the early expansion of the west axis that Wenmiao's Yimen (namely Dachengmen) and its Jialang are 'connected with eaves and ridges', forming a 'super porch' that runs across the entire Xuegong building complex. It not only sewed the two axes of Wenmiao and Xianxue and embraced the giant Panchi, but 'vertically drawing' the magnificent and pure Dachengdian courtyard to the north. This key stroke laid the foundation of today's Xuegong textbook-like, unmodified cut pattern of Song style, and aroused the viewer's praise that Langwus were stately standing and Yimens were like couples of wings. It is suspected that during the reconstruction of Danghushuyuan in the Qing Dynasty, the east Jiawu of Limen was demolished, but its grand looking had not been reduced.

文中四项主要改造，都是为了横向舒展开整座文庙建筑群的面阔，以达到州级尺度，并与先期扩建的泮池相称：一、大成殿增设两侧挟殿及前轩（这也可从侧面验证嘉定县城隍庙大殿形制的『存古』）；二、两庑各十二间分别退后重建；三、改造仪门挟廊；四、建并列的棂星门三座——文庙的尺度规制就此奠定——其中大成殿两庑各十二间，甚至东路州学部分的明伦堂五间、前轩三间的尺度形制，都神奇地留存至今。只待百年后的明天顺四年（一四六○年），大成殿被整合为一庞然巨殿，最终完成这一苍劲乐章的谱写。

尤值得一提的是，这次扩建后期，增建东路州学仪门（即礼门）五间时，因再无腾挪余地，且为免参差琐碎，极可能索性与前期拓展的西路州文庙仪门（即大成门）、挟廊『连檐通脊』，形成横贯整个州学宫建筑群的『超级门廊』，不只缝合庙、学双轴，抱拥巨形泮池，且向北『纵引』出恢宏纯净的大成殿廊院。这关键性一笔，奠定了今日学宫教科书般的、宋魄未改的斩截格局，让此后明永乐二十一年（一四二三年）的观者有『廊庑秩秩，仪门翼翼』的赞美——疑清代改建当湖书院时拆除了礼门之东挟屋，但尚不至令其减色。

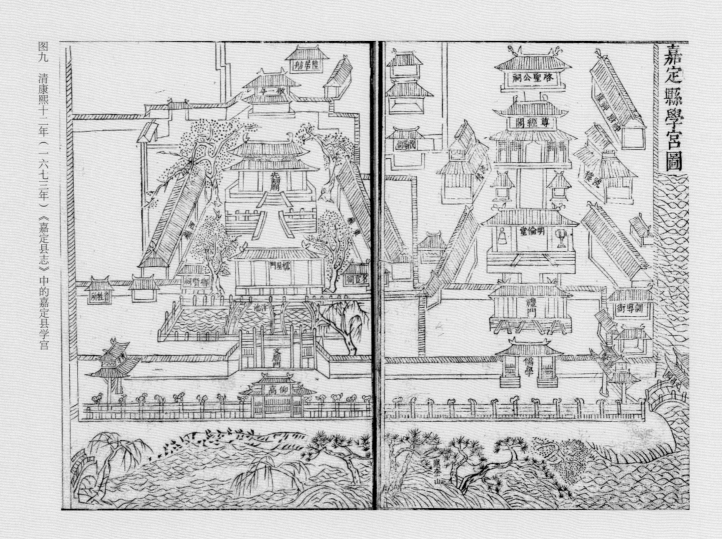

嘉定縣學宫圖

Fig.9  The map of Jiading County Xuegong in *Jiadingxianzhi*, 1673

Later, 'in 1390… Three stone bridges called Panqiao was built.' The current three bridges on Panchi have a relaxed arch and a strong waist, which seems to preserve the appearance of this construction. The east Panqiao purely built of Qing Stone seems to be a relic at that time, while the other two bridges seem to be in rebuilt 1879 with Huagang Stone. On the middle Panqiao, the imperial stone with a cloud and dragon pattern that raised above the ground is extremely vivid and exquisite, as if it were a three-dimensional version of the Southern Song Dynasty Chen Rong's Jiulngtu, full of the flavor of the Song and Yuan Dynasties, which were extremely rare in China. Comparing with Yang's text, 'Shizhou and Danchi of several Chi long', we don't know whether this was the imperial road stone with the cloud and dragon pattern, and were used by later generations. The Qing Stone 'Changxishi' (six in total) on the outer side of three Panqiaos were decorated with Kunmen pattern, with flowers in the circle, in which the Song style still exists and is more convergent and seems to be a relic of the early Ming Dynasty.

The broad and heavy Qing Stone platform in front of Dachengdian is in line with Yang's description of 'The original two *Langwus* were demolished. The base of the architecture was heightened'. This platform and the part of the stone-imitating-wood fence on the south side of the platform (whose Xunzhang is pumpkin shape), the cirrus cloud head is connected toWangzhu, and the cloud head of the double pillars of Lingxingmen (which was inherited in1373 when Lingxingmen was rebuilt), may be the relics of this period of reconstruction. However, the fence in front of Dachengdian may even be earlier.

此后『洪武二十三年（一三九〇年）……甃泮池建石梁三』，今存泮池三桥均拱势舒缓，收腰有力，似保存着这次建造的面貌，而纯青石构筑的东泮桥似乎犹是当时遗物，其余两桥则似在清末光绪五年（一八七九年）的重修中以花岗石翻砌。而中泮桥上的『剔地起突』云龙纹御路石极尽生动精湛，仿佛是南宋陈容《九龙图》的三维版本，宋元气息极为浓郁，为国内罕见。对照杨文中『石甃丹墀若干丈』，不知是否即（包括）此云龙纹御路石，而为后世沿用。泮池三桥外侧的青石『长系石』（共计六根）端部以壶门花纹修饰，其内团花宛转，宋风犹存而较为收敛，亦似为明初遗物。

而大成殿前宽阔厚重的青石月台则符合杨文『撤旧两庑，高筑基址』的描述。此月台与台上南侧的部分石仿木勾栏（寻杖为瓜棱形）、卷云头连身望柱，还有棂星门明间双柱的云头『洪武六年（一三七三年）重建棂星门时沿用』，均可能为这一期重建的遗物。勾栏甚至可能时代更早。

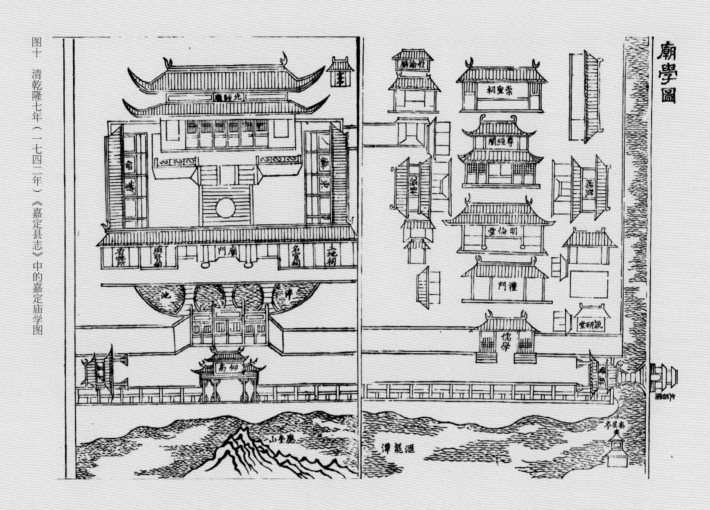

图十 清乾隆七年（一七四二年）《嘉定县志》中的嘉定庙学图

Fig.10  The map of Jiading County Xuegong in *Jiadingxianzhi*, 1742

## 'Grand Hall' Era in the Early Ming Dynasty: the Hierarchical Temple Group and the Authoritarian 'Grand Hall'

In the early Ming Dynasty, Jiangnan area had declined for a hundred years in unprecedented confinement. However, in 1460, although the political situation of the north area was uncertain, the tide of revival had sprouted in Jiangnan. This year, presided over by LONG Jin, the prefect of Jiading County having taken the position for three years, who came from Jishui County, Jiangxi Province, a large-scale renovation of the Jiading Xuegong, especially Dachengdian and its corridors and courtyard Gallery was carried out. This renovation Integrated the several expansions in the late Yuan Dynasty into a complete whole and laid the foundation for the current appearance of Jiading Xuegong. The most important of all, is the integration of 'Jiadian and Qianying', which mentioned in *Yang's text* and combined version of Dachengdian with ancient style, into a 'grand' hall with a double eaves, gable and hip roof, three-room width, eight-purlin depth, and Fujie around the hall, showing the new style of 'Grand Hall' in the early Ming Dynasty.

If we say that the unprecedentedly freedom and prosperity of the Song Dynasty had once extend and develop the group art of Chinese architecture to a certain extreme that baosha and platform in the front of the hall, Jiawu and Jialang on both sides, and Weilang and back hall behind in one go often went out of large architecture, then in the early Ming Dynasty, in addition to the convergence of the system and the concentration of power, which opened the 'grand men' era of politics, also opened the 'gran hall' era in architecture. That is, to a certain extent, it had played down the overall logic of the previous building groups that set the stage, the climax, the consistent flow, and the endless aftertaste, and the intention was to strengthen the independent expression of a single building. Fengtiandian of the Forbidden City in Beijing in the early Ming Dynasty, which was nearly 100 meters wide and nearly 50 meters deep, was the best example. In Japan, the change behaved as the transformation from 'Qingdianzao' to 'Zhuadianzao'. Later in 1474, when the Zunjingge on the east axis of Jiading Xuegong was built, 'just built three rooms and two sidedoors' was still an ancient technique that focuses on hierarchy rather than absolute volume, which may also have outstanding achievements considering of the size of Dachengdian.

明前期的『巨殿时代』：层次化的殿组与威权化的『巨殿』

明前期的江南，曾在空前禁锢中衰颓百年，然而在北国政局阴晴不定的天顺四年（一四六〇年），复兴的春潮或已在江南萌动。这一年，由到任三载的嘉定知县、江西吉水人龙晋主持，对已降为县级的嘉定学宫，尤其大成殿廊院进行了大规模翻建，将元后期以来的次第拓展最终整合为一浑然整体，奠定了学宫的今貌——其中最重要的，是将杨文中拥有『挟殿、前楹』，风格存古的组合版大成殿，整合为一重檐歇山顶、面阔三间、进深八架、副阶周匝的『庞然』巨殿，展现出明前期的『巨殿』新风。

如果说，空前自由而繁荣的宋代，曾将中华建筑的组群艺术，延续、发展到某种极致，凡大型建筑，往往前出抱厦、月台，侧拥挟屋、挟廊，后曳尾廊、后殿，连绵一气；那么明代前期，除了制度收敛、权力集中，开启政治的『巨人时代』之外，也一度开启了建筑的『巨殿时代』。即在一定程度上淡化了此前建筑群组烘托铺垫、高潮迭起、一气贯注、余韵不尽的整体逻辑，而着意于强化单体建筑的独自表述——明初面阔近百米，进深近五十米的北京故宫奉天殿就是最好的实例。在日本，则似表现为由『寝殿造』向『主殿造』的变化。此后成化十年（一四七四年）兴建嘉定学宫东路尊经阁时，『为楹者三，扶者二』，则仍是注重层次而非绝对体量的存古手法，或许亦有突出大成殿体量的考虑。

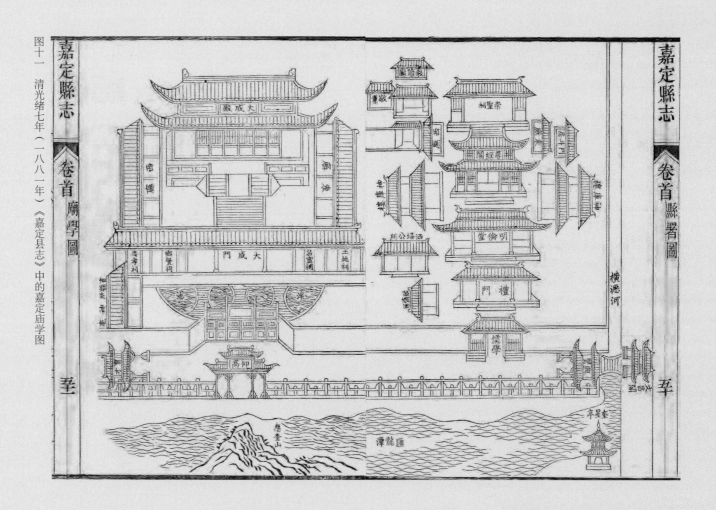

图十一　清光绪七年（一八八一年）《嘉定县志》中的嘉定庙学图

Fig.11  The map of Jiading County Xuegong in *Jiadingxianzhi*, 1881

# 

Dachengdian of the Jiading Wenmiao in this period, whose middle room was more than 6 meters wide and side rooms were more than 5 meters, was wide and sparse. Its gable and hip roof was more than 10 meters high, without shrinking the gable wall. It felt that it was deep and heavy, and the atmosphere is magnificent. Not only the upper eaves dominated the position, but also could effectively take into account indoor ventilation and shelter from wind and rain, similar to the main hall of Xuanyuangong in Dongting Dongshan, Suzhou, Jiangsu Province and architecture that of the same period in Japan, which should be a common technique in Jiangnan area at that time. Inside, the rainbow beams flied across and rose up high, Tuofeng roamed, and Minban spread out. Moreover, the columns of Fujie had an obvious Cejiao while the middle ridged purlin had an significant *Shengqi*. The ancient style was fierce, and it seemed that more of the original appearance remained, similar to Dachengdian of the Suzhou Wenmiao during 1465-1478 of the Ming Dynasty. According to the documents, the 'reconstruction' 1684 and 1877 seemed to have only replaced part of the beams. On the contrary, in the repairs after 1949, the upper eaves Jinwen gongyanbi of Ming style were replaced with the late patterns.

This integrated reconstruction may also enable Dachengmen of Wenmiao and its Jialang to better connect with Limen of the Xianxue, completing the final form of the 'super gate'. The pillars of Dachengmen and the two Langwus that exist today have been replaced in the late Qing Dynasty, but the beams still has the early style.

Mingluntang of Xianxue on the east axis 'collapsed and was rebuilt' in 1426, and it can be regarded as a reconstruction. However, in this historic reconstruction, it 'had not been finished yet and just did some repairs'. The strong and exquisite beams in the early Ming Dynasty and huge Qing Stone Fupen pillars, as well as the fact that the ground of the main house is higher than the front baosha, seemed preserving more of the original production at the end of the Yuan Dynasty. The reconstruction in 1678 and 1876 should be regarded as big repairs. However, baosha in the front used to be gable and hip roof, while today's flush gable roof is suspected to be modified during the 'reconstruction' in the late Qing Dynasty.

这一时期的嘉定文庙大成殿明间面阔逾六米，次间逾五米，开间阔大，面目疏朗。其歇山屋顶总高十余米，不作收山，愈觉其深沉厚重，气象宏远，不只上檐统率地位明显，且能有效兼顾室内通风与遮蔽风雨，与元末明初之江苏苏州洞庭东山轩辕宫大殿和日本同期建筑相似，应为当时江南常用手法。其内部则虹梁飞跨、上昂翘耸，驼峰流转，皿板棋布，且副阶檐柱侧脚明显，正脊『生起』显著。古风烈烈，似较多留存着初建时的面貌，而与明成化年间（一四六五至一四七八年）的苏州府文庙大成殿较为相近——文献中康熙二十三年（一六八四年）及光绪三年（一八七七年）的『重建』似乎只是抽换了部分梁架，反倒是一九四九年后的修缮中，将上檐的明风锦文拱眼壁换成了晚期纹样。

这次整体化重建，也可能使县文庙大成门及其挟廊与县学礼门更好地贯通一气，完成了『超级门廊』的最后定型。今存大成门及两庑部分柱础已为晚清更换，但梁架仍存早期样式。

东路县学明伦堂在此前明宣德元年（一四二六年）的『坠而复端』，可视为重建，而在这次历史性重建中『尚完未竣，则葺之而已』。今仍存浑厚精湛的明早期梁架和硕大的青石覆盆柱础，其正屋地面较前部抱厦升高，似较多存留着元末创建初制。清初雍正元年（一六七八年）和清末光绪二年（一八七六年）的重建应视为大修，惟前部抱厦似原为歇山顶，今日之硬山顶疑系清末『重建』时修改。

图十二　嘉定文庙大成殿明凤梁架，由徐瑞彤绘制

Fig.12　The Ming Style beam of Jiading Wenmiao Dachengdian, drawn by XU Ruitong

The pillars of Limen in front of Mingluntang is already in the late Qing style, but the beams still have a Ming style. It should be that the original system and even part of the original components were preserved in the reconstruction in 1879. This gate, together with the main house of Mingluntang, Dachengmen of Wenmiao, Jialang and the original first gate of Danghushuyuan, may be all overhanging gable roof, but were all changed into flush gable roof in the later reconstruction.

Just two years before this reconstruction, LONG Zhixian, a diligent and capable talent, had just presided over the reconstruction of Jiading County Chenghuang Temple, and also initiated the dredging of the Wusong River, leaving a lot of spatial memory in the city.

明伦堂前部礼门脊柱下亦为硕大的青石覆盆柱础，似同存元末创建初制，梁架仍存明风。应系晚清光绪五年（一八七九年）的重建中存留了部分原制。此门与明伦堂正屋、文庙大成门、挟廊、当湖书院头门原均可能为悬山屋顶，而为后期改建为硬山顶。

就在这次重建两年前，龙晋知县刚刚主持重建了县城隍庙，还首倡疏浚了吴淞江，一代孜孜干才，在这座城市留下了诸多空间印记。

图十三　嘉定县学明伦堂明风梁架，由徐瑞彤绘制

Fig.13　The Ming Style beam of Jiading Xianxue Mingluntang, drawn by XU Ruitong

## The Multi-generation Landscape and Geographical Integration in the Late Ming Dynasty

Jiading Xuegong emerged in the low-lying wetland in the south of the city at the beginning of the Southern Song Dynasty. It had been suffering from hurricanes during the Song and Yuan Dynasties for hundreds of years. 'In the corner of the coast, there were frequent hurricanes, and houses were destroyed.' 'It had been years suffering from the rain and the wind.' 'Buildings were frequently repaired and frequently collapsed, lasting only one or two decades.'

Apart from the hurricane, the east part of Xuegong and the west bank of Hengli had suffered from foundation settlement and bank collapse for a long time, and they had to rely on soil to contend with repeated embankment construction. For example, when *Yang's text* talking about building Mingluntang at the end of the Yuan Dynasty, 'the base was low, people were recruited to transport the soil, and the foundation was completed within two months... The stone bank was more than fifty Zhang.' In 1535, 'Xianxue was in the east of the river, and was shot by the tide frequently, so the stone embankment was built forty Zhang east of the gate'. In 1603, 'twenty Zhang of the east stone embankment was rebuilt'.

When the coast moved eastward and the wind disaster was farther away, and the temple was gradually expanded in the late Yuan Dynasty and early Ming Dynasty, gradually becoming solemn and finalized, the active response to the natural landscape environment and the aesthetic transformation in the name of divinity were put on the table.

The south part of Xuegong was originally an open area in the south of the city. It was so-called 'when the grand hall was open, it could accommodate thousands of horses'. The south part was an unknown slope, so Dangshanlou was built during 1249-1251. It must be located in the south of Xuegong and the building acted as a mountain, forming a level of dialogue between Xuegong and the natural scene. However, two hundred years later, the true sense of landscape shaping had been diligently built for a hundred years and has gone through five steps.

# 明后期的叠代山水理景与地脉整合

嘉定学宫崛起于南宋建城之初的城南低洼湿地，宋元数百年间，一直苦于飓风时袭——『矧地滨海隅，飓风数作，层甍颓败』『雨虐风伐，盖已有年』『栋宇乍修乍圮，不能支十年、二十年之久』。

飓风之外，学宫东部、『横沥』西岸则长期苦于地基沉降和坍岸，只能靠垫土与反复筑堤抗衡。如《杨文》中元末筑明伦堂时『基地低洼，募人运土，不二月而基成……石堤东岸五十余丈』。明嘉靖十四年（一五三五年）『学故东临于河，每为潮水啮射，乃修石堤四十丈于门之东偏』，明万历三十一年（一六○三年）『重甃学东石岸二十丈』。

待到海岸东移，风灾渐远，庙貌经元末明初的拓展，渐渐庄严而定型之际，对自然山水环境的主动因应，和以神性之名的审美化改造就摆上了桌面。

学宫南部原为城南旷地，所谓『明堂开敞，可容万马』，其南部为无名土坡，所以宋淳祐九年至十一年（一二四九至一二五一年）曾建有『当山楼』，想必横亘宫南，以楼当山，形成学宫与自然景象对话的层次——而两百年后，真正意义上的山水塑造则孜孜百年，历经五步。

图十四 嘉定学宫当湖书院大门明代梁架，由徐瑞彤绘制

Fig.14 The Ming Dynasty beam of Jiading Xuegong Danghushuyuan Gate, drawn by XU Ruitong

Firstly, heap the mountain to shadow the temple, and gather momentum by curved water. In 1460, apart from the establishment of Xuegong, the original soil slope in the south of Xuegong was added, Yingkuishan (Yingkuishan or Siyishan) was piled up, and Liuguang Temple in the south and Hengli River in the east were shadowed, in order to make it tortuously circumnavigate the mountains, gather momentum, and then return to the north stream, avoiding straight dash to the east bank of Xuegong. The image was roughly between 'shadow the mountain' and 'Anshan' (using the mountain as the table for study). After this renovation, 'there were mountains in the front, with the pines and cypresses lush, and the rivers flowing' in Xuegong area. Moreover, the result of Imperial Examination had been brand new. Four years later, in 1464, Jiading County won two Jinshi in the examination, and in1469 and 1475, both won three Jinshi. In the era of 'cultural capitalism', such immediate spatial incentives and psychological hints inspired the enthusiasm for landscape management for the next century.

But at the beginning, the pattern had not been relaxed, the mountains had a tendency to force Xuegong, and the rivers had many twists and cannot fully spread the water surface. Therefore, it was needed to ease the tide, soften the shape of the mountain, expand the potential energy of the scene, and integrate into the urban water network, laying the foundation for the subsequent transformation.

Secondly, south move the peak and sort out the landscape into five 'table'. Decades later, the artificial earthen mountain gradually collapsed and lost its power, and the call to reinvigorate the ground and consolidate the results was revived. In 1506, under the leadership of the central 'Xunan' Rao Tang, who comprehensively inspected and urged local work, Yingkuishan was moved to the south and heightened, and was horizontally separated into five peak imitating penholders, making people thinking of Shizilin Garden in Suzhou in the late Yuan Dynasty and Jingshan in Beijing in the early Ming Dynasty. A 'luminous' strange stone was moved to the middle peak, to strengthen momentum and lighten the stars.

一、隆山屏寺，曲河蓄势。明天顺四年（一四六〇年）学宫定制之余，于学宫之南增培原有土坡，堆掇应奎山（映奎山、四宜山），南屏留光佛刹，东屏『横沥』激流，使之迂曲绕山而过，盘桓蓄势，再归故道北流，避免直冲而过学宫东岸。其意象大致在『屏山』和『案山』（以山为学宫桌案）之间。这次改造后，学宫『前有土山，松柏郁然，而河水萦之』，科考战绩焕然一新——四年后的天顺八年（一四六四年）嘉定县一科两进士，成化五年（一四六九年）、十一年（一四七五年）均是一科三进士，在『文化资本主义』的时代，这样立竿见影般的空间激励和心理暗示成果，激发了此后百年的山水经营热忱。

但草创之始，格局未舒，山有逼宫之势，河多扭曲之形，未能充分展开水面，以缓解潮势，柔化山形，扩张景象势能，融入城市水网——为后续的改造留下了伏笔。

二、分峰南移，理景为案。数十年后，人工土山渐渐坍圮失势，振作地脉、巩固成果的呼声再起。正德元年（一五〇六年），在全面巡查督促地方工作的中央『巡按』饶榶主导下，应奎山被南移加高，且横分五岭，拟形笔架，成为学宫前的庞然巨案，亦让人想到元末苏州狮子林、明初北京景山的山形。其『中峰』前更移来『夜光』奇石，点睛壮势，辉耀奎星。

图十五　上海嘉定县文庙仰高坊与两侧的勾栏叉子南临汇龙潭的样貌，摄于一八七〇年代

Fig.15  The appearance of Yanggaofang and Chazi Goulan on both sides in Shanghai Jiading County Wenmiao, facing Huilongtan in the south photographed in 1870s

The south wall of the depressive path space in front of the Xuegong had also been transformed into an open wooden fence naturally, and Yingkuifang was embedded at the south end of the central axis of Wenmiao, perhaps facing the luminous stone. The three-hundred-year Xuegong had finally been faced and concentrated on the landscape, as if the classic layout of the literati garden that the main building 'across the water faced the mountain' was first completed. At that time, literati gardens in Jiangnan area were about to boom that Suzhou Zhuozheng Garden, Jiading County Gongshi Garden (Qiuxiapu Garden), Yi Garden (Guyi Garden), Songjiang Yi Garden, and Shanghai Yu Garden were about to arise. The transformation also began to have a conscious sense of perspective. Gongbu Shangshu GONG Hong, the first generation of the winner of the examination after Yingkuishan was piled up, Jinshi in 1478, and the first owner of Qiuxiapu Garden once said, 'because of the power of the universe, the high mountains and the deep water, the beautiful earth, and the submerged spirituality… The result of the Imperial Examination was prosperous. Wasn't it because of this mountain?'

Two years later, Jiading County really won three Jinshi in the examination in 1508.

Thirdly, move the mountain to the southwest and the temple to the South Gate. Only three years later, in 1509, another 'Xun'an' LI Tingwu once again advocated to move Yingkuishan as a whole to the southwest and further heighten it to make the spatial relationship between it and Xuegong more relaxed and calmer. And two new pavilions were built to enrich the city's skyline. So far, the volume of Yingkuishan had reached its extreme, but it seemed to have overly frustrated the landscape and spatial logic, and its horizontal stretch, which implied political authority, making its subsequent 'effectiveness' less obvious.

In 1517, Liuguang Temple, which had been carrying a century of resentment, was moved out of the South Gate and the original site was changed to Lianchuanshuyuan. Yingkuishan finally relieved and was able to focus on the landscape and poetry.

学宫前原本郁闭的夹道空间南墙，也在这次改造中顺理成章地易为开敞的木栏，并在其中的文庙中轴南端嵌建应奎坊，或许正与夜光石相对——三百年学宫至此终得一气贯注，直面山水，朝飞暮卷，凝神畅意，仿佛主体建筑『隔水面山』的文人园林经典布局初成——当时正值江南造园热潮蓄势欲发之际，苏州府拙政园、嘉定县龚氏园（秋霞圃）、猗园（古猗园）、松江府城颐园、上海县豫园们即将粉墨登场。这次改造也初具自觉的理景意味——第一代应奎山堆掇后的科场赢家、成化十四年（一四七八年）进士、秋霞圃首任主人、工部尚书龚弘热切地期待道：『由是乾象坤势，山高水深，土膏毓秀，灵化潜融……科目之盛，孰不曰自此山欤？』

两年后的正德三年（一五〇八年），果然又是一科三进士。

三、山移西南，寺移南门。仅仅三年后的正德四年（一五〇九年），又一任巡按李廷梧再度倡导，将应奎山整体向西南迁移并进一步加高，使其与学宫间的空间关联更为舒缓从容，山上并新筑两亭，丰富了城市的天际线——至此应奎山体量已臻极致，但似令山水景象与空间逻辑顿挫过度，有失横向的舒展，隐含着政治威权意味，其后续『实效』亦不明显。

正德十二年（一五一七年），背负百年怨愤的留光寺被迁出南门，原址改为练川书院，应奎山水如释重负，得以专注于景象与诗情。

图十六　宾兴桥与文昌阁，摄于一八七〇年代

图十七　宾兴桥与文昌阁，摄于一八七〇年代

图十八　汇龙潭西北角新渠与育才坊旧影

图十九　一九三六年秋，上海友声旅行团在孔庙汇龙潭畔，侯黄二先生纪念碑前留影，左前为辅文山

Fig.16　Bingxin Bridge and Wenchangge, photographed in 1870s
Fig.17　Bingxin Bridge and Wenchangge, photographed in 1870s
Fig.18　An old photo of Xinqu and Yucaifang in the north-west corner of Huilongtan
Fig.19　In the autumn of 1936, the Shanghai Yousheng Tour Group took a photo in front of the Monument of Mr. Hou and Mr. Huang near Huilongtan of Wenmiao, in which Fuwenshan was in the left front

In 1540, in the southeast corner of Xuegong and on the east bank of Hengli, a small 'Fuwenshan' was added, which seemed to be the insignificant remaining vein of Yingkuishan, accompanied by the saying that 'the towering of the southeast makes the culture prosperous', which also strengthened the east space limitation in front of Xuegong. The next year, Jiading County also won three Jinshi in the examination. In 1549, Wenchangde was built across the road at Qingyun Bridge (Binxing Bridge) in front of Fuwenshan, and 'Longmen' plaque was hung high, making the space scene and imagery in front of Xuegong more full and layered. The next year, Jiading County still won three Jinshi in the examination.

Fourthly, cut the mountain, dig the pool, and move water gate to pass the river. When Yingkuishan was heightened for three times and Liuguang Temple was no longer a 'threat', local scholars might gradually realize that a horizontally linear water network city like Jiading had been already 'too many mountains than water'. If people arrogantly rised lonely peaks and vigorously vibrated the ground, maybe it was effective for a while but may not last long enough. Therefore after the climax of 1583 with winning four Jinshi in the examination, the city opened his eyes and combed the neighboring ground veins from the city or even the 'regional' scale, and switched from mountain worship to waterscape management, merging the mountain into the water and leading the mountain with water.

In 1588, Zhixian XIONG Mi added the 'Huilong' water gate to the east of Jiading's south water gate (which still exists so far), deliberately welcoming the water from Tai Lake and Wusong River into the city from the southeast, and guiding the circulation, called 'Shuimingtang' (Huilongtan) between Xuegong and Yingkuishan. Then the water went north through the city and into the Li River. At the same time, the height and volume of Yingkui Mountain had been greatly reduced to create a undulating and smooth horizontal landscape scene, in order to resonate with the horizontally stretched Xuegong building complex. 'And because of the south water coming from the north like arrow, and not being the accurate direction of southeast, so the water gate was moved to the east… Then it met the waters of Zhenze and Wusong, meandering and flowing. The Mingtang waters flowed, and were vast and blue. It was the origin of the gods and the prosperity, so called it Huilongtan, and the name of the water gate was also called Huilong.' The construction of the entire urban spatial context was then merely driven by the pool.

247

嘉靖十九年（一五四〇年），在学宫东南角、『横沥』东岸增筑小型『辅文山』，仿佛应奎山的渺渺余脉，附会了学宫『巽方高耸则文运昌盛』之说，也强化了宫前空间的东侧限定。翌年再度一科三进士。嘉靖二十八年（一五四九年），更在辅文山前的青云桥（宾兴桥）头跨路而建文昌阁，高悬『龙门』匾额，令宫前空间景象、意象更饱满而有层次。翌年仍获一科三进士。

四、削山凿潭，移关通江。当应奎山三度增高，留光寺『威胁』不再，地方士人或亦渐觉嘉定这样的横向线性水网城市已经『土胜于水』，如妄起孤峰、强振地脉，固能收效一时，却未必能够持久。于是在万历十一年（一五八三年）一科四进士的高潮后，放开眼界，自城市甚至区域尺度，对邻近地脉加以整体梳理，由山景崇拜转向水景经营，融山入水，以水带山。

万历十六年（一五八八年），知县熊密在嘉定南水关（至今犹存）以东添设『汇龙』水关，刻意迎纳太湖和吴淞江之水自东南方入城，引导其在学宫与应奎山之间回旋为『水明堂』（汇龙潭），再向北穿过城市，注入浏河。同时大幅削减应奎山的高度与体量，打造起伏舒缓、流畅浑成的横向山水景象，以求与横向舒展的学宫建筑群共振——『又以迤南之水箭激而北来，非正巽方也，故徙其关而东之……而下合震泽、吴淞之水蜿蜒迤迆而纳之，明堂万脉宗流，洸洋一碧，恍乎薄日月而撼烟涛，是神物之所由兴也，故命之曰汇龙潭，而关之名亦曰汇龙』——由区区一潭带动了整个城市空间脉络的营造。

图二十一　魁星阁下的石砌驳岸与潭中荡舟的游人，摄于一九三〇年代

图二十　晚清气息十足的魁星阁端立于嘉定庙、学东南水口，远处可见宾兴桥、文昌阁、仰高坊、当湖书院门等，摄于一九三〇年代

Fig.18  The late Qing style Kuixingge stood at the south-east water gate of Jiading County Wenmiao and Xianxue, and in the distance Binxingqiao, Wenchangge, Yanggaofang, and the Gate of Danghu Shuyuan can be seen, photographed in 1930s

Fig.19  The stone embankment under Kuixingge and the tourists rowing in the pond, photographed in 1930s

Fifthly, dredge the canal in view directions and gather the water as a dragon. In 1603, At the same time that Zhixian HAN Jun was repairing the halls of Xuegong, he further focused on the waterscape of Huilongtan and carved the water in front of Xuegong. On the east side of Yingkuishan, the old river Hengli which had been blocked by mountains since 1460 was restored, and North Yangshubang in the northeast, South Yangshubang in the southeast, and Tangjiabang flowing into Huilongtan were welcomed. On the west side of Yingkuishan, the old river of Yenujing was restored in the southwest, and a new canal was drilled in the northwest to pass Yenujing, and to welcome the two rivers into the lake. Therefore, the mountain was surrounded by water on all sides, while Xuegong faced the mountain across the water (tan), and the pool was surrounded by five waters, which was the trend of 'five dragons playing with the pearl', and finally cross Bingxing Bridge, vent from 'Hengli' and Fuahua Tower to the north.

At the same time, the volume of mountain was further reduced, the pool was expanded, and the horizontal integration of the landscape and Xuegong was strengthen, just like 'Nan Zong' landscape scrolling. This was also in sync with the gradual shift from the vertical Zhang Nanyang style to the horizontal Zhang Lian style of the contemporary garden-making technique.

Finally, Kuixingting (later called Kuixingge) was built at the south of Wenchangde and southeast of Huilongtan to defend the embankment. Therefore, the entire layout of the building complex was laid out in a 'facing mountain across the water' style, just like a large-scale literati garden in the late Ming and early Qing Dynasties, while Kuixingting at the location where Biguangting in Jiading Qiuxiapu Garden, the Qintai in Songjiang Yi Garden, and the Zhiyujian in Jichang Garden in Wuxi.

Five years later in 1608, the city center on the north side of the Xuegong, the Fahua Tower of the Southern Song Dynasty at the intersection of Lianqi River and Hengli was rebuilt as a city 'brush', while Huilongtan was like the 'inkstone' of the city, echoed by the image and interspersed as a corridor of sceneries.

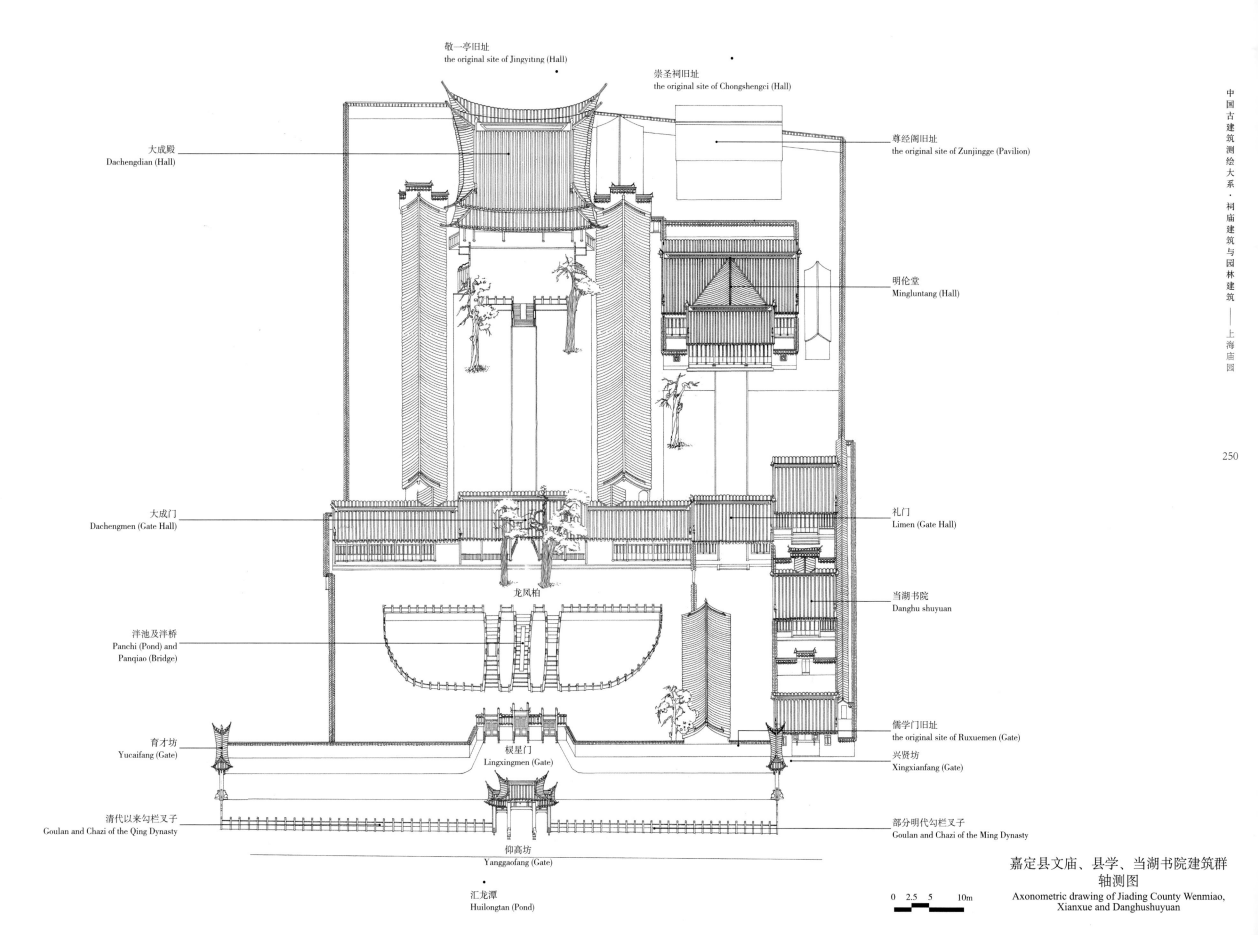

敬一亭旧址
the original site of Jingyiting (Hall)

崇圣祠旧址
the original site of Chongshengci (Hall)

尊经阁旧址
the original site of Zunjingge (Pavilion)

大成殿
Dachengdian (Hall)

明伦堂
Mingluntang (Hall)

大成门
Dachengmen (Gate Hall)

礼门
Limen (Gate Hall)

龙凤柏

当湖书院
Danghu shuyuan

泮池及泮桥
Panchi (Pond) and
Panqiao (Bridge)

育才坊
Yucaifang (Gate)

儒学门旧址
the original site of Ruxuemen (Gate)

棂星门
Lingxingmen (Gate)

兴贤坊
Xingxianfang (Gate)

清代以来勾栏叉子
Goulan and Chazi of the Qing Dynasty

部分明代勾栏叉子
Goulan and Chazi of the Ming Dynasty

仰高坊
Yanggaofang (Gate)

汇龙潭
Huilongtan (Pond)

嘉定县文庙、县学、当湖书院建筑群
轴测图
Axonometric drawing of Jiading County Wenmiao,
Xianxue and Danghushuyuan

0   2.5   5      10m

And eight years later in 1616, Wenchangge of Xuegong was moved slightly east to Fuwenshan, forming a barrier and high landmark in the southeast of the Xuegong, which implied a mountain meaning and echoes Kuifeng. Finally, this landscape chapter changed from emphasizing the mountain to emphasizing the water, from vertical to horizontal, from blocking to leading, from static to dynamic, from isolated to grand, and completed nearly two hundred years of unremitting writing.

The plan of the main part of the Jiading Academy, the platform foundation, the column foundation, and a considerable part of the large wooden structure are still relics from the middle and late Yuan Dynasty to the early Ming Dynasty. The layout of the three Paifangs and the 'Chazi' and fence in the leading part was established during 1506-1521, and there were also some original remains. The south and southeast corner of the garden area was finalized in the late Ming Dynasty, of which although the buildings were newly rebuilt, there are still many Ming Dynasty stone structures in Binxing Bridge. The entire historical relics and artistic value of Jiading Xuegong and its garden are really enough to disregard Jiangnan area.

Living in a prosperous, deeply Confucianized and firstly citizenized Jiangnan area, Jiading, with the heart of 'holding firmly', hundred years of continuously polished the core Confucian space of the city, and finally recorded in the history together with celebrities and heroes in the late Ming Dynasty and magnates and scholars in the Qing Dynasty, becoming an immortal example of urban spiritual space creation.

又八年后的万历四十四年（一六一六年），学宫文昌阁被微微东移至辅文山上，形成学宫东南的屏障与高标，隐含山意，呼应奎峰，最终使这一山水篇章由重山而重水，由纵向而横向，由屏堵而导引，由静态而动态，由孤立而宏大，完成了近两百年的不懈书写。

嘉定学宫主体部分的平面布局、台基、柱础，以及相当部分大木结构，犹为元中后期至明前期的遗物；其前导部分的三坊及叉子勾栏的布局，则确立于明正德年间（一五〇六至一五二一年），亦有部分初始原物遗存；而其南部及东南隅山水苑区，则定型于明晚期，虽建筑均为新近重建，但宾兴桥尚存诸多明代石构——其整体历史遗存与艺术价值，实足以傲视江南。

身居富庶繁华、深度儒化而又率先市民化的江南海隅，嘉定一城以『固守』自持之心，对城市核心儒教空间不断打磨，百年未倦，并最终共着明末侯黄英烈，清代钱王巨儒而长垂史册，成为城市精神空间营造的不朽范例。

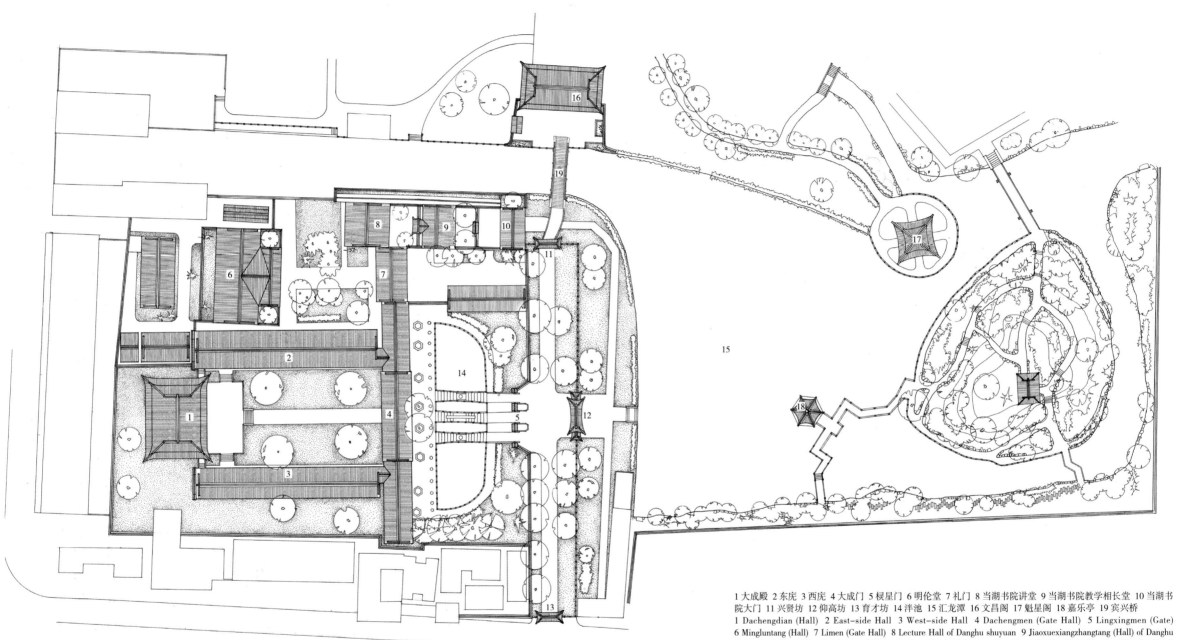

1 大成殿 2 东庑 3 西庑 4 大成门 5 棂星门 6 明伦堂 7 礼门 8 当湖书院讲堂 9 当湖书院教学相长堂 10 当湖书院大门 11 兴贤坊 12 仰高坊 13 育才坊 14 泮池 15 汇龙潭 16 文昌阁 17 魁星阁 18 嘉乐亭 19 宾兴桥

1 Dachengdian (Hall) 2 East-side Hall 3 West-side Hall 4 Dachengmen (Gate Hall) 5 Lingxingmen (Gate)
6 Mingluntang (Hall) 7 Limen (Gate Hall) 8 Lecture Hall of Danghu shuyuan 9 Jiaoxuexiangzhangtang (Hall) of Danghu
shuyuan 10 Gate of Danghu shuyuan 11 Xingxianfang (Gate) 12 Yanggaofang (Gate) 13 Yucaifang (Gate) 14 Panchi (Pond)
15 Huilongtan (Pond) 16 Wenchangge (Pavilion) 17 Kuixingge (Pavilion) 18 Jialeting (Pavilion) 19 Binxingqiao (Bridge)

嘉定县文庙、县学、当湖书院、汇龙潭、应奎山建筑群屋顶平面图
Roof plan of Jiading County Wenmiao, Xianxue, Danghushuyuan, Huilongtan and Yingkuishan

0  6  12  24m

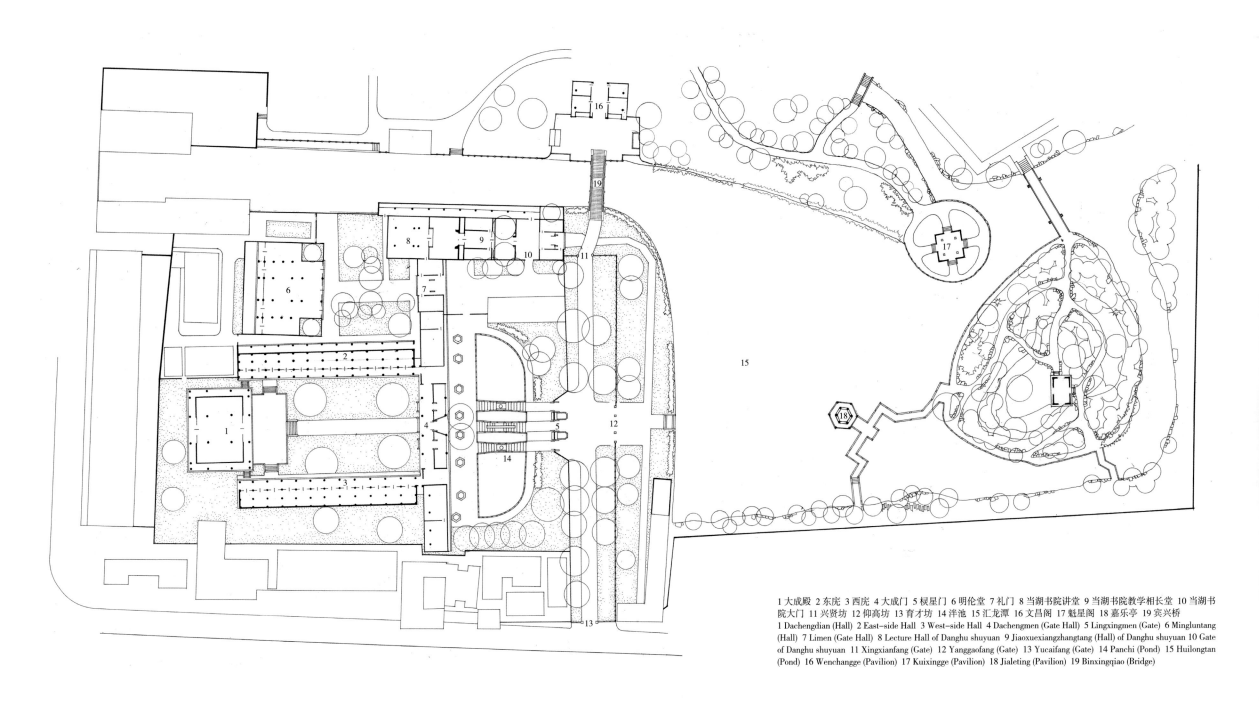

1 大成殿 2 东庑 3 西庑 4 大成门 5 棂星门 6 明伦堂 7 礼门 8 当湖书院讲堂 9 当湖书院教学相长堂 10 当湖书院大门 11 兴贤坊 12 仰高坊 13 育才坊 14 泮池 15 汇龙潭 16 文昌阁 17 魁星阁 18 嘉乐亭 19 宾兴桥

1 Dachengdian (Hall) 2 East-side Hall 3 West-side Hall 4 Dachengmen (Gate Hall) 5 Lingxingmen (Gate) 6 Mingluntang (Hall) 7 Limen (Gate Hall) 8 Lecture Hall of Danghu shuyuan 9 Jiaoxuexiangzhangtang (Hall) of Danghu shuyuan 10 Gate of Danghu shuyuan 11 Xingxianfang (Gate) 12 Yanggaofang (Gate) 13 Yucaifang (Gate) 14 Panchi (Pond) 15 Huilongtan (Pond) 16 Wenchangge (Pavilion) 17 Kuixingge (Pavilion) 18 Jialeting (Pavilion) 19 Binxingqiao (Bridge)

0  6  12  24m

嘉定县文庙、县学、当湖书院、汇龙潭、应奎山建筑群一层平面图
Ground floor plan of Jiading County Wenmiao, Xianxue, Danghushuyuan, Huilongtan and Yingkuishan

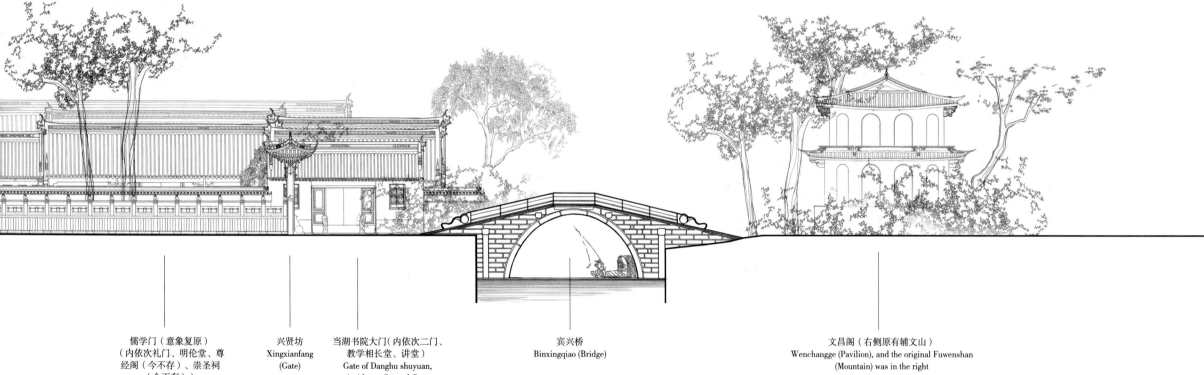

儒学门（意象复原）
（内依次礼门、明伦堂、尊
经阁（今不存）、崇圣祠
（今不存））
Ruxuemen (Gate) by
imagination, inside are Limen
(Gate Hall), Mingluntang (Hall),
Zunjingge (Pavilion, destroyed)
and Chongshengci (Hall,
destroyed)

兴贤坊
Xingxianfang
(Gate)

当湖书院大门（内依次二门、
教学相长堂、讲堂）
Gate of Danghu shuyuan,
inside are Second Gate
Jiaoxuexiangzhangtang (Hall)
and Lecture Hall

宾兴桥
Binxingqiao (Bridge)

文昌阁（右侧原有辅文山）
Wenchangge (Pavilion), and the original Fuwenshan
(Mountain) was in the right

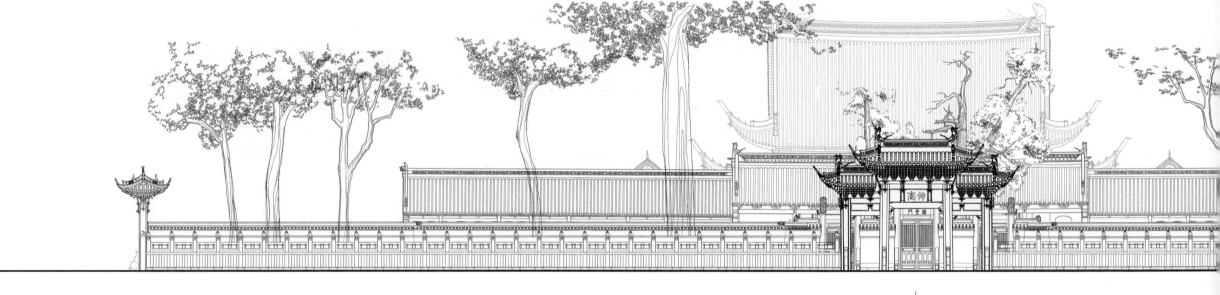

育才坊
Yucaifang (Gate)

仰高仿（内依次棂星门、泮池泮桥、
大成门、大成殿）
Yanggaofang (Gate), inside are
Lingxingmen (Gate), Panchi (Pond) and
Panqiao (Bridge), Dachengmen (Gate Hall)
and Dachengdian (Hall)

0    1    2        5m

注：文昌阁系根据历史照片示意复原

**嘉定县学宫南立面图**
South elevation of Jiading Xuegong

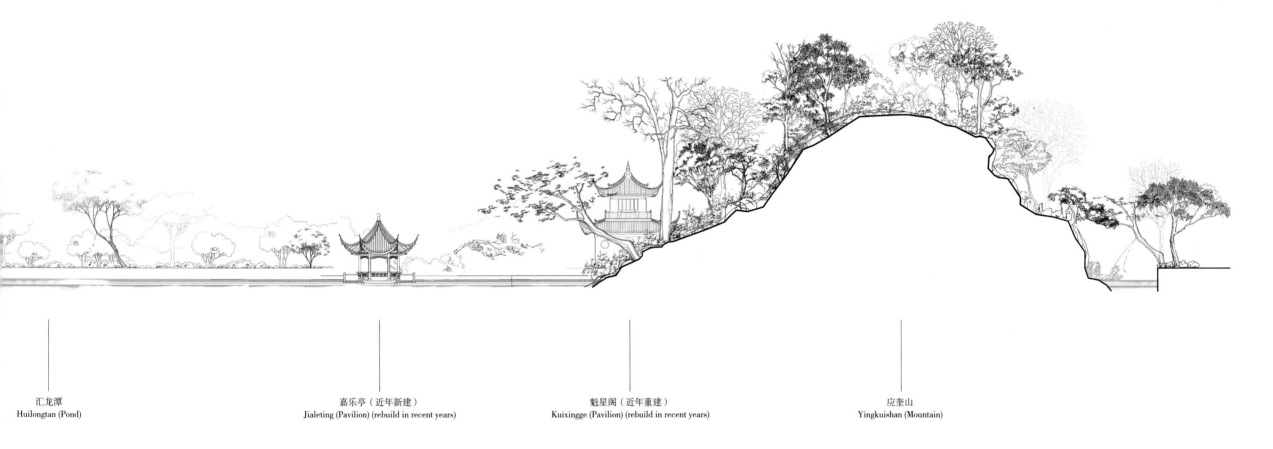

汇龙潭
Huilongtan (Pond)

嘉乐亭（近年新建）
Jialeting (Pavilion) (rebuild in recent years)

魁星阁（近年重建）
Kuixingge (Pavilion) (rebuild in recent years)

应奎山
Yingkuishan (Mountain)

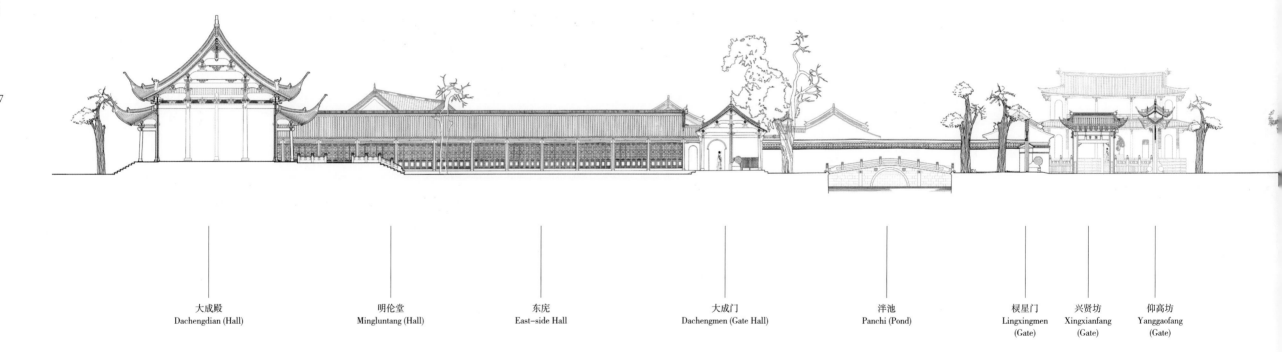

| 大成殿 | 明伦堂 | 东庑 | 大成门 | 泮池 | 棂星门 | 兴贤坊 | 仰高坊 |
|---|---|---|---|---|---|---|---|
| Dachengdian (Hall) | Mingluntang (Hall) | East-side Hall | Dachengmen (Gate Hall) | Panchi (Pond) | Lingxingmen (Gate) | Xingxianfang (Gate) | Yanggaofang (Gate) |

注：文昌阁系根据历史照片示意复原

嘉定县文庙中轴剖面图
Section of Jiading Wenmiao's main axis

0  10  20      50m

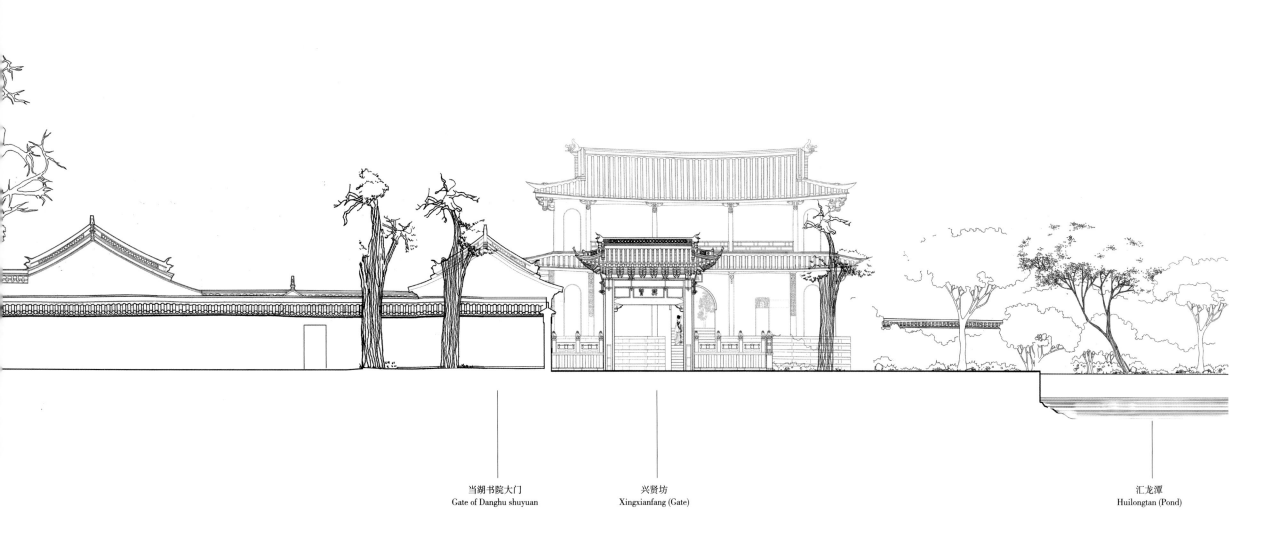

当湖书院大门
Gate of Danghu shuyuan

兴贤坊
Xingxianfang (Gate)

汇龙潭
Huilongtan (Pond)

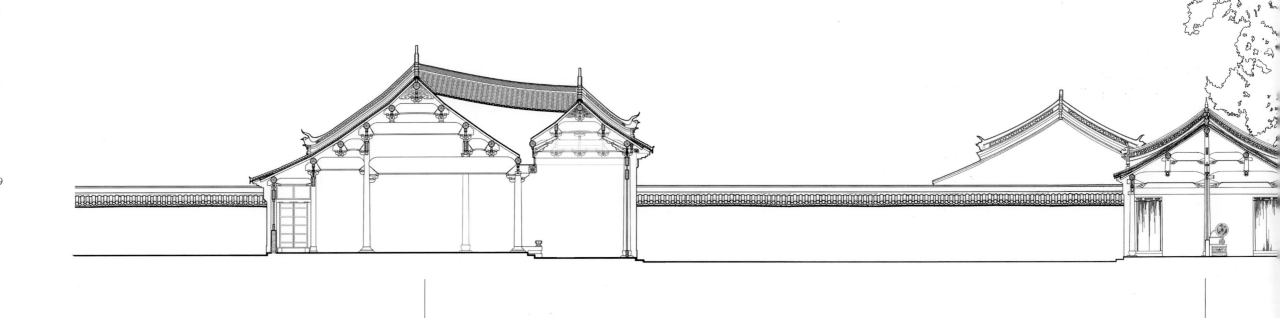

明伦堂（其北尊经阁、崇圣祠已毁）
Mingluntang (Hall), Zunjingge (Hall)
and Chongshengci (Hall) in the north of
Mingluntang have been destroyed

礼门（其南首进儒学门已毁）
Limen (Gate Hall), Ruxuemen (Gate Hall)
in the South of Limen has been destroyed

注：文昌阁系根据历史照片示意复原

0  1  2          5m

嘉定县学中轴剖面图
Section of Jiading Xianxue's main axis

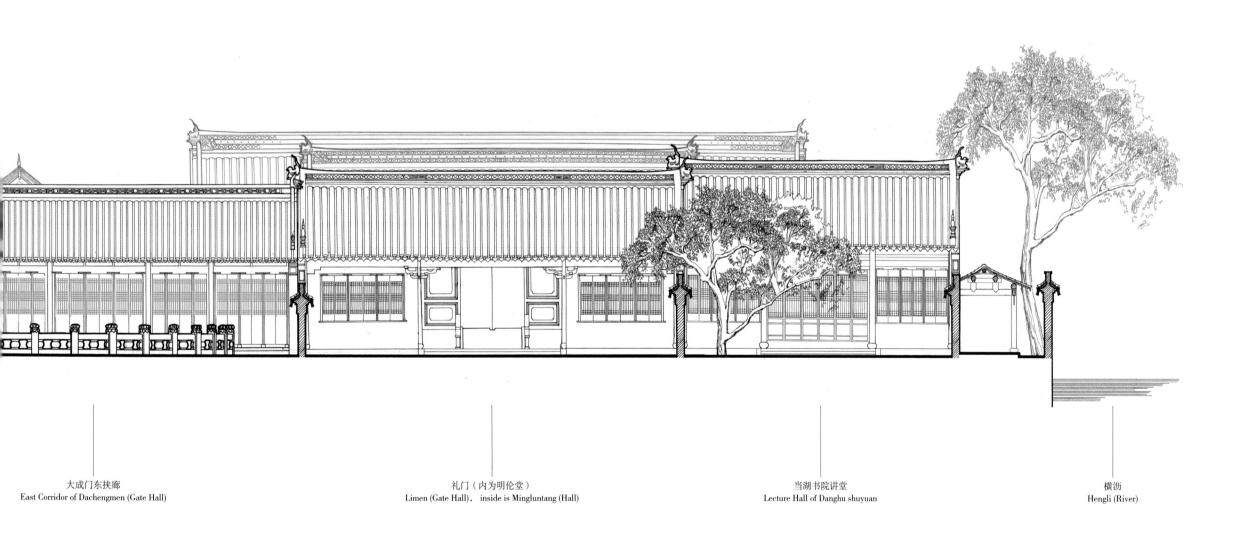

大成门东挟廊
East Corridor of Dachengmen (Gate Hall)

礼门（内为明伦堂）
Limen (Gate Hall)，inside is Mingluntang (Hall)

当湖书院讲堂
Lecture Hall of Danghu shuyuan

横沥
Hengli (River)

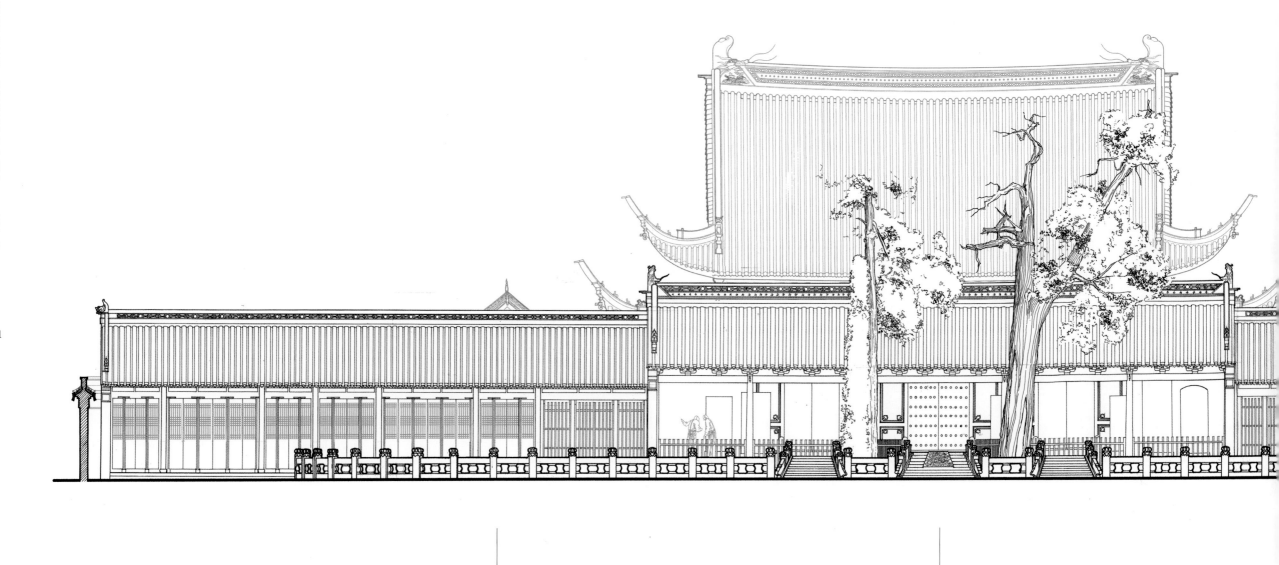

大成门西挟廊
West Corridor of Dachengmen (Gate Hall)

大成门（内为大成殿）
Dachengmen (Gate Hall)，inside is Dachengdian (Hall)

0　1　2　　　　5m

大成门及县学礼门等南立面
South elevation of Dachengmen and Limen

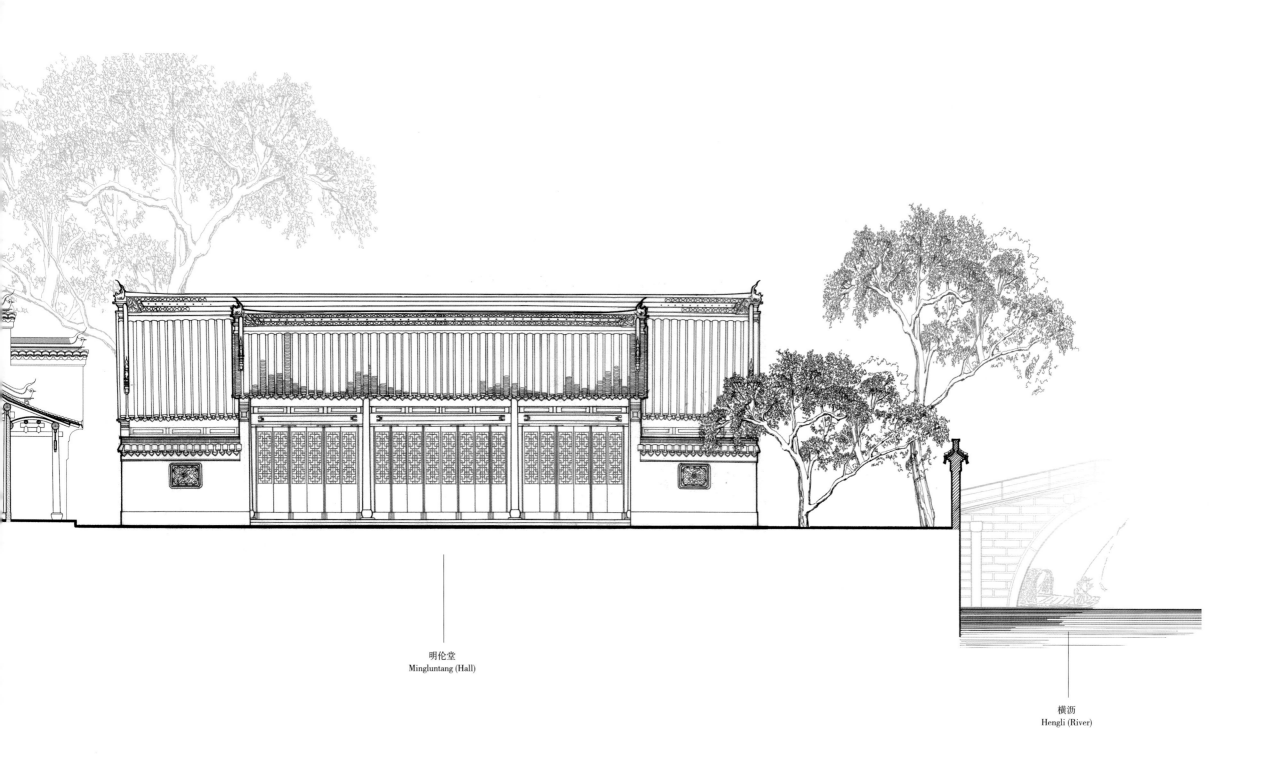

明伦堂
Mingluntang (Hall)

横沥
Hengli (River)

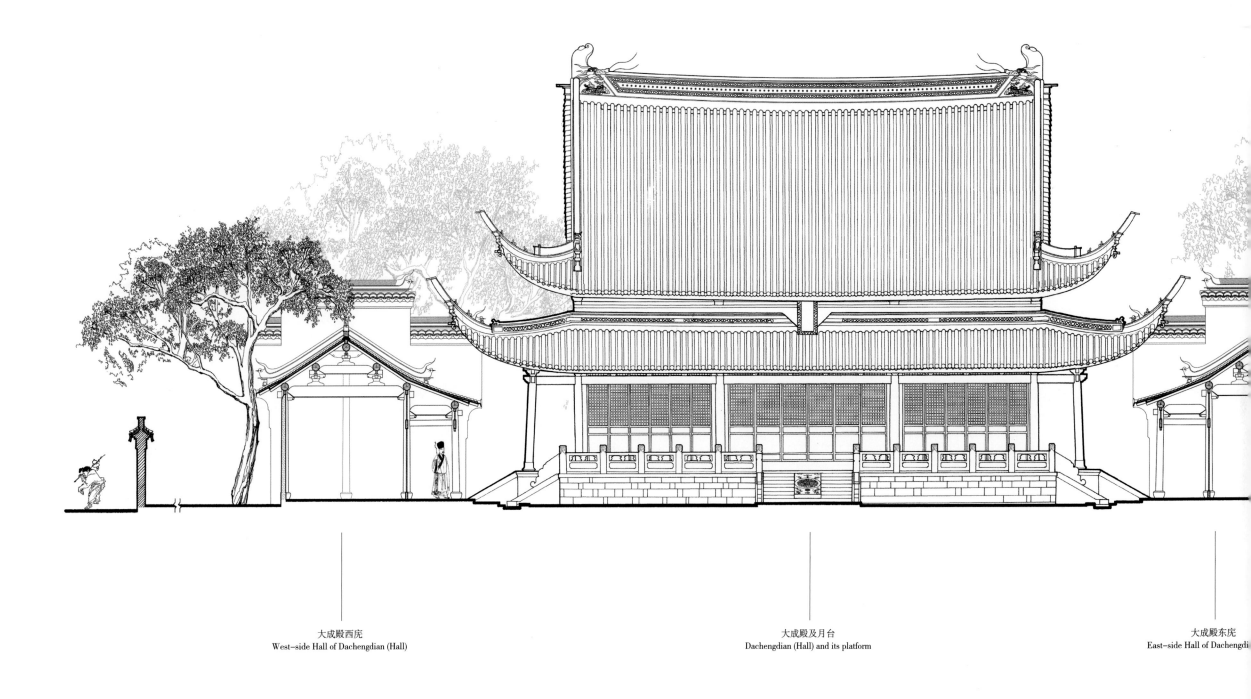

大成殿西庑
West-side Hall of Dachengdian (Hall)

大成殿及月台
Dachengdian (Hall) and its platform

大成殿东庑
East-side Hall of Dachengdi

注：大成殿两庑内四界梁架被天花遮挡未能测绘，按大成门梁架绘制

## 大成殿及县学明伦堂等南立面
South elevation of Dachengdian and Mingluntang

0 1 2 5m

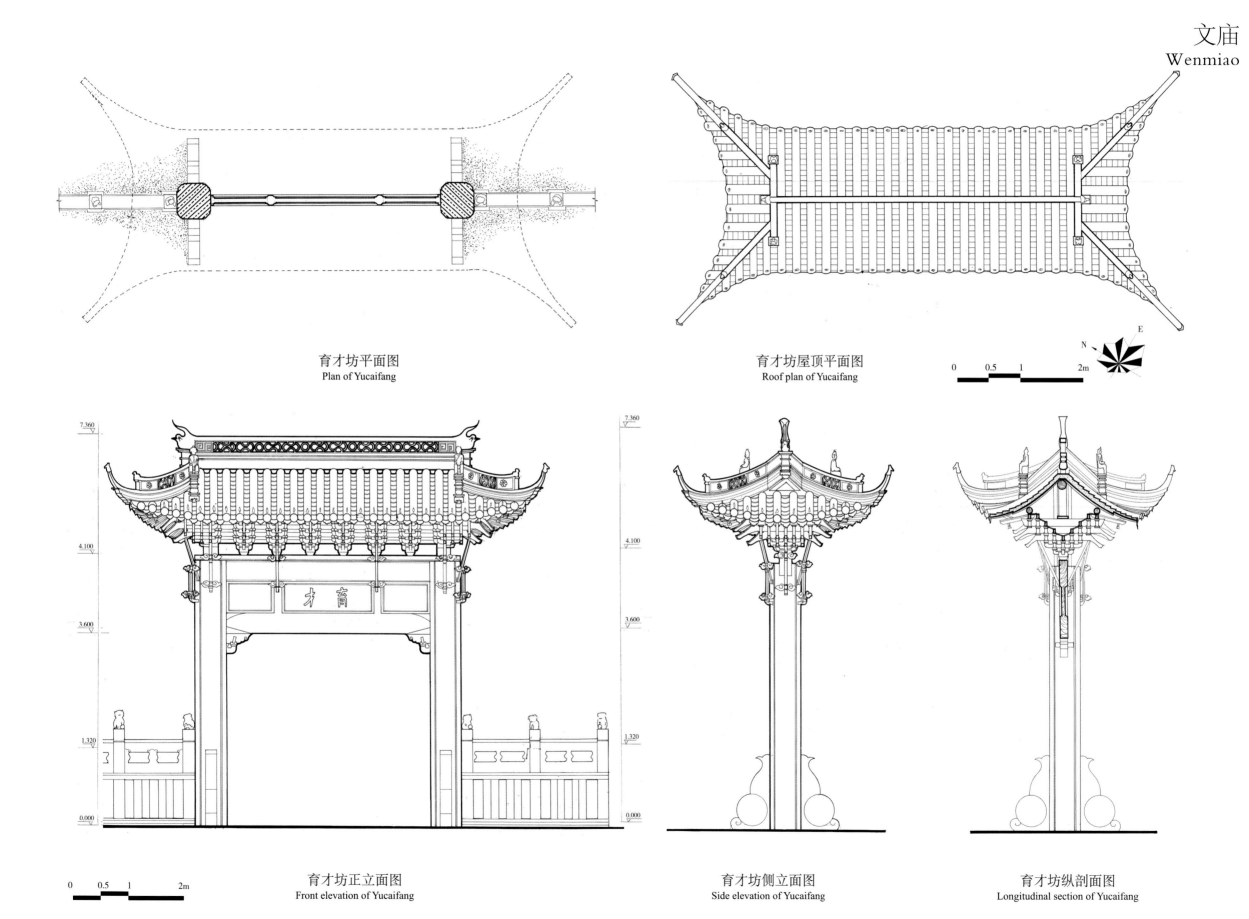

育才坊平面图
Plan of Yucaifang

育才坊屋顶平面图
Roof plan of Yucaifang

N E

0    0.5    1         2m

育才坊正立面图
Front elevation of Yucaifang

0    0.5    1         2m

育才坊侧立面图
Side elevation of Yucaifang

育才坊纵剖面图
Longitudinal section of Yucaifang

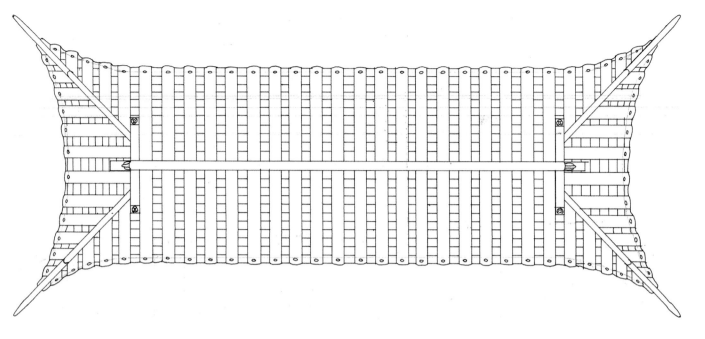

兴贤坊屋顶平面图
Roof plan of Xingxianfang

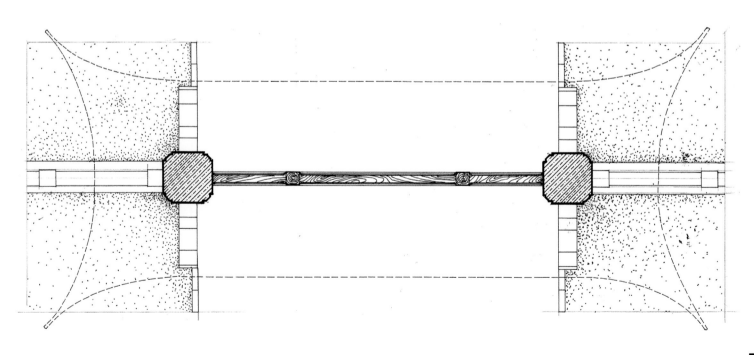

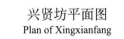

0 0.25 0.5 1m

N E

兴贤坊平面图
Plan of Xingxianfang

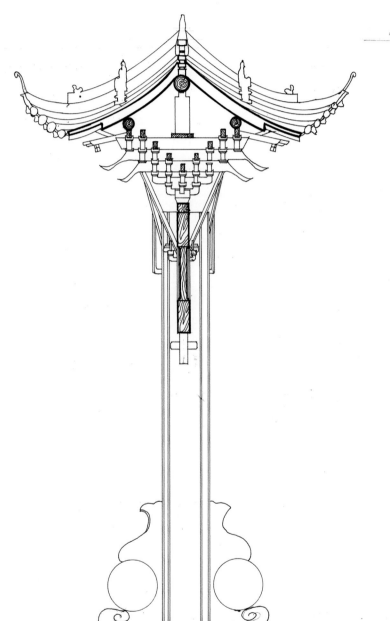

兴贤坊纵剖面图
Longitudinal section of Xingxianfang

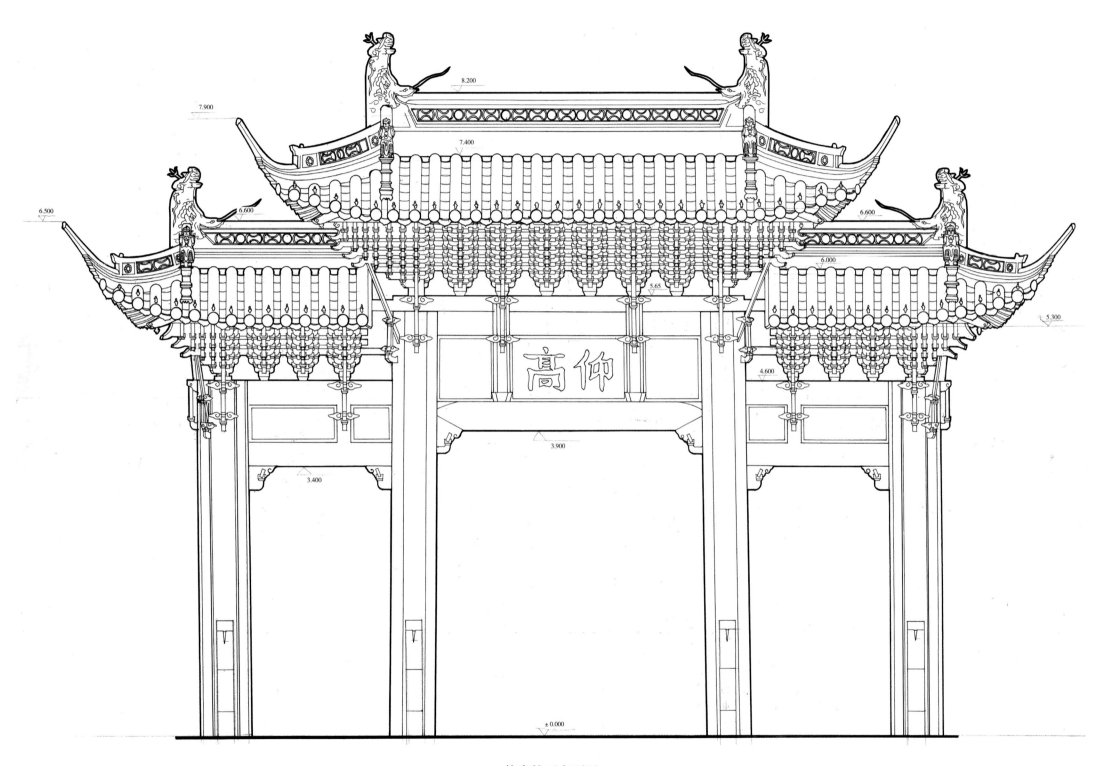

仰高坊正立面图
Front elevation of Yanggaofang

0　0.25　0.5　　1m

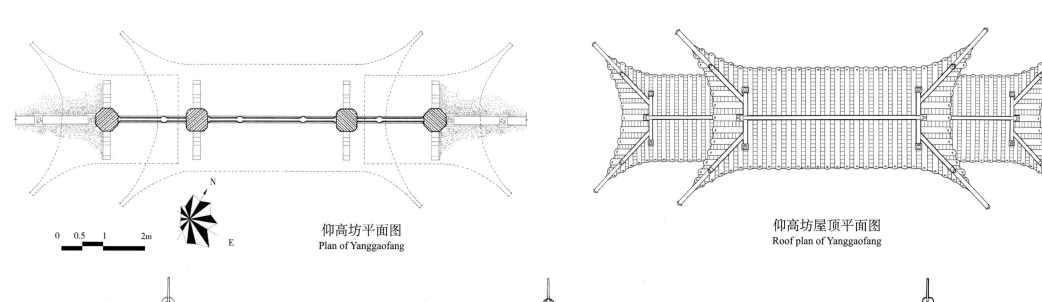

仰高坊平面图
Plan of Yanggaofang

0 0.5 1 2m

仰高坊屋顶平面图
Roof plan of Yanggaofang

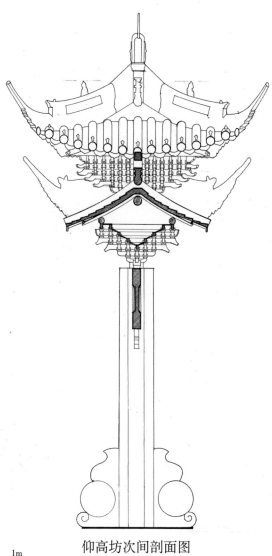

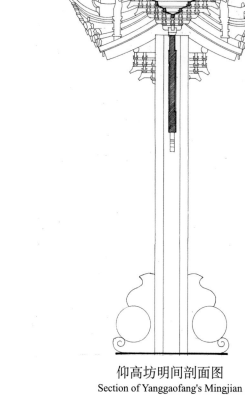

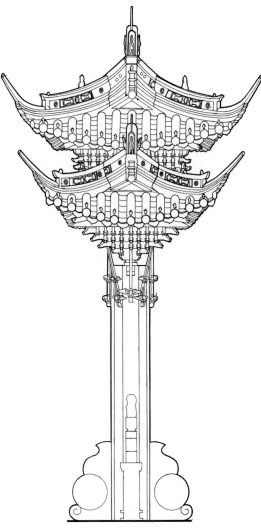

0 0.25 0.5 1m

仰高坊次间剖面图
Section of Yanggaofang's Cijian

仰高坊明间剖面图
Section of Yanggaofang's Mingjian

仰高坊侧立面图
Side elevation of Yanggaofang

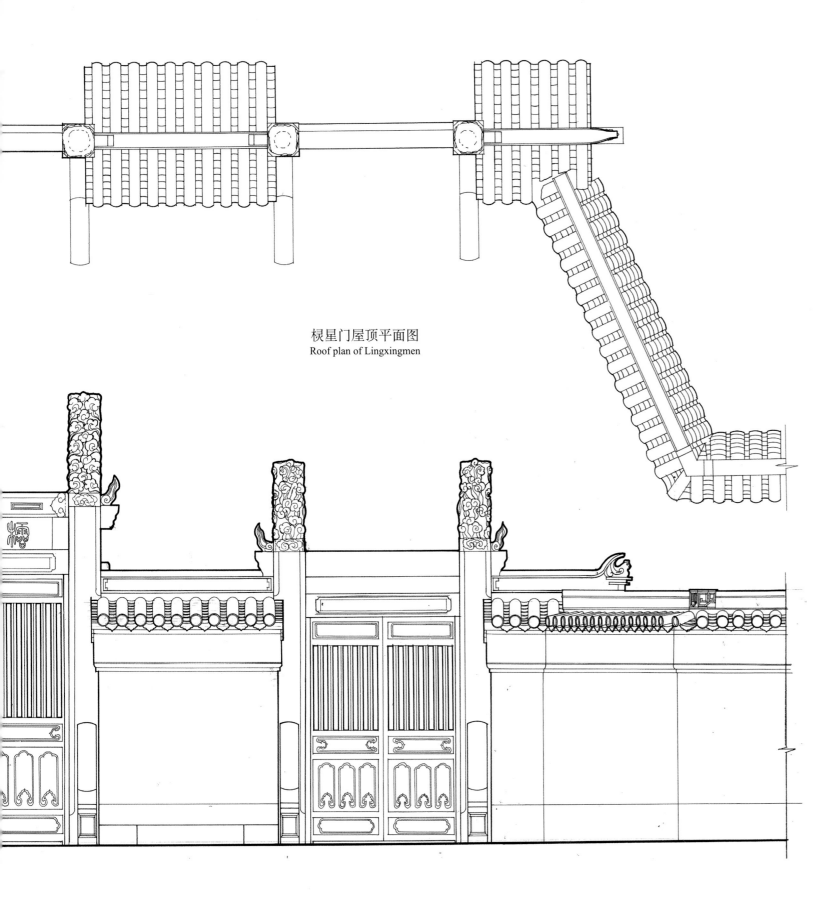

棂星门屋顶平面图
Roof plan of Lingxingmen

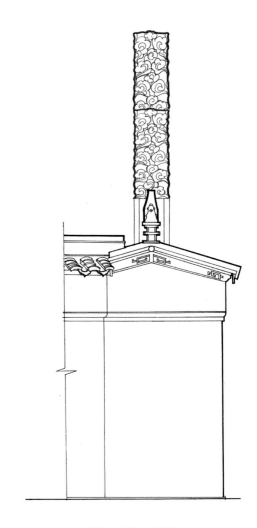

棂星门侧立面图
Side elevation of Lingxingmen

0　0.25　0.5　　1m

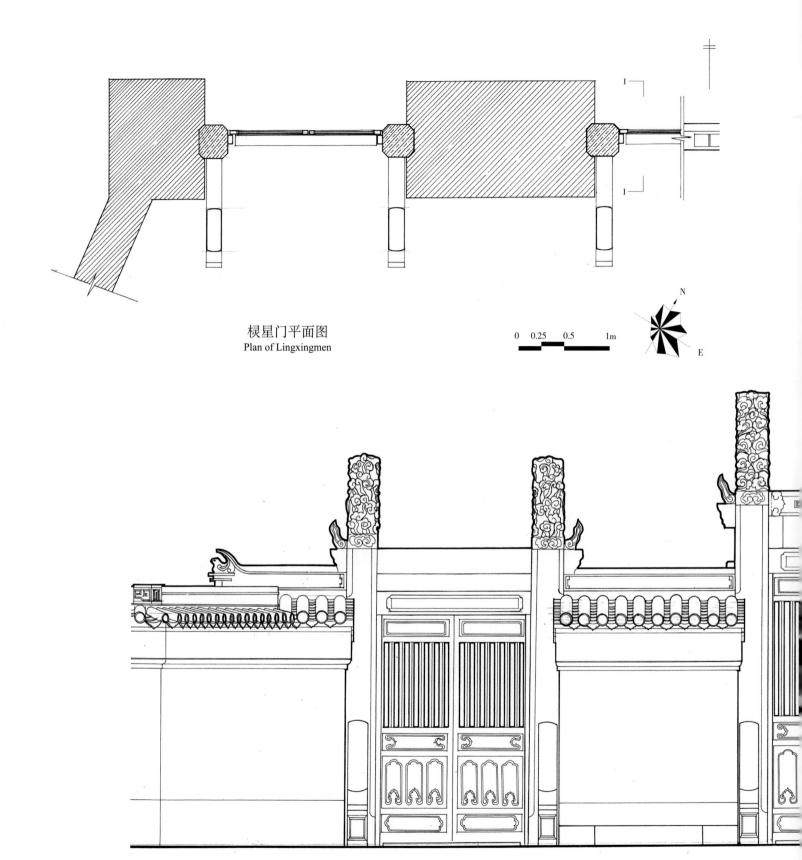

棂星门平面图
Plan of Lingxingmen

0  0.25  0.5    1m

棂星门 1—1 剖面图
Section 1-1 of Lingxingmen

棂星门正立面图
Front elevation of Lingxingmen

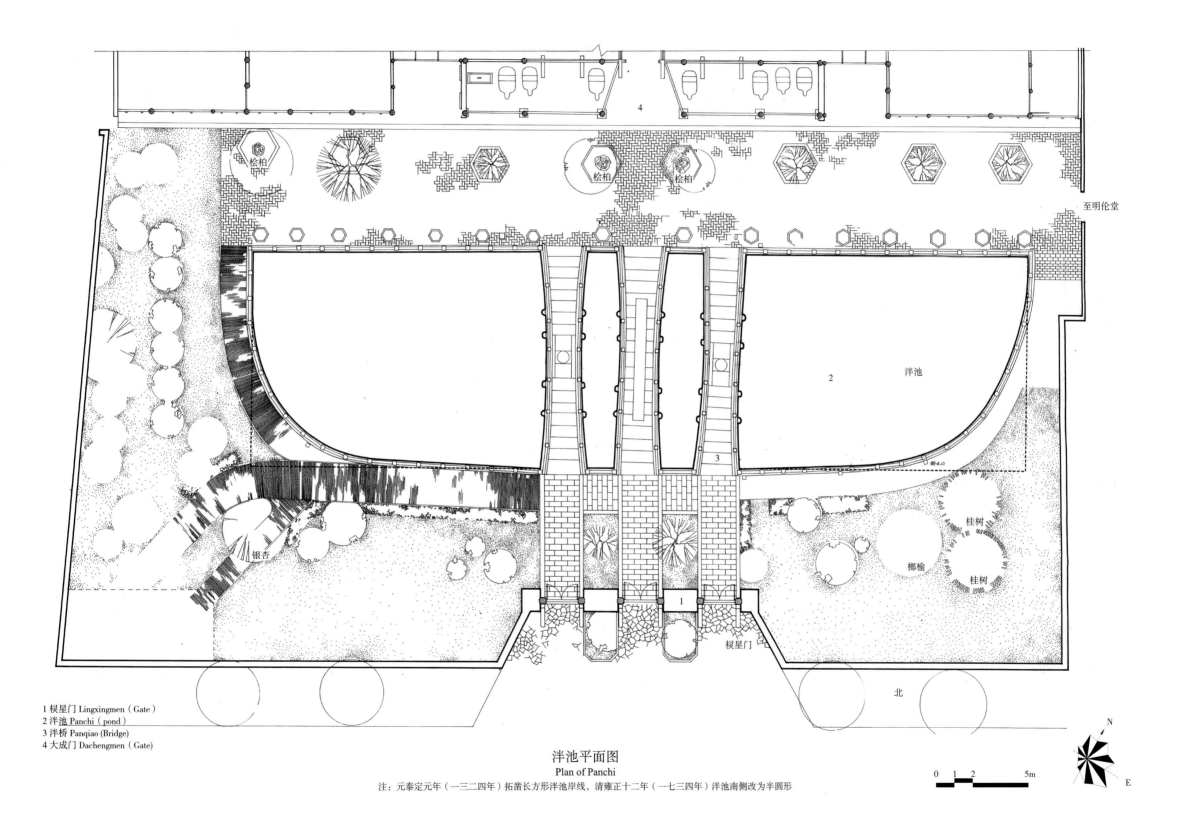

至明伦堂

桧柏

桧柏

桧柏

4

2 泮池

3

银杏

桂树

椰榆

桂树

1

棂星门

北

1 棂星门 Lingxingmen（Gate）
2 泮池 Panchi（pond）
3 泮桥 Panqiao (Bridge)
4 大成门 Dachengmen（Gate)

泮池平面图
Plan of Panchi
注：元泰定元年（一三二四年）拓凿长方形泮池岸线，清雍正十二年（一七三四年）泮池南侧改为半圆形

0 1 2    5m

N

E

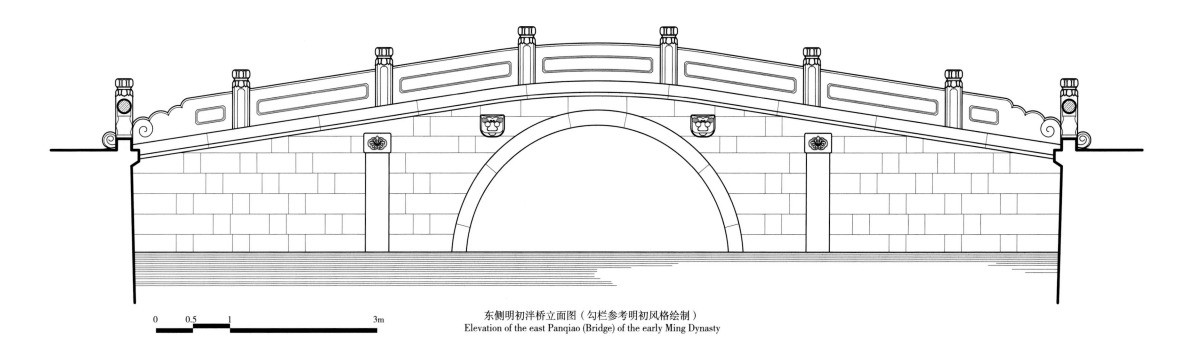

0    0.5    1                    3m

东侧明初泮桥立面图（勾栏参考明初风格绘制）
Elevation of the east Panqiao (Bridge) of the early Ming Dynasty

0    0.25    0.5    1m

泮桥云龙纹御路大样图（元明）
Yulu of cloud and dragon pattern on Panqiao (Bridge)

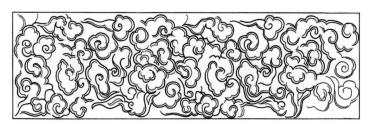

长系石大样图（中晚清）
Changxishi of the middle and late Qing Dynasty

长系石大样图（明初）
Changxishi of the early Ming
Dynasty

排水井盖大样图
Manhole Cover

0    0.1    0.2              0.5m

泮桥立面及详图
Elevation and Detail of Panqiao

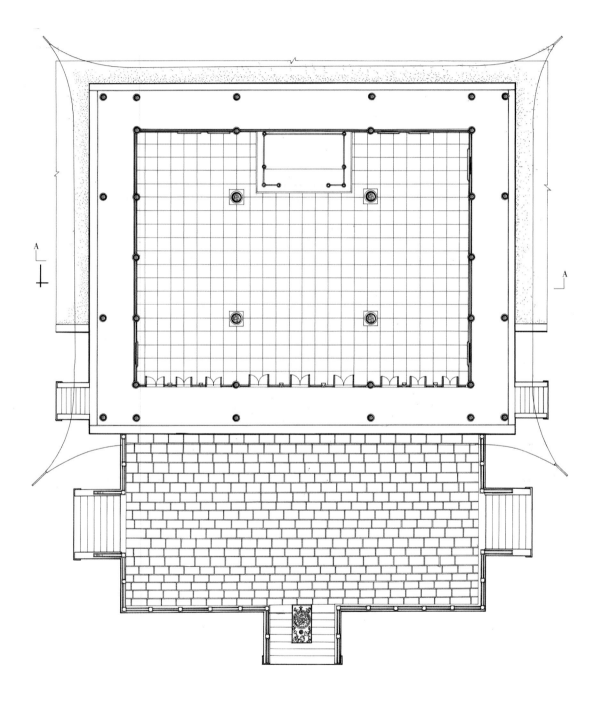

大成殿平面图
Plan of Dachengdian

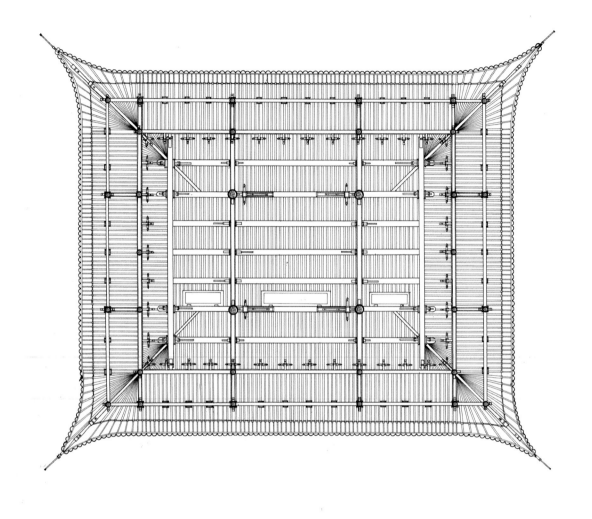

大成殿屋架仰视图
Bottom view of Dachengdian's beams

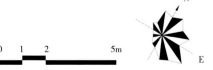

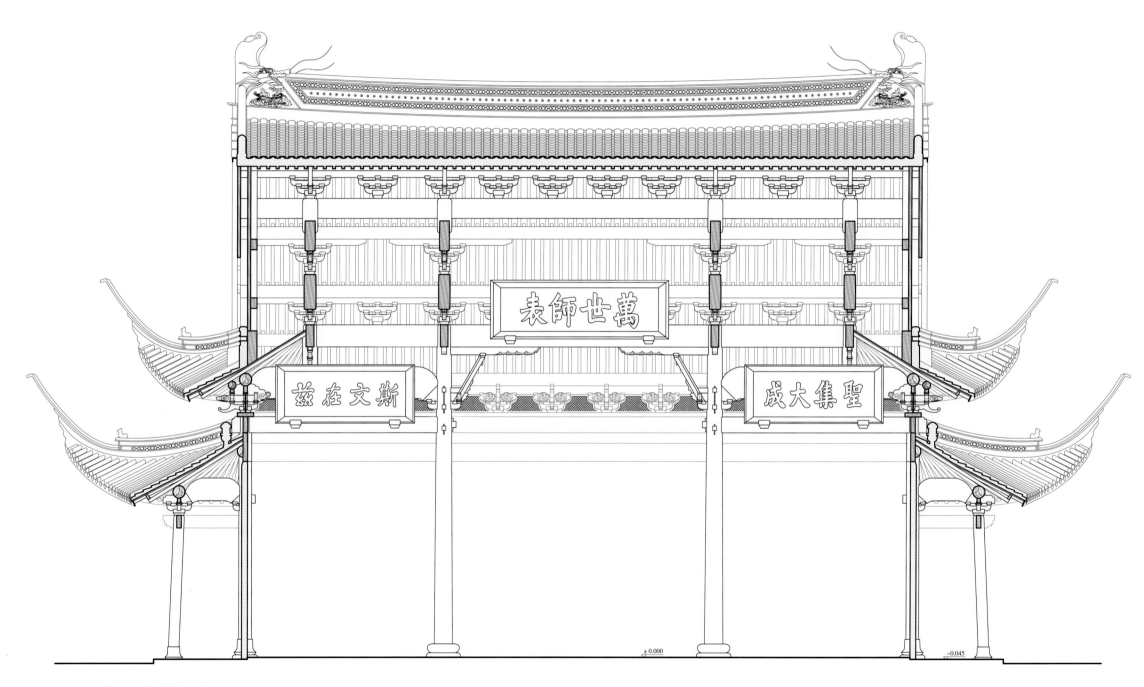

萬世師表

斯文在兹

聖集大成

±0.000

-0.045

注：山花板内侧不可见部分系推测所得

0    1    2                    5m

大成殿纵剖面图
Longitudinal section of Dachengdian

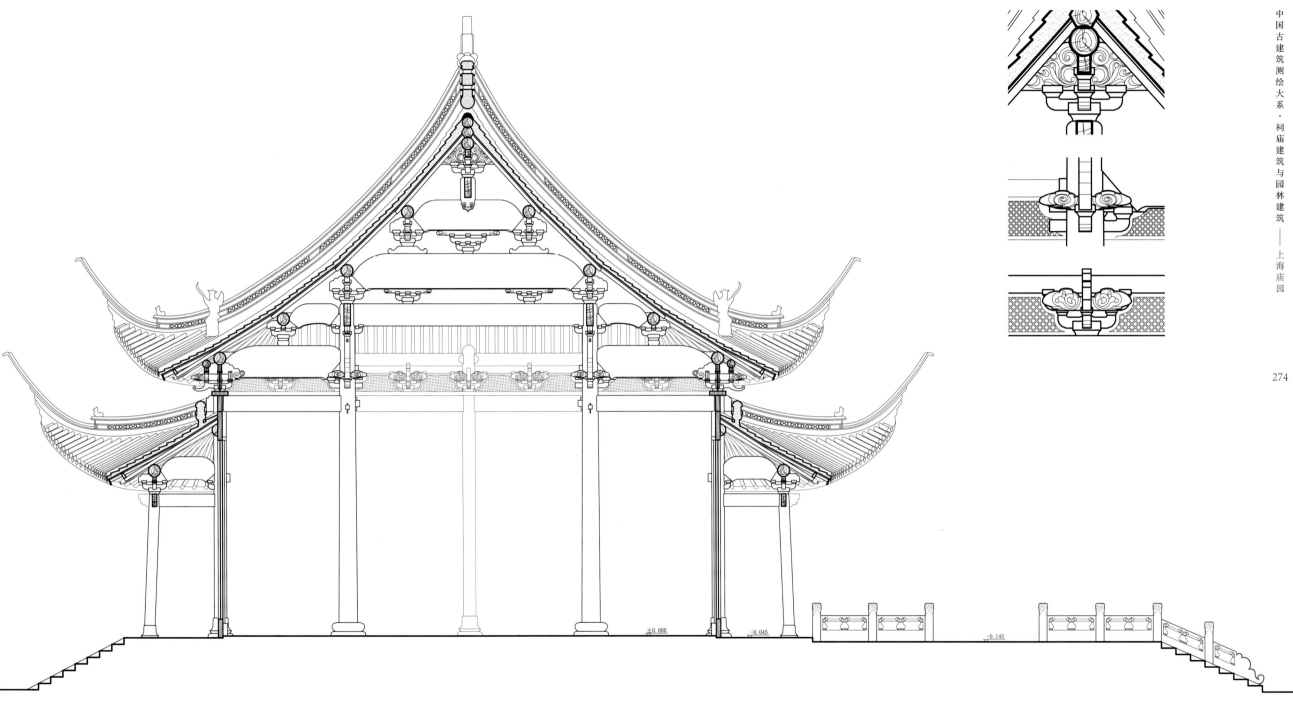

±0.000

−0.045

−0.145

大成殿正贴剖面图
Section of Dachengdian's Zhengtie

0    1    2        5m

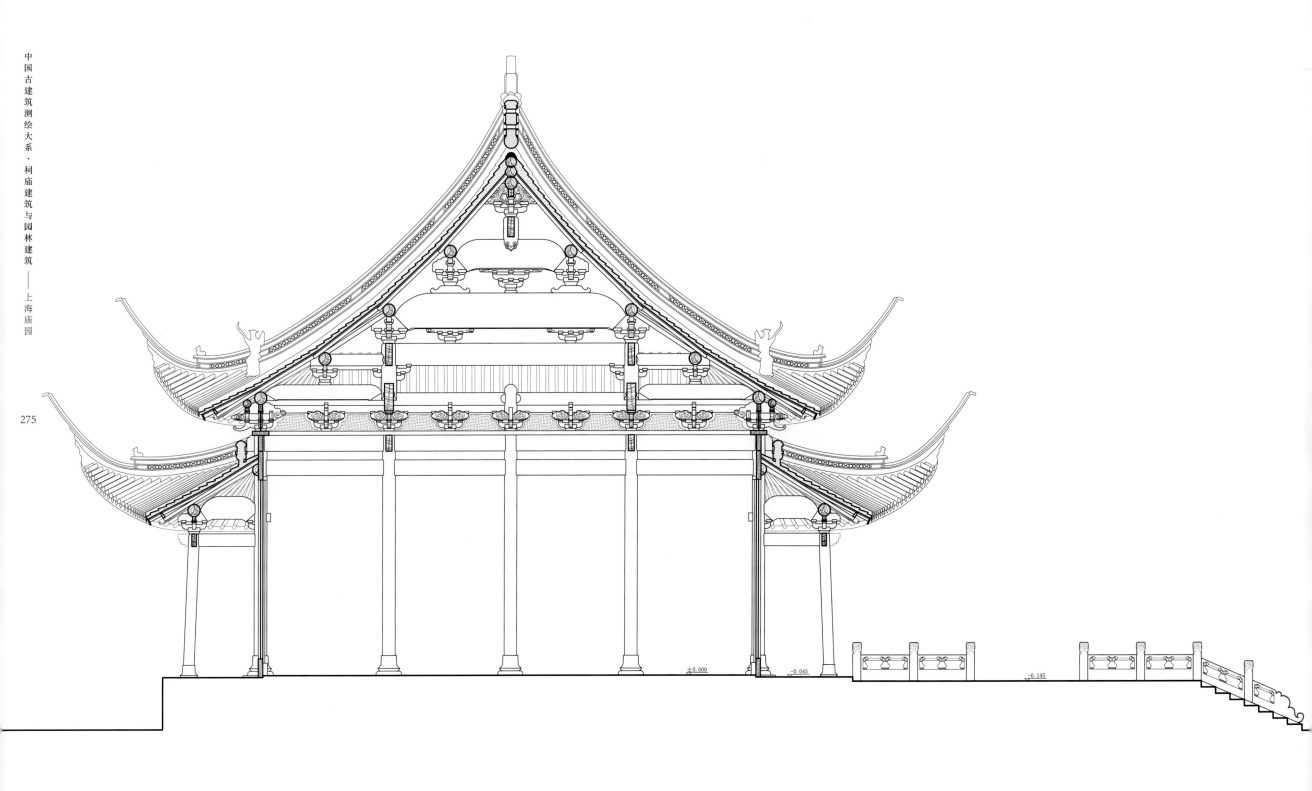

±0.000    -0.045    -0.145

0    1    2    5m

大成殿边贴剖面图
Section of Dachengdian's Biantie

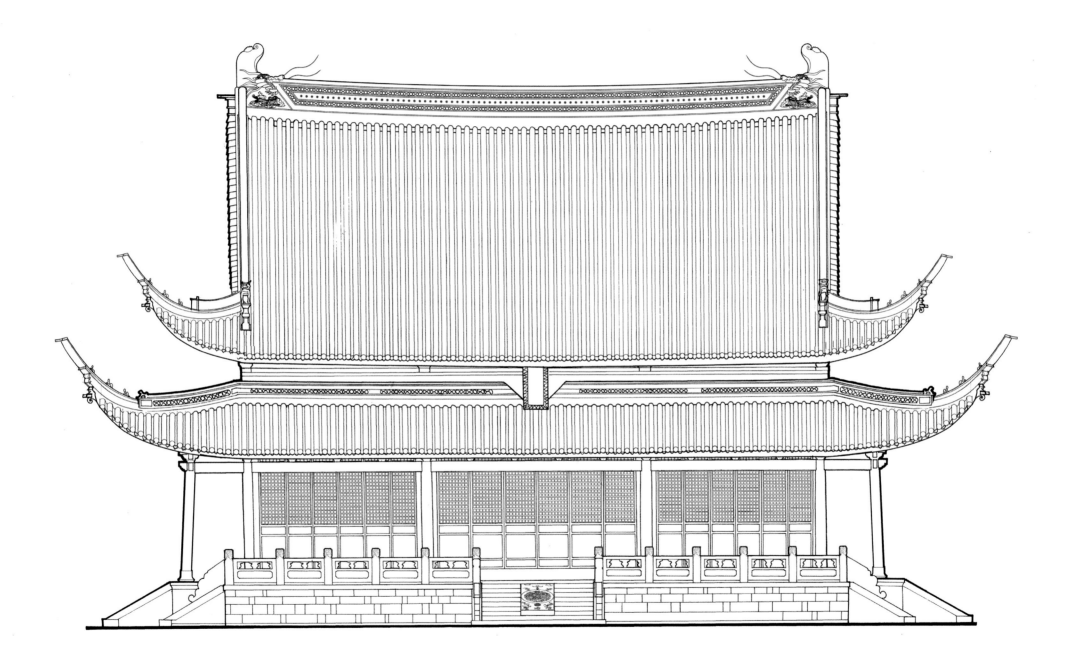

大成殿正立面图
Front elevation of Dachengdian

0  1  2      5m

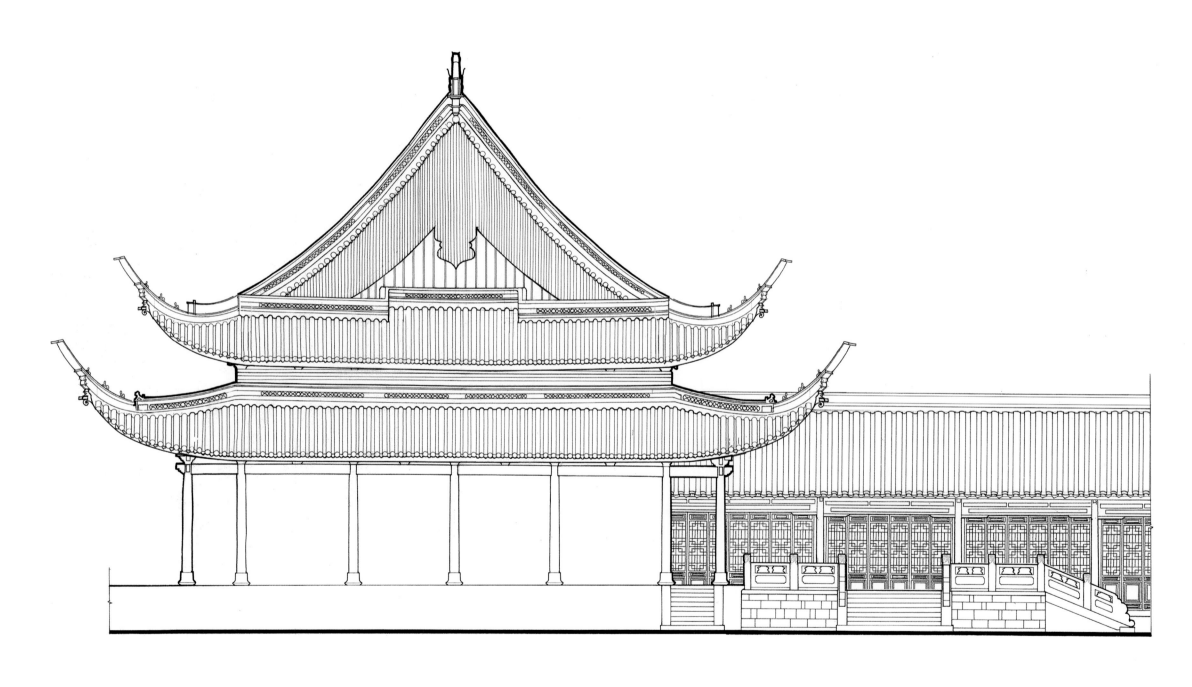

大成殿侧立面图
Side elevation of Dachengdian

0 1 2 5m

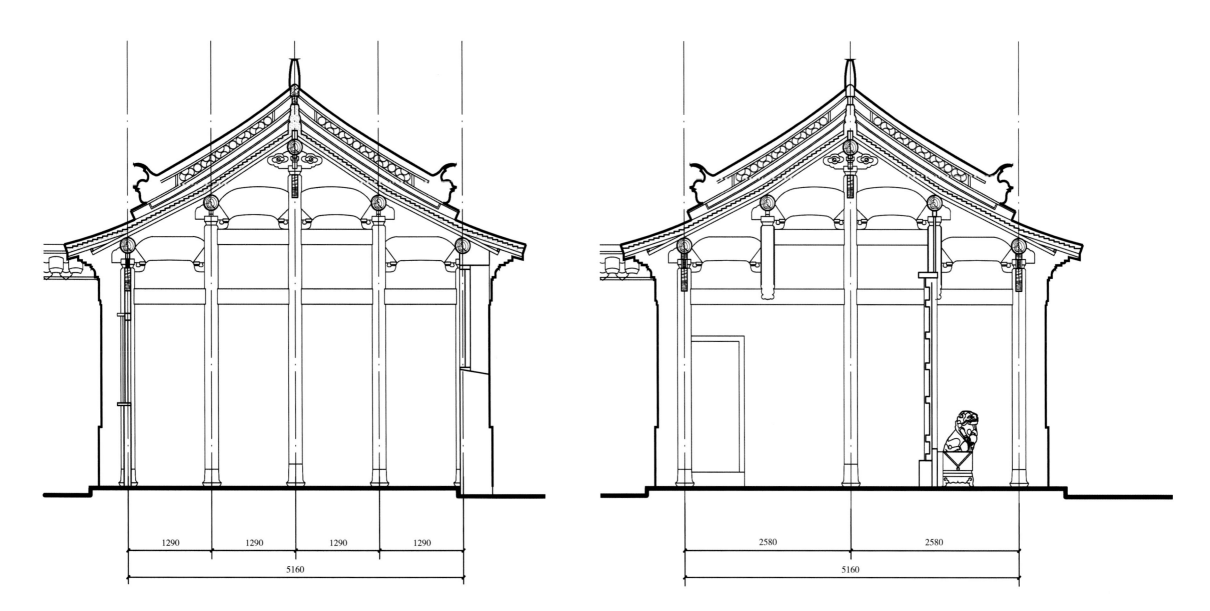

当湖书院头门（疑为原训导署门）边贴剖面图
Section of Danghu-Shuyuan Toumen's Biantie

当湖书院头门（疑为原训导署门）正贴剖面图
Section of Danghu-Shuyuan Toumen's Zhengtie

0    0.5    1    2m

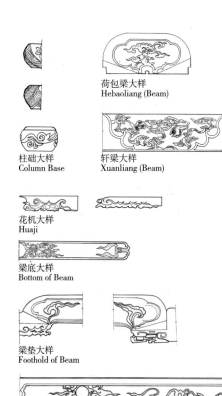

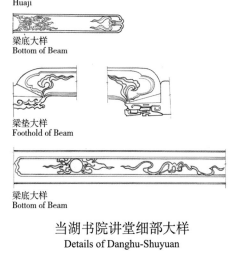

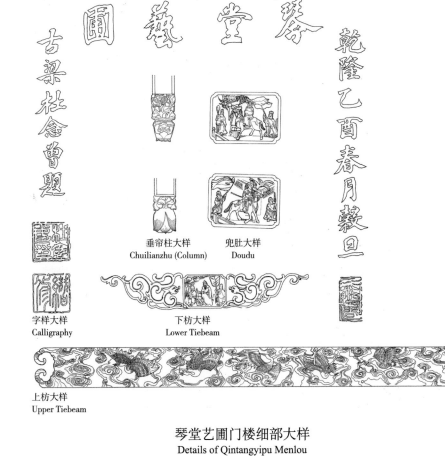

琴堂艺圃门楼立面图
Elevation of Qintangyipu Menlou

0  0.5  1  2m

当湖书院讲堂细部大样
Details of Danghu-Shuyuan

琴堂艺圃门楼细部大样
Details of Qintangyipu Menlou

荷包梁大样
Hebaoliang (Beam)

柱础大样
Column Base

轩梁大样
Xuanliang (Beam)

花机大样
Huaji

梁底大样
Bottom of Beam

梁垫大样
Foothold of Beam

梁底大样
Bottom of Beam

垂帘柱大样
Chuilianzhu (Column)

兜肚大样
Doudu

字样大样
Calligraphy

下枋大样
Lower Tiebeam

上枋大样
Upper Tiebeam

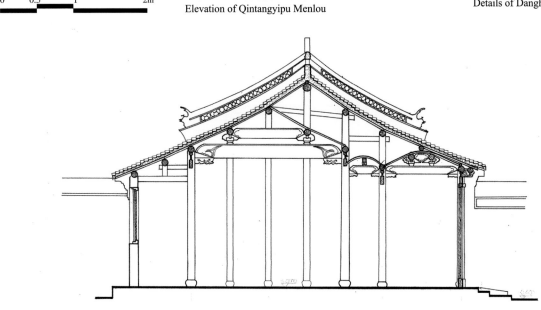

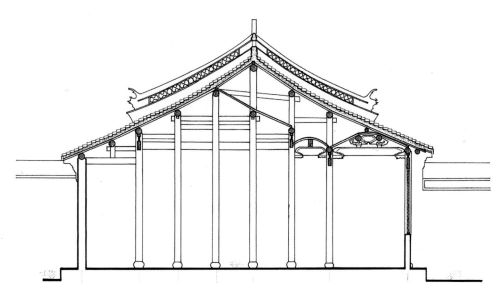

当湖书院讲堂正贴
Section of Danghu-Shuyuan's Zhengtie

0  1  2  5m

当湖书院讲堂边贴
Section of Danghu-Shuyuan's Biantie

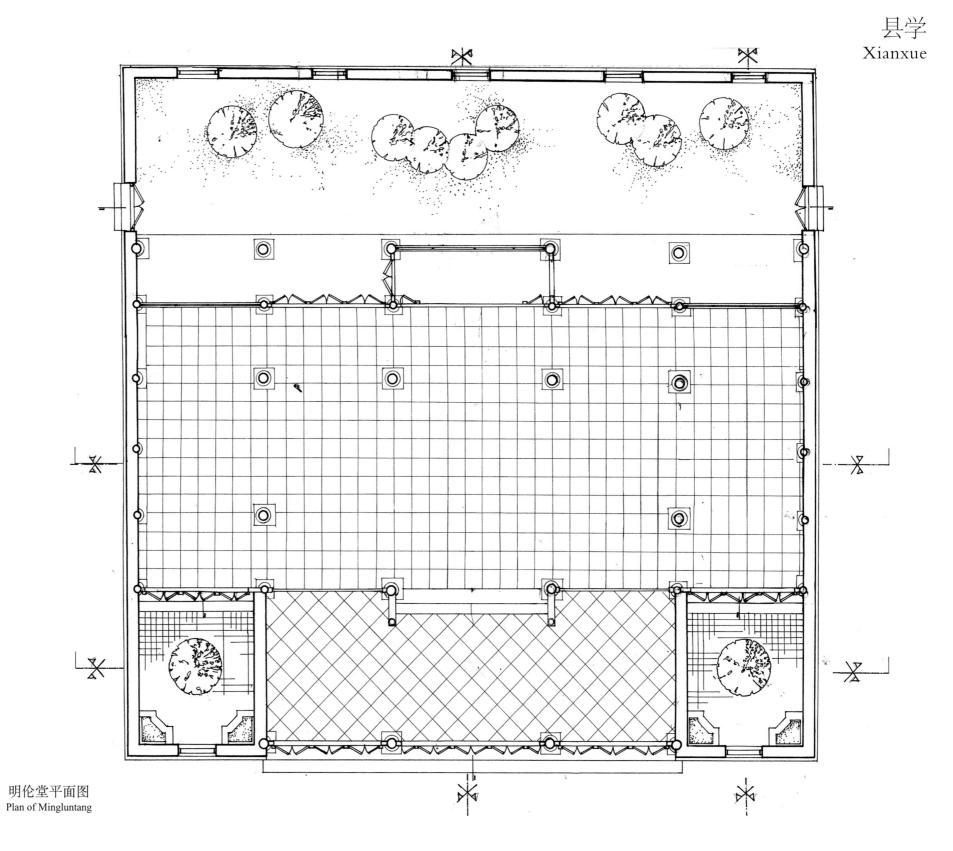

N

E

0    0.9    1.8    3.6m

明伦堂平面图
Plan of Mingluntang

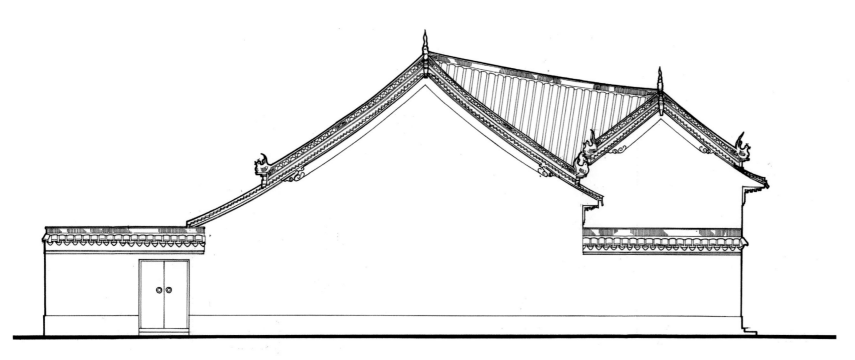

明伦堂侧立面图
Side elevation of Mingluntang

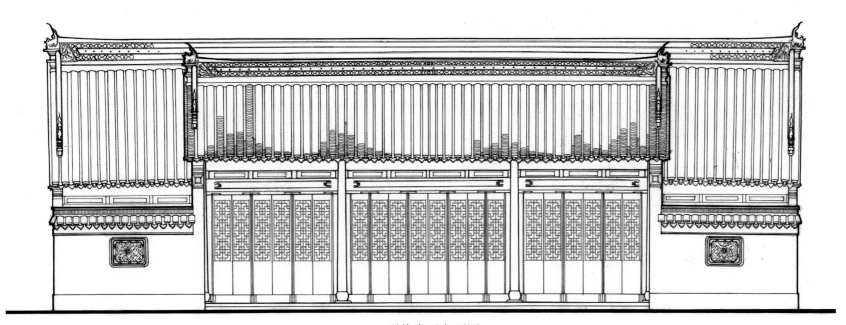

明伦堂正立面图
Front elevation of Mingluntang

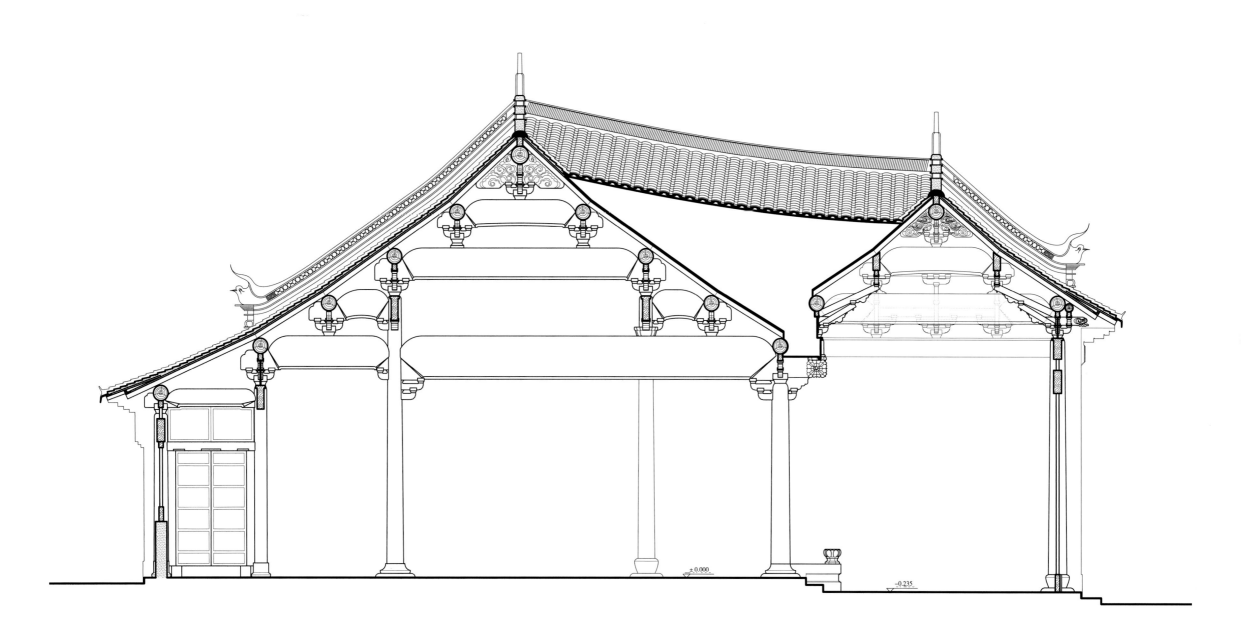

±0.000

−0.235

明伦堂正贴剖面图
Section of Mingluntang's Zhengtie

0　　0.5　　1　　　　　　　　　3m

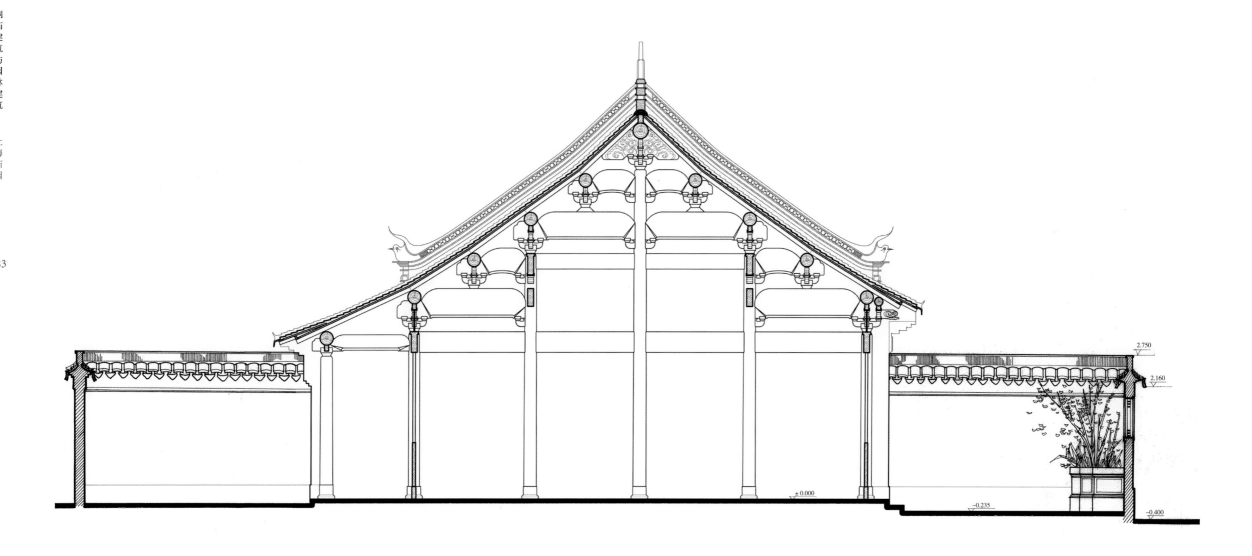

2.750

2.160

±0.000

−0.235

−0.400

0　　1　　2　　　　　　　　5m

明伦堂边贴剖面图
Section of Mingluntang's Biantie

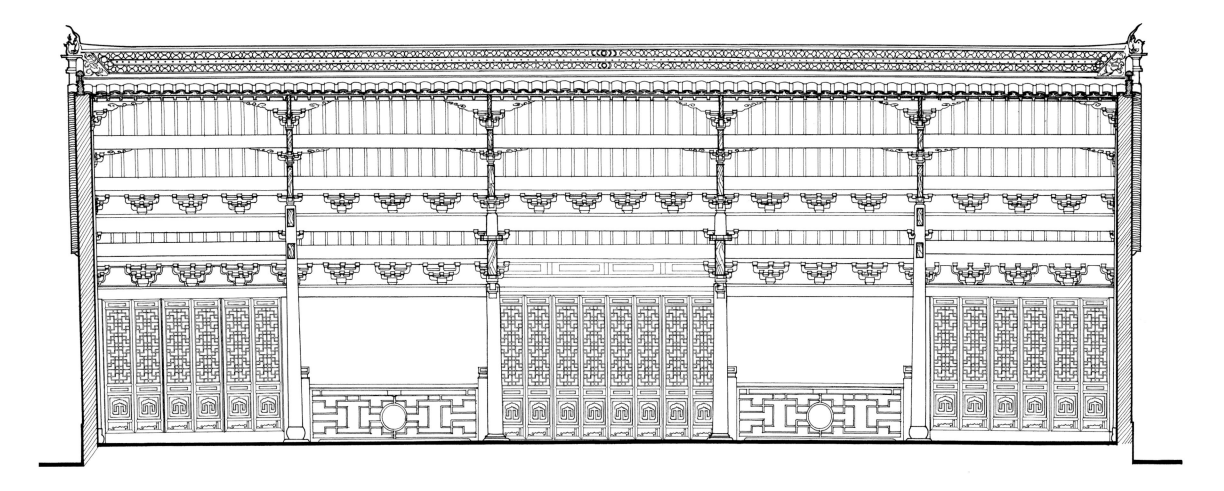

明伦堂纵剖面图
Longitudinal section of Mingluntang

0　0.9　1.8　3.6m

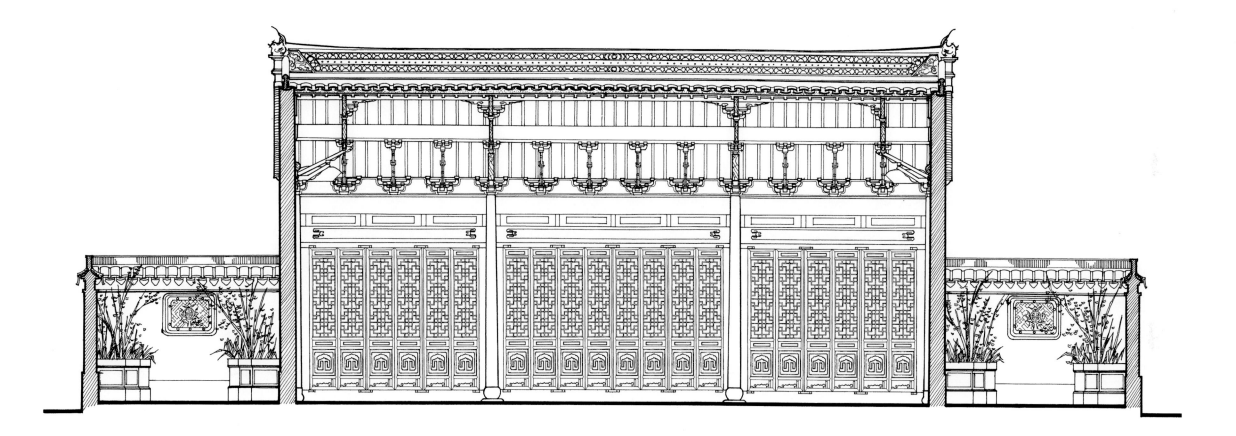

明伦堂抱厦纵剖面图
Longitudinal section of Mingluntang's Baosha

0    0.9    1.8         3.6m

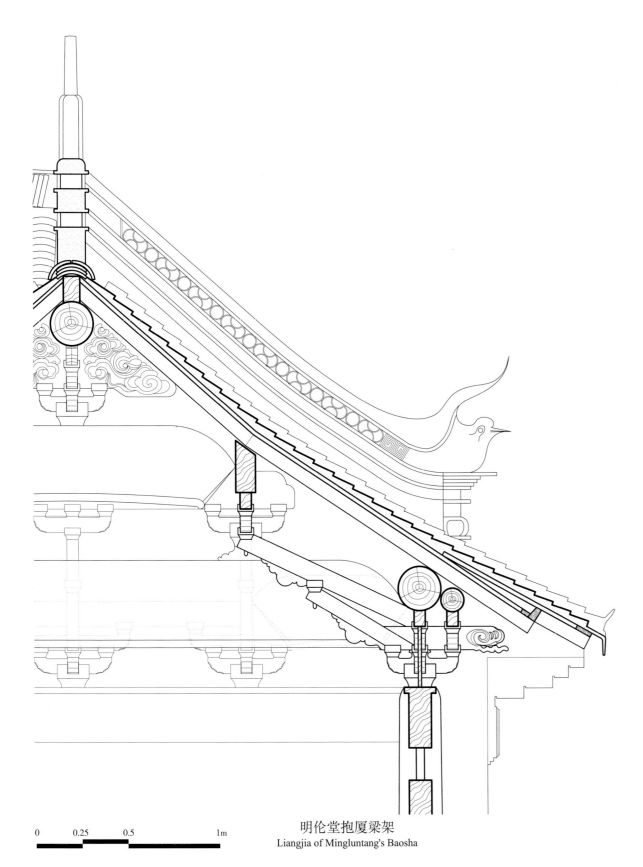

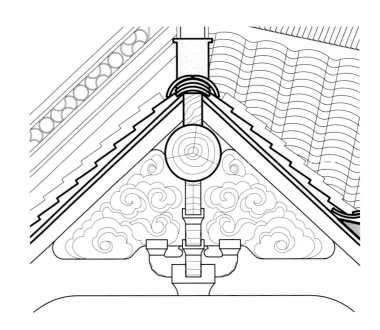

明伦堂正屋山雾云详图
Shanwuyun of Mingluntang's Zhengwu

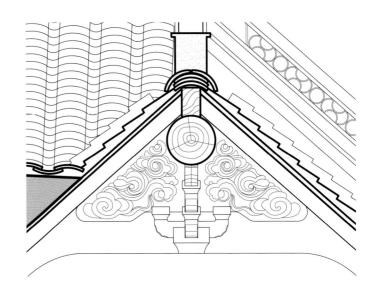

明伦堂抱厦梁架
Liangjia of Mingluntang's Baosha

0    0.25    0.5    1m

明伦堂抱厦山雾云详图
Shanwuyun of Mingluntang's Baosha

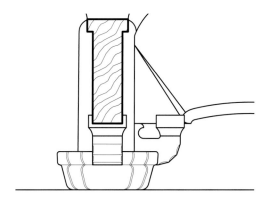

明伦堂正屋近仰莲状平盘斗立面和平面
Elevation and Plan of Mingluntang Zhengwu's Pingpandou

0    0.1    0.2              0.5m

明伦堂后廊靴形柱礩立面和平面
Elevation and Plan of Mingluntang's Xuexingzhuzhi

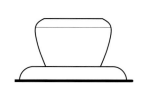
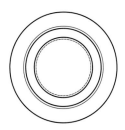

明伦堂前部荸荠形石鼓立面和平面
Elevation and Plan of Mingluntang's Biqixingshigu

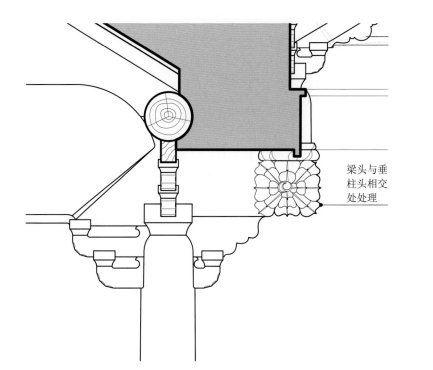

梁头与垂
柱头相交
处处理

明伦堂正屋与抱厦屋顶交接处
Connection of Mingluntang's Zhengwu and Baosha

0        0.25        0.5              1m

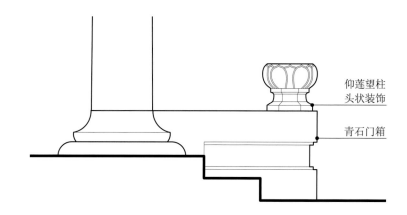

仰莲望柱
头状装饰

青石门箱

明伦堂正屋与抱厦台阶交接处
Connection of Mingluntang's Zhengwu and Baosha

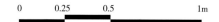

0        0.25        0.5              1m

0  0.5  1                    3m
礼门横剖面图
Cross-section of Limen

0      0.25      0.5m
礼门"组装"抱鼓石立面图
Elevation of Limen's Baogushi

中国古建筑测绘大系·祠庙建筑与园林建筑——上海庙园

注：嘉乐亭系近年新建

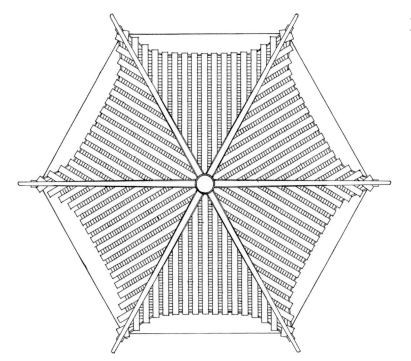

汇龙潭嘉乐亭屋顶平面图
Roof plan of Huilongtan Jialeting

汇龙潭嘉乐亭平面图
Plan of Huilongtan Jialeting

汇龙潭嘉乐亭瓦当大样
Wadang of Huilongtan Jialeting

汇龙潭嘉乐亭滴水大样
Dishui of Huilongtan Jialeting

汇龙潭嘉乐亭立面图
Elevation of Huilongtan Jialeting

0 0.5 1 2m

0 0.5 1 2m

0 0.02 0.04 0.1m

0 0.5 1 2m

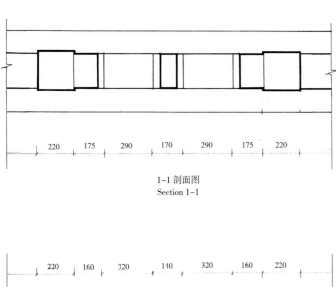

1-1 剖面图
Section 1-1

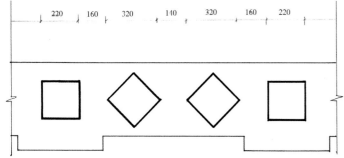

2-2 剖面图
Section 2-2

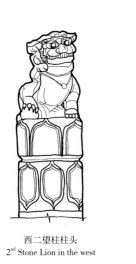

3-3 剖面图
Section 3-3

勾栏正立面图
Front Elevation of Goulan

0    0.2    0.4              1m

西一望柱柱头
1st Stone Lion in the west

西二望柱柱头
2nd Stone Lion in the west

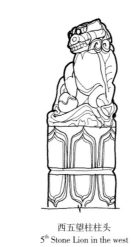

西三望柱柱头
3rd Stone Lion in the west

西四望柱柱头
4th Stone Lion in the west

西五望柱柱头
5th Stone Lion in the west

西六望柱柱头
6th Stone Lion in the west

西七望柱柱头
7th Stone Lion in the west

西八望柱柱头
8th Stone Lion in the west

注：仰高仿东侧勾栏西首第一、第七根望柱柱头石狮疑为明正德元年原物

嘉定县庙学前勾栏与望柱石狮大样图
Goulan and Wangzhu Stone Lions in front of Jiading county Wenmiao and Xianxue

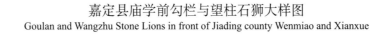

0    0.1    0.2              0.5m

291

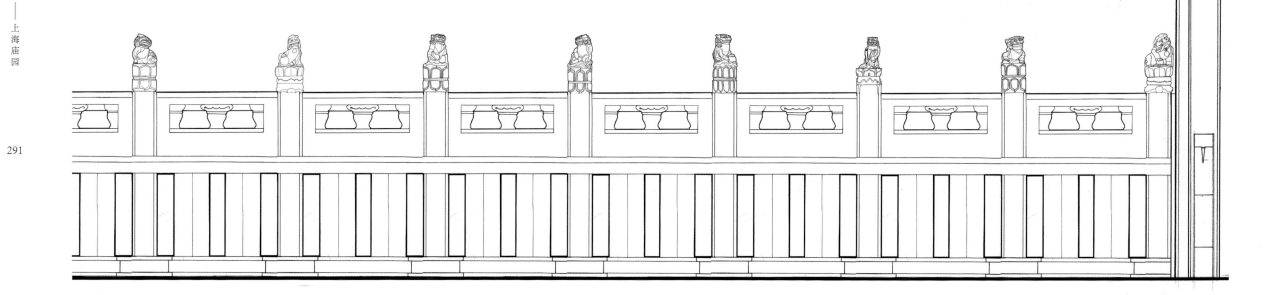

仰高仿东侧勾栏北立面图
North elevation of Goulan in the east of Yanggaofang

0　　0.2　　0.4　　　　　　　1m

**Item of survey:** Songjiang County Shantang and Gardens (Zuibaichi Garden)

**Address:** No64, South Renmin Road, Songjiang District, Shanghai

**Age of construction:** Early and middle Shunzhi Emperor of Qing Dynasty

**Site area:** 12.1 mu

**Competent organization:** Management Office of Shanghai Zuibaichi Garden

**Survey organization:** College of Architecture and Urban Planning, Tongji University

**Time of survey:** 1999

测绘项目：松江府城善堂及附园（醉白池）

地　　址：松江区人民南路六十四号

始建年代：清顺治中前期

占地面积：十二·一亩

主管单位：上海市醉白池公园管理处

测绘单位：同济大学建筑与城规学院

测绘时间：一九九九年

# Songjiang County Shantang and Gardens (Zuibaichi Garden)

松江府城善堂及附园（醉白池）

# Introduction

Songjiang Shantang and its garden (Zuibaichi Garden) are located a few miles south-west of the west gate (Guyangmen) of the original Songjiang Prefecture. In the old days, it was near to the secluded 'Yushutou' in the east, and faced the famous temple Chaoguosi and 'Yilanlou' (which was abandoned today) across 'Changqiaohe' (a tributary of Xiuzhoutang River, which had been filled up nowadays) and Changqiao Street. And the other three sides were fields and villages. A few miles away on the north side, you will find 'Junzhi Street' (now Zhongshan Road) and Xiuzhoutang city area that traversed the city from east to west and connected with Shuicicang in the west. A few miles along the west of Xiuzhoutang River, you will find Guanjialou in Yi Garden that located in deep courtyard, spreading horizontally.

The spatial group includes two parts, Shantang including several houses in the south-east and Zuibaichi Garden in the north and west. The overall spatial structure may have been laid at the beginning of the Qing Dynasty. It was a villa built in the south-western suburbs of the city by Gu Dashen (whose style name is Zhenzhi, and pseudonym is Jianshan, 1620-1674 or 1675) as a Jinshi, chief of the Ministry of Industry, water conservancy, and landscape painter in 1652. Among them, Shantang (the original residential space, including the warehouse in the south of the pond) covers an area of about 3.7 mu, while Zuibaichi Garden covers an area of 8.4 mu (including Xuehaitang courtyard of 1.1 mu and excluding the newly-built Yulanyuan, Shangluyuan and the outer garden).

# 导　言

松江府城善堂及附园（醉白池）位于原松江府城西门（谷阳门）外、西南方数里。

旧日东临幽僻的榆树头，并隔长桥河（秀州塘南支流，已填）、长桥街与名刹超果寺一览楼（今废）的峻拔身姿相望；其余三面则为野田村舍——北侧数里外，就是东西横贯府城、西接水次仓城的郡治大街（今中山路）与秀州塘一线城市带，沿秀州塘西行数里，就是俏踞深院、盈盈横展的颐园观稼楼了。

该空间群落包括东南部的善堂数进及西、北部的附园醉白池两部分，其整体空间结构可能初奠于清初，为清顺治九年（一六五二年）进士、工部主事、水利家、山水画家顾大申（字震雉，号见山，一六二〇至一六七四年或一六七五年）兴筑于府城西南郊的别业。其中善堂（原住居空间，含池南仓廒）占地约三·七亩，附园醉白池占地约八·四亩（含雪海堂院落一·一亩，不含新建玉兰院、赏鹿苑与外园）。

As a rare 'village land' waterscape garden in Jiangnan area, with the image of rural wild ponds as the main scenery, this garden was built to honor BAI Juyi from the Tang Dynasty, HAN Qi from the Song Dynasty (who built the Zuibaitang on the water of a private residence) and even DONG Qichang from the Ming Dynasty (whose pseudonym is Sibai) and other sages, notably taken he textual description in BAI Juyi's 'Chishangpian' as the basis for the layout and artistic purpose. Although it was built in the early Qing Dynasty, it contains more of the artistic genes and techniques that were plain and comfortable before the late Ming Dynasty. Although it has been included into the local charity agency for a long time, it has maintained part of the literati garden after additional construction, enclosure and decoration.

The wide Simianting in the north-east side of the pond in the garden is connected to the back building of the former residence in the east, and keeps a considerable distance from the main water surface in the south. There are still Qing Stone base and boot-shaped pillars from the late Ming and early Qing dynasties. Its Juzhe is gentle and the cornice is low, having a picturesque sense. Although the internal beams had been changed, it still corresponds to the east and west of Guanjialou in Yi Garden. It is a precious sample of the early architectural style and location logic of the gardens in Jiangnan area.

作为江南地区难得的、以乡村野塘意象水体为主景的村庄地水景园，此园因追慕唐人白居易、宋人韩琦（曾于私宅池上建醉白堂）甚至明人董其昌（号思白）等先贤而建，显著地以白氏《池上篇》中的文字描述为布局依据与艺术旨归。虽建于清初，却更多包含着晚明以前平淡舒朗、质朴简易的艺术基因与手法特征；虽长期归入地方善堂，历经添建、围合与增饰，仍保持着部分文人园风貌。

园中池北偏东的横长四面厅东接曾经的住宅后楼，南与主体水面保持相当距离，仍存留着明末清初的青石台基与靴形柱礩，举折平缓、檐口低压，极饶画意。虽内部梁架改易失色，仍堪与颐园观稼楼东西相应，为江南园林早期建筑风貌与位置逻辑的珍贵样本。

## Changes in Water Space: Condensed Thousand Years of Chinese Gardening History

This garden actually located in the coastal wetland settlement where the legendary Dongwu LU Xun family lived since the Three Kingdoms. In the east, there was the fish pond and home of LU Mao, LU Xun's younger brother. A few miles to the northeast, it was passed down as the villa (Puzhao Temple after Tang Dynasty) of LU Ji and LU Yun, who were LU Xun's grandsons and great writers. Along the way, ponds and pavilions were beautiful, the splendor overflowed, the cranes were singing and fishes were leaping, and the sound were heard in the sky. It combined the functions of production, residence,leisure, and sightseeing, creating the initial melody of space and cultural spirit for Songjiang.

In the Northern Song Dynasty, the official dotted pavilions and terraces in Beizhu and Zhongzhou here, and made them famous with conscious scene construction, becoming a public recreation area in the south-west suburbs of the city, known as Mao Lake, West Lake, and Fangsheng Pond. The past glory of one family was magnified to the entire city. On the lake is the Guyang Garden built by ZHU Zhichun, a Jinshi in 1091, which maintained the spatial hierarchy.

In the Yuan Dynasty, the lake quickly blocked due to hydrological changes. The newly built city wall of Songjiang City in the late Yuan Dynasty also passed by the lake with its powerful authority. As a result, the lake outside the city which were integrated into the city had gradually become flat ground as city markets. 'Living here today, there was no relic of the past to be found'. Hundred mu of lake and Lihe Beach in the city became the 'Old West Lake' in the literature and the precious waters of urban sceneries (the remaining water was filled in in the 1870s).

# 池湖之变：浓缩的千载中华造园史

此园实居于三国以来，传说中的东吴陆逊家族聚居的滨海湿地聚落中，其东侧即传为陆逊之弟陆瑁的养鱼池与故宅，东北行数里，则传为陆逊之孙、文学家陆机、陆云的别业（唐以后为普照寺）——一路或许池亭迤逦，华彩流溢，鹤鸣鱼跃，声闻于天，兼具着生产、居住与休闲、景象等功能，为松江一地谱就了空间与文化精神的初始旋律。

北宋时官方在这里的北渚、中洲点缀亭台，以自觉的景象营造使其名胜化，成为城西南近郊的公共游赏区，人称瑁湖、西湖、放生池。昔日的一门风流被放大至整座城市。湖上并有北宋元祐六年（一○九一年）进士朱之纯兴筑的谷阳园，保持着空间的层次。

元代此湖因水文改变而迅速淤塞，元末新筑的松江府城城垣又挟着强悍的威权，不由分说裂湖而过，于是城外湖面除融入城濠者外，渐成平陆市肆，『今比栉而居，昔之遗迹无复可寻矣』，城内的盈盈湖面百亩和唳鹤一滩，则成为文献中的『旧西湖』和珍贵的城市景象水域（十九世纪七○年代其残水被填没）。

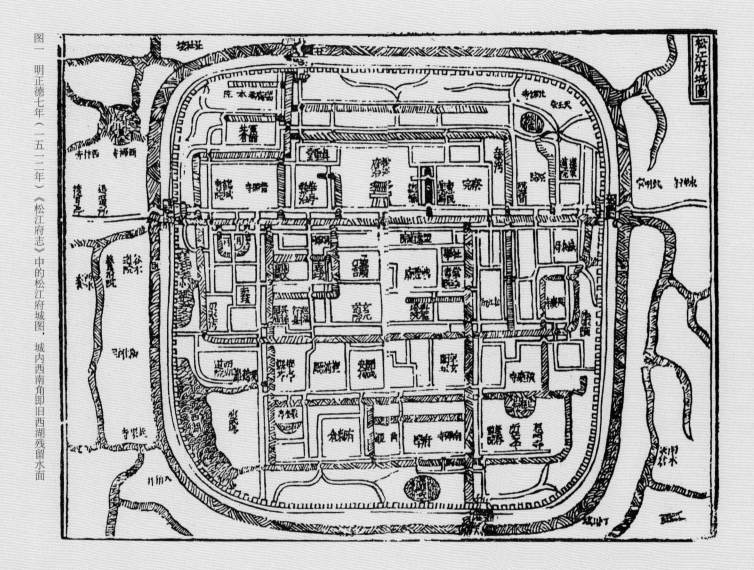

图一 明正德七年（一五一二年）《松江府志》中的松江府城图，城内西南角即旧西湖残留水面

Fig.1 The map of Songjiang Prefecture, *Songjiangfuzhi*, in 1512, the remaining the old West Lake is in the south-west corner of the city.

Fortunately, the Huating Cranes, which has been flying for thousands of years, still lingers in and out of the city wall to the west of Chaoguo Temple and the site of 'Zuibaichi' in the waterland. During the Qianlong period of the Qing Dynasty, ZHANG Minghe's *Gushui Jiuwen* recorded: 'Zuibaichi used to be Dong Sibai's drinking and chanting place'. It proves that in the late Ming Dynasty, it was the place where DONG Qichang (1555-1636), who was Jinshi in 1689 and a master of calligraphy and painting, held parties or meetings. However, it wasn't mentioned in *Zuibaichiji* written by HUANG Zhijun (1668-1748), who was Jinshi in 1721 and an opera artist, which made the relationship between Zuibaichi and DONG Sibai become obscure.

Fortunately, in the early and middle Shunzhi period of the Qing Dynasty, the 'perfect garden owner' GU Dashen, who was good at painting, rich in wealth but was trapped in official career, finally met this land.

The center of the Jiangnan painting circle in the late Ming Dynasty had actually moved east from the TANG Yin's 'Wumen' to DONG Qichang's Songjiang. The interpenetration and blending of landscape painting and landscape gardening seems to have reached its peak. In the late Ming Dynasty, most of the figures in the Songjiang painting circle were good at gardening too. After all, they only needed to respect and avoid the threshold of ZHANG Nanyang's 'high-tech school', adhere to the 'Nanzong' painting style advocated by DONG Qichang, and use the south of the Yangtze River as the prototype to build horizontal landscapes by soil slopes and rocks. For example, GU Zhengyi, the owner of Zhuojin Garden outside the East Gate, 'was especially good at decorating bamboo and stone'. The construction of Sunjia Garden (Dongguo Caotang) was also urged by the owner SUN Kehong. As the stamina of the Songjiang painting circle, GU Dashen continued to take on the spirit of his predecessors at the time of dynasty change and civilization collapse. With his skills and enthusiasm, he turned this land into an ancient, simple and remote village water landscape garden, leaving a fresh, elegant and endless return page for the colorful gardening history in the late Ming and early Qing dynasties.

好在流风千载的华亭鹤唳，尚缭绕不绝于城垣内外，直至超果寺西、水田野塘中的醉白池址。清乾隆年间章鸣鹤《谷水旧闻》称：『醉白池尝为董思白觞咏处』。证其晚明时曾为万历十七年（一六八九年）进士、一代书画宗师董其昌（一五五五至一六三六年）的诗酒文会之地，但清康熙六十年（一七二一年）进士、戏曲家黄之隽（一六六八至一七四八年）作于乾隆初年的《醉白池记》却未言及此，令这清池一泓与董思白的关系变得明灭隐约。

好在其后不久的清顺治中前期，精于绘事、饶于资财，却又困于仕途的『完美园主』顾大申终与这片流风千载的土地坎坷相逢了。

明末的江南画坛重心，实已由唐寅们的『吴门』东移向董其昌们的松江，山水画与山水园的互渗、交融似也达到顶峰。晚明松江画坛人物即多擅造园，毕竟只要敬避避张南阳一脉的『高技派』门槛，坚守董其昌所倡导的『南宗』画风，以江南地貌为原型，营造土坡点石，曲池萦带的横向山水即可。如东门外濯锦园主人顾正谊『缀竹石尤为擅长』；孙家园（东郭草堂）的兴筑亦得园主孙克弘指点——顾大申作为松江画坛后进，于朝代更迭、文明蹉跌之际，赓续起前辈风流，技之所长，情之所钟，一时将这片土地打造为高古简远、若不经意的村庄地池池塘型水景园，为明末清初缤纷繁丽的造园史留下了清新雅淡、余韵不绝的回归一页。

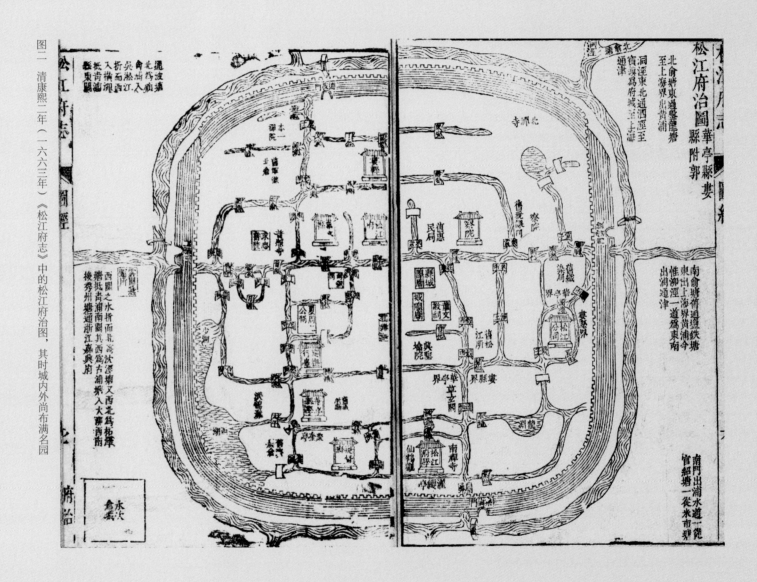

图二 清康熙二年（一六六三年）《松江府志》中的松江府治图，其时城内外尚布满名园

Fig.2  The map of Songjiang Prefecture *Songjiangfuzhi*, in 1663, at that time, both inside and outside the city, there was still a lot of famous gardens.

After many years of eunuchs, GU Dashen, a technical and literary bureaucrat, reluctantly recalled the beauty of his homeland and said, 'My hometown was outside the seclusive place, and I enjoyed the life in the suburbs in the west of the city…Bamboos covered the path and the tide rose in the square pond…The house was surrounded by crop fields. In autumn, the color was beautiful. We watched the harvest in the fields, as if calling me back'. 'It must be in the autumn of 1657', if the scene described and HUANG Zhijun's 'Zuibaichiji' later fit together, it should be the same place, and perhaps there was no such high-profile name as 'Zuibaichi' at that time. GU Dashen later bought the XU family's Xishe Caotang and other gardens, but it seems that this 'adolescent' hometown experiencing wind and rain was the place where the wandering soul returns.

In the early Qianlong period more than sixty years after GU Dashen's death, the garden was returned to GU Sizhao, who was the former Xundao of Dantu County, after several changes of ownership. However, the overall layout may not change much. In the middle Qing Dynasty, this garden did not belong to Chenghuang Temple of the city like other famous gardens in Shanghai. Instead, the owner of the garden changed several times. It was purchased by Yuyingtang attached to Chaoguo Temple in the east og the river during the Jiaqing and Daoguang periods and became its west hall to manage rent collection and accounting. For more than a hundred years, this garden had undergone additional construction until the final form of the late Qing Dynasty. In the meantime, the west and south sides of the pond are gradually enclosed by the building. Pond and shore treatment is becoming more and more complex, creating illusion of scale The 'Banshan Banshui Banshuchuang' pavilion that protrudes above the pond, especially makes the original pond transformed into a flat lake. The name of the big and small lake pavilions on the front line of the east of the pond also proves this point.

With a thousand years of history and the immortal water space, the garden experienced the 'Fish Pond' that was full of fun in the Three Kingdoms period to the 'West Lake' that was consciously retouched and opened to the public in the Northern Song Dynasty, and from the wild pond in the countryside where the literati garden returned to the nature in the late Ming and early Qing Dynasties to the public garden that was repeatedly enclosed and re-decorated of civil society in the late Qing Dynasty. The changes in water space of Zuibaichi Garden has become a microcosm of the history of Chinese gardening for thousands of years.

此后多年的宦游劳顿中，技术型而兼文艺范官僚顾大申会时时无奈地忆及故园的美好：『我家五湖表，考槃城西郊……修篠掩芳径，方塘疏回潮……绕宅被禾黍，原隰秋纷骚，刘获观西成，田里欢想招』『其时必在顺治十四年秋』，所描述景物与稍后黄之隽的《醉白池记》若合符节，应即一地，或许当时尚未有『醉白池』这样稍嫌高调的名称。顾大申后来尚曾购买徐氏的西佘草堂等其他园亭，但似乎这座『青春期』的风雨故园才是游子的魂归之地。

顾大申去世六十余年后的乾隆早期，此园于数次易主后归于曾任丹徒县训导的顾思照，但格局应变化不大。清中叶时，此园并未像上海地区现存其他名园那样归属邑庙，而是又经数度易手，于嘉庆、道光年间为附设于河东超果寺的育婴堂购得，成为其专管收租入账的西堂——百余年间，此园历经添建，直至清末定型。其间池西、南两面渐被建筑群落围合，池岸处理渐趋复杂而生尺度幻觉，『半山半水半书窗』一亭挑临池上，尤令原本的清塘一泓向平湖一望转化，池东一线亭轩的大、小湖亭之称也旁证了这一点。

千载胜迹，不灭水光，从三国时天趣烂漫的盈盈『鱼池』，到北宋时历经自觉润饰、面向公众的脉脉『西湖』；从明末清初文人园天趣回归的乡村野塘，到晚清市民社会公众园林的反复围合、增饰和重新湖景化——醉白池一隅的千载池、湖之变，也成为中华千载造园史的精微缩影。

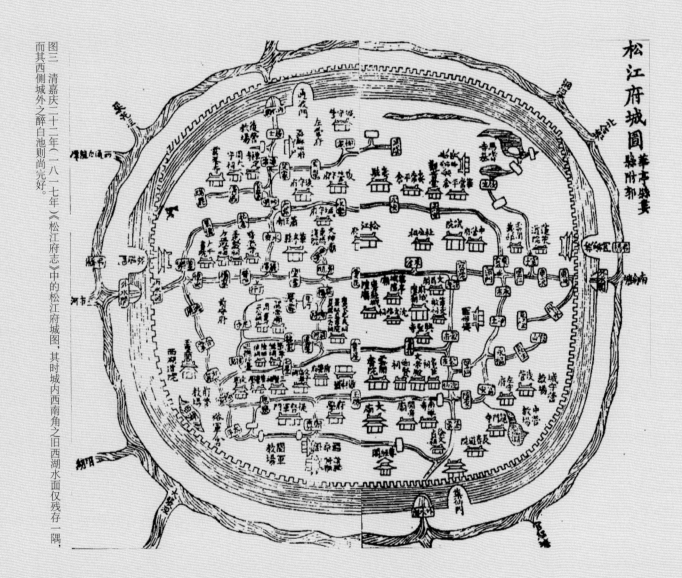

图三 清嘉庆二十二年（一八一七年）《松江府志》中的松江府城图，其时城内西南角之旧西湖水面仅残存一隅，而其西侧城外之醉白池则尚完好。

Fig.3 The map of Songjiang Prefecture *Songjiangfuzhi*, in 1817, at that time, only a part of the old West Lake in the south−west corner of the city remained, while Zuibaichi outside the city on its western side was still in good condition.

## Classics and Return: From 'Facing the Fields across the Water' to 'the Pavilion and the Tree as the Focus'

HUANG Zhijun has left us a valuable record of the appearance of Zuibaichi Garden in the early years of Qianlong period in the Qing Dynasty. 'The pond is square and long, about three or four mu, locates at the right side of the house. In the east of the pond, there is a big elm tree for at least 200 years old. A pavilion is under the tree and locates near the water, so people call it Laoshuxun, and a corridor passes through the pavilion. The fields in west of the pond are connected to the residence and are enclosed by bamboos. The bamboos are sparse and can be watched through. In the south of the pond there are two or three families, with windows closed and women washing, just like a painting. The hall is in the north of the pond, and its four sides are all open. The north and the west of the hall are surrounded by the bamboos and stones. The bamboos and stones also travel around the pond from the north to east. The north land of the garden can also hold houses to plant flowers and it is also enclosed by bamboos…'

From this article and combined with the current situation, we can infer the overall spatial appearance of the garden in the early Qing Dynasty.

1. Garden was in the west and residences were in the east, borrowing scenery from the wall of residences. The overall spatial relationship between the garden and the house (the later Shantang) of the complex may never change. In the early Qing Dynasty, the handover interface between houses and gardens in the east of the pond was used, with the tall and undulating west wall of the residences as the background, and the small corridor leaning against the wall as the foreground, forming the core picture and scene of the whole garden. In the east of Caixiachi in Wangshi Garden in Suzhou, the west 'Guanyindou' gable of the residence was also used as the background of cloud-shaped mountain meaning. Yipu Garden retreated the tall gables as far as possible to make it feel like a remote mountain, which is the most appropriate choice.

典则与回归：从隔池面野到轩树点睛

黄之隽为我们留下了清乾隆初年醉白池面貌的宝贵记录：『池方而长，可三四亩，据宅之右。池东有老榆槎桠，二百年物，轩荫其下，临流而坐，曰：老树轩，贯以长廊，池西畝亩连瓦，限之以篱，篱疏可眺也，池南两三人家，窗户映带，妇孺浣汲，望若画图。池北堂临之，敞其四面，堂北与西竹石环列，又北则池尾绕而东。又北有隙地可构屋莳花木，亦篱限其外焉……』

由此文结合现状，可推断清初该建筑群之整体空间面貌。

一、西园东宅，借景宅垣：该建筑群之园、宅（后为善堂）空间整体关系或许从未改变，且清初即利用池东的宅园交接界面，以住宅高大起伏的西墙为背景，以倚墙的小型轩廊为前景，构成全园核心画面与景象——苏州网师园彩霞池东，亦同样借住宅观音兜西山墙为云形山意的背景；艺圃则尽可能将高大山墙退远安置，俾成缥缈远山，拿捏最为恰当。

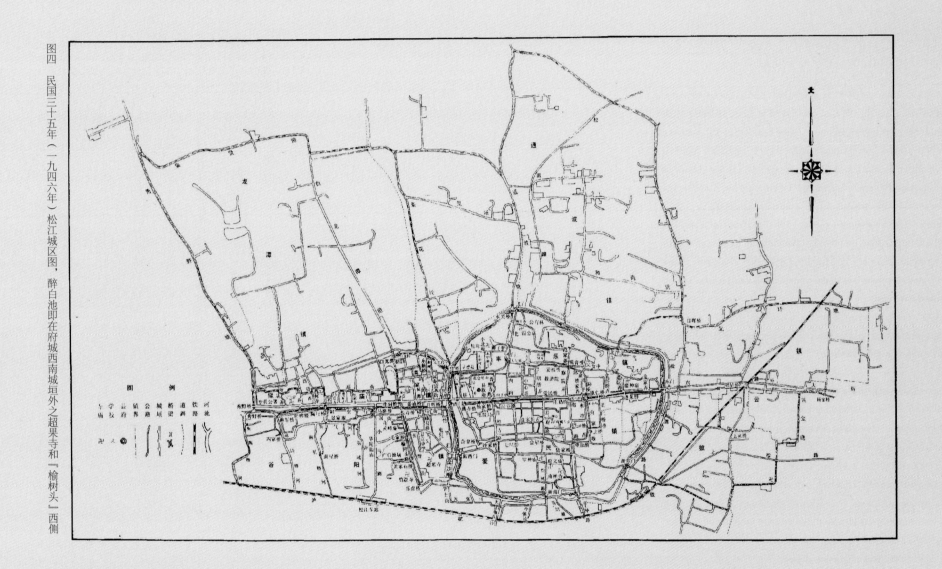

Fig.4　The map of Songjiang City and its surrounding area in 1946, in which Zuibaichi Garden is located in the south of Chaoguo Temple and Yushutou outside the south-west city wall

2. The garden faced the fields across the water and leaned against the bamboos and stones. The south bank interface of the pond in the garden did not cut the hills or slopes as usual, to form the scenery with Simianting on the north side of the pond as a classic scene of the mountain across the water. Instead, it was completely opened to the farmhouse in the south of the garden, in order to fully integrate with the countryside scenery and fully extend the garden space, fully reflecting the characteristics of the village gardening. At that time, the south of the pond was far away from the city area of 'Junzhi Street', so the scene was the most natural. Among them, the water pavilion in the southwest corner of the pool called 'Banshan Banshui Banshuchuang', was suspected to be a river port, where people nearby can get water and wash clothes, and it was opposite to Laoshuxuan, helping to create a vivid picture. Or this pond was originally publicly owned, because once the road to draw water was blocked, the life of people would feel inconvenient.

The northern side of the garden was relatively the nearest market place, which was completely covered by bamboos and stones, forming the thick green background of the whole garden and the 'Profound Realm' of deep and secluded canals, and making the southern side more clear and refreshing. The northern wall was made of painting bamboo fences instead of solid walls. The pastoral scenery on the west side of the garden must be slightly worse than the south side, but it was also enclosed by bamboo fences to form a moderate level of separation. And the east of the garden was the residential wall. Therefore, the entire garden's viewing sight can be concentrated on the picturesque countryside on the south side, and the combination of a tree and a pavilion was the precise foreground and focus of the picture.

3. The garden had square pond and long canal, forming the stratification of the square and the strength. The whole garden took the water as the main scene, which was close to the square shape, the shoreline was flat and wide, the water surface was slightly lower and bright, and there were no bridges, islands, rocks, or bays, constituting a rural and wild pond back to realism. The north-west corner flowed out a long canal, which traveled north to east, forming the tail of the pond. The square and strong turning point echoed the square pond's simplicity, and outlined the meaning of endless space. This exquisite water also made Simianting 'trapped' in the nearly square peninsula surrounded by bamboos and water, forming an excellent plan echoing relationship with the square hall, which may not have been changed for hundreds of years.

二、隔池面野，背倚竹石：园中池塘南岸界面未按惯例掇山或起坡，以成池北四面厅对景，形成经典的隔水面山之局，而是完全向园南野田农舍打开，作敞开式借景处理，以求与乡野风光的透彻融合和园林空间的充分延展，充分体现了村庄就地造园的特征——想必当时池南因远离郡治大街一线城市带，景象亦最为自然入画。其中池西南角水亭半山半水书窗一带位置疑原为河埠，可供邻人汲水、浣洗，与老树轩相对，助成生动画面——或者此塘本为公有，一旦阻断汲水道路，邻人生活即感不便。

园北侧相对最近市肆，则以竹石完全遮蔽，形成全园厚重的绿色背景与深径幽渠的奥境，愈显南侧之清旷，并以画意竹篱代替实墙作北垣。园西侧田园想必风光亦较南侧稍逊，则亦以竹篱围合，作一适度的分隔层次；园东则为住宅墙面——故全园观赏视线能集中于南侧之如画田园，而以一树一轩组合为精约前景与画面焦点。

三、方池长渠，方劲叠映：全园以盈盈一水为主体景象，水体近方，岸线平整显豁，水面稍低下而明亮，略无桥、岛、矶、湾分隔，构成回归写实的乡间野塘。其西北角分出长渠，向北复向东，形成池尾。方劲的转折，呼应着方塘的朴拙，勾勒出空间不尽之意。这一精湛水形也令四面厅『身陷』水竹环绕的近方形半岛，与方正的厅形产生极好的平面呼应关系，很可能数百年未改。

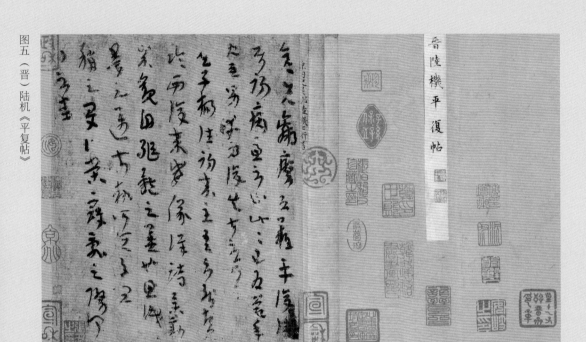

图六 （北宋）赵佶《瑞鹤图》

图五 （晋）陆机《平复帖》

图七 （明）董其昌《山水卷》

Fig.5  LU Ji (Jin Dynasty)'s *Pingfutie*
Fig.6  ZHAO Ji (Northern Song Dynasty)'s *Ruihetu*
Fig.7  DONG Qichang (Ming Dynasty)'s *Shanshuijuan*

The bridge on the north-west corner of the pond across the canal and diagonally opposite Simianting was well positioned, simple and powerful, and quite close to the Ming style. Unfortunately, it was converted into a mediocre Taihu Stone arch bridge in 1963 in order to imply the image of water.

It was speculated that in the early Qing Dynasty, the tail of the pond in the north-east corner may continue to communicate with the Changqiao River to the east. However, the branch canal that straight out of the canal to the north-west corner of the whole garden was not mentioned in Huang Zhijun's article. The branch canal may have connected to the west tributary of the Changqiao River, and this tributary still existed in the Republic of China.

4. The main hall hided itself from the pond and connected to the residences. The main hall of the whole garden, Simianting, used low cornices to reduce its sense of scale, and used horizontal and long proportions to enhance the sense of speed and direction. And it connected to the back building of the residence in the east, like wandering near the water, escaping from the central axis of the residence. At the same time, it tried to retreat from the south side of the pond and hided into the bushes to avoid interference with the scale of the main landscape space. Four approaches were used, and one hall was hidden. Its location and form may not have been changed for nearly 400 years. In addition to successfully hiding its own volume, people inside can calmly visit fields, ponds and villages in the south, enjoying the bamboos, stones and secluded canals and experiencing the speed of low-key sightseeing.

Similar blanking treatments for main halls are still common in Jiangnan gardens from the middle and late Ming to the early Qing. For example, between Leshoutang of Yu Garden and the main water surface, the 'big platform, which was several ren deep and double in width, was built by stones, with strange rocks on the left and right, hidden in the shape of rocky slopes and valleys, with famous flowers and rare trees included'. Today this 'big platform' is already half in the garden and half outside the garden, and both have a generous and tolerant image. Jingmiaotang of Nanjing Zhan Garden, Shanguangtanyingguan of Jiading Qiuxiapu Garden, Kanlou of Songjiang Yi Garden, and Zhusutang of Shanghai Rishe Garden, their relationship with the main water surface is also similar.

池塘西北角跨渠斜对四面厅的板桥位置精当，质朴有力，颇近明风，惜一九六三年改建为平庸造作、暗示湖泊景象的湖石拱桥。

推测清初时，东北角池尾可能继续向东与长桥河相通。而此渠向全园西北角直出的支渠并未为黄文言及，该支渠可能曾连通向长桥河之西侧支流，民国时此支流仍存。

四、主厅消隐，退池接宅：全园主要厅堂——四面厅以低矮檐口减小自身尺度感，以横长比例强化速度感与方向感；且东接住宅后楼，仿佛凌波微步，自住宅中轴横逸而出；同时尽量退离南侧池面，隐入树丛，避免对主体景象空间的尺度干扰——四管齐下，一堂深隐，其位置及形态亦可能近四百年未改。

于成功消隐自身体积之余，更可从容南顾野塘村舍，周览竹石幽渠，体验低调纵目之快。

类似的主要厅堂消隐处理尚多见于明中晚期至清初的江南园林。如豫园乐寿堂与主体水面间的「广庭，纵数仞，横倍之，甃以石如砥，左右累奇石，隐起作岩峦坡谷状，名花珍木，参差在列」。今日此广庭已半在园内，半在园外，均觉宽绰有容。南京瞻园静妙堂、嘉定秋霞圃山光潭影馆、松江颐园看楼、上海县日涉园竹素堂与主体水面间的关系也与之类似。

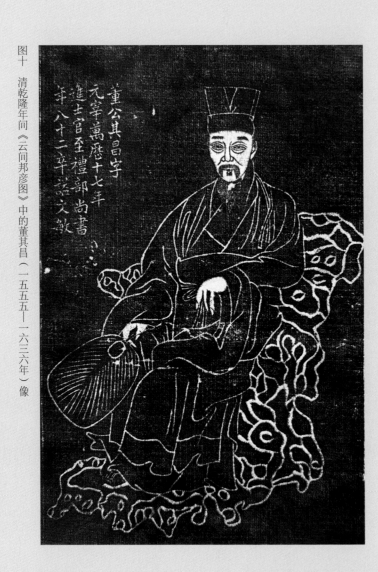

图十 清乾隆年间《云间邦彦图》中的董其昌（一五五五—一六三六年）像

图九 韩琦（一〇〇八—一〇七五年）像

图八 白居易（公元七七二—八四六年）像

Fig.8 The portrait of BAI Juyi (772−846)
Fig.9 The portrait of HAN Qi (1008−1075)
Fig.10 The portrait of DONG QICHANG (1555−1636) in *Yunjianbangyantu* of Qianlong period in the Qing Dynasty

5. The garden took the pavilion and the tree as the focus, to create a secluded scene. In old days, in the east of the pond, Laoshuxuan was set up as the focal point of the whole garden at the intersection of the residential gables and the 200-year-old elm. The location of the pavilion and the tree was suspected to be 'Dahuting' at the end of the existing Qing Stone revetment. Its layout resembled that of Ruyuting of Suzhou Yipu Garden and Zhiyujian of Wuxi Jichang Garden, and the latter was also covered by maple trees on Hebutan. In the middle Qing Dynasty, during the GU Sizhao era, a corridor might be added to reach this pavilion from the Simianting. A pavilion and a corridor together cut the volume of the west gable of the residence, making it better integrate into the landscape.

In the early Qing Dynasty, a secluded environment was set up in the bamboo and stone area in the north of the garden to add another level beyond the focus of the scene. When Letianxuan was rebuilt in the Republic of China, there was originally a thatch pavilion, which seems to be closer to the appearance of the early Qing Dynasty, and can be roughly comparable to Yuou Courtyard of Yipu Garden. Nowadays, the bamboo forest is sparse, with a huge pavilion, and the elegant touch is no longer elegant, which is somewhat inferior.

BAI Juyi had written in *Chishangpian* that 'a residence of ten mu should have a garden of five mu, a pond and some bamboos… It should possess a hall and affiliated houses, pavilions and bridges, boats and books, and wine and delicious food… Enjoying the garden, I would live all my life here.' The bamboo pond is clear and open, and the simple strokes are quiet, like a space monologue of Zuibaichi Garden in the early Qing Dynasty.

五、轩树点睛，幽境别构：旧日池东于住宅山墙与两百年老榆的交织部位设置全园景象焦点老树轩，轩、树位置疑在今存青石驳岸结束处的大湖亭——其格局恍似苏州艺圃乳鱼亭、无锡寄畅园知鱼槛，后者亦有鹤步滩上枫杨与之掩映成画。清中叶顾思照时代或添建廊道，从四面厅达此轩。一轩一廊削减了住宅西山墙的体量感，令其更好融入园景。

清初并于园北竹石深处别设一幽境，以增加景象焦点之外的另一层次。民国改建乐天轩时此处原存一茅亭，似更接近清初面貌，大致可比照艺圃浴鸥小院。当下则竹林稀疏，一轩庞然，逸笔不逸，有所逊色。

白居易《池上篇》诗云：『十亩之宅，五亩之园，有水一池，有竹千竿……有堂有序，有亭有桥，有船有书，有酒有肴……优哉游哉，吾将终老乎其间』——竹池清旷，简笔幽怀，仿佛清初醉白池之空间独白。

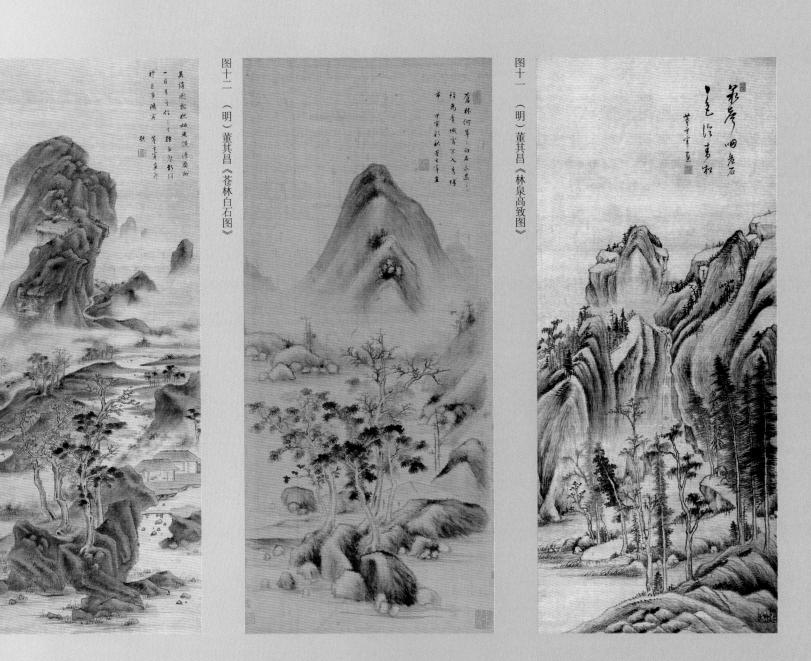

图十三 （明）董其昌《吴绢图》

图十二 （明）董其昌《苍林白石图》

图十一 （明）董其昌《林泉高致图》

Fig.11  DONG Qichang (Ming Dynasty)'s *Linquangaozhitu*
Fig.12  DONG Qichang (Ming Dynasty)'s *Canglinbaishitu*
Fig.13  DONG Qichang (Ming Dynasty)'s *Wuxiaotu*

## Inversion of Subject and Object: Internalization of the Scene in a Dual Context

In the middle Dynasty, the garden had been changed hands for several times after GU Sizhao's era, until it was put under the management and maintenance of public Shantang for more than a hundred years. The times were turbulent and the world changed, and the famous garden that had been evacuated from generation to generation was unable to stand alone.

On the one hand, being in the 'post' era of the late classical period had made its scene space increasingly introverted, dense and decorative.

On the other hand, the coming 'first' wave of the new era pushed its space to open to the city, try to share part of the city's secular functions, meet the public's leisure gatherings and other needs, and move towards verticalization, secularization and 'urbanization'. The change in direction was similar to the changes of other 'Shanghai style gardens' like Xu Garden and even CHANG-SU-HO Garden in the same period.

Under the dual context, the landscape and space logic of the building complex had undergone tremendous changes.

### 1. Subject and guest inversion

The basic scene pattern of the main hall in the garden with Simianting hidden in the north bank, backed by bamboos and stones, 'facing the fields across the water' disappeared with the degradation of the field outside the garden and the closure of the scene surrounding the pond. After the fields in the south of the pond reluctantly transformed into a warehouse, bamboos in the north of the pond were becoming more and more flourish, and the pavilions and corridors on the east and south sides of the pond are connected and the shoreline was regularized, the east bank of the pond leaning against Shantang was naturally upgraded to the first viewing subject, while the south bank where the water corridor leaned against the wall and went straight was second.

And from the water pavilion in the south-west of the pool to the west bank of the big sloping tree and Laoshuxuan, and even from the north-west estuary of the pond to the north bank where the hundred-year camphor and the Taihu Stone Mudantai set off

主客倒置：双重语境下的景象内化

清中叶顾思照时代后的数度易手，直至归入公共善堂管理和维护的百余年间，时代风浪迭起，世潮变迁，一代萧疏散淡的名园亦未能独善。

一方面，身处古典后期的『后』时代，使其景象空间日益内向化、繁密化与装饰化。

另一方面，新时代将至的『先』浪潮，又推促其景象空间向城市打开，尝试分担部分城市世俗功能，满足公众的休闲聚会等多方面需求，向纵向化、世俗化与城市化方向转变，类似同时代『海派园林』徐园甚至张园们的变化。

双重语境下，该建筑群的景象空间逻辑发生了巨大变化。

## 一、主客倒置

园中主要厅堂四面厅深隐北岸，背倚竹石、隔池面野的基本景象格局随着晚清园外田野景象的退化和池周空间的闭合而不复成立。在池南田园无奈地变身为仓廒，池北竹树日益繁茂，以及池东、南两侧亭廊贯通，岸线规整化之后，背倚善堂的池东岸自然升级为第一观赏主体，一线水廊倚墙直行的南岸则次之。

而自池西南水亭至老树横斜的卧树轩一带的池西岸，甚至复自池西北水口至数百年香樟与湖石牡丹

图十四 （清）顾大申《松桧鬱图》

图十五 （清）顾大申《松柏飞泉图》

图十六 （清）顾大申《仿董源山水》

Fig.14  GU Dashen (Qing Dynasty)'s *Songguiyutu*
Fig.15  GU Dashen (Qing Dynasty)'s *Songbaifeiquantu*
Fig.16  GU Dashen (Qing Dynasty)'s *Fangdongyuanshanshui*

each other, there were trees and rocks along the way, and the tree shade were over the head, which became the most important viewing interface in the garden. Especially, the most dramatic change of the role in the north of the pond, was the only 'host and guest inversion' seen.

There were two more specific response changes. One was that a moon gate was set up under the high wall on the south bank and in the middle of the water corridor, from which people can look back at the main scenery of the whole garden from the warehouse courtyard. And the old tree and the slope across the bank, had become a 'facing slope across the water layout, with a sense of tension and ritual, which can be contrasted with the extremely elegant and double moon gates under the high walls of Yuou Courtyard of Suzhou Yipu Garden. The second was that in the north bank space, which was originally densely sheltered by bamboo forests and darkened by a dark path, a viewing gallery facing the moon gate on the south bank was opened, pointing to the area of Letianxuan in the north of the garden, obtaining a considerable depth of scene.

## 2. Two halls being established together

The main hall of the whole park was originally Simianting with low tiled eaves and a long and horizontal body. It was particularly picturesque among the gardens in the south of the Yangtze River, and its location was also typical and appropriate. But after all, the size was compact and it was difficult to meet the needs of the dense public activities in the late Qing Dynasty. Therefore, in the late Qing Dynasty, a functional space 'Chishang Caotang' was built on the west side of the Simianting, which was juxtaposed with Simianting in a similar shape and orientation, with only a slight step back. With its steep roof, majestic volume and high standing posture, it had naturally become the new main hall of the whole garden. The original Simianting was infinitely and embarrassingly reduced to a room overflowing to one side. The two halls were connected between the east and the west, so that the shadow space of the bamboos and stones in the north of the garden was over-shaded. Fortunately, in summer, all doors and windows can be removed from the hall, and it became a veranda on the water. At this time, its volume was maximized to be blurred and transparent, and for a while, the bamboos and stones in the north of the garden reappeared, which made people feel back to the original garden. In winter, the doors and windows were wrapped, the line of sight was blocked, and the space faded.

The huge Xuehaitang in the west of the garden was built in the same year as the quasi

台掩映的池北岸，一路树石逶迤、绿荫张天，反成园中最主要的被观赏界面，尤以池北岸的角色转换最

为戏剧化，是仅见的主客倒置。

更具体的因应变化有二：一是南岸高墙下、水廊中央辟设月门一道，可自仓院回望全园主要景象空间，隔岸老树陂陀，竟成隔水面坡之局，颇具张力与仪式感，可与苏州艺圃浴鸥小院高墙下极尽飘逸之重重月门相对照。二是原本以竹林密密遮蔽、幽径暗通的北岸空间中被辟出正对南岸月门的视廊，指向园北乐天轩一带，获得相当景深。

## 二、双厅并起

全园主要厅堂原为瓦檐低小，长身横逸的四面厅，于江南诸园中尤具画意，位置亦较为典型妥帖，但毕竟面积精约，难以满足晚清繁密公共活动之需。故清末又在其西侧兴建功能性空间『池上草堂』，以相似形态、朝向与四面厅并列，仅稍稍退后。其屋面陡峻，体量雄巨，昂然高峙，自然成为全园新的主要厅堂，原四面厅则无限尴尬地沦为其向一侧溢出的挟屋空间。两堂东西连理，令园北竹石荫翳的景象空间被过度遮蔽——好在此堂夏日可撤去全部门窗，成为水上凉堂，此时其体量被最大限度虚化、透明化，一时园北竹石重现，令人恍归千载池上；冬日则门窗包裹，视线阻断，空间失色。

园西庞巨的雪海堂系与池上草堂同年构建的准『礼堂』空间，因别处一院，故未对核心景象空间产

图十七　雪海堂前明洪武年间（一三六八至一三九八年）青石对狮之雄狮，今不存，疑为原松江府城隍庙前石狮，由徐瑞彤绘制

Fig.17  The Qing Stone male lion of 1368−1398 in front of Xuehaitang, which is destroyed and suspected the stone lion in Songjiang Chenghuang Temple, drawn by XU Ruitong

'auditorium' space of Chishangcaotang. Because it was located in another courtyard elsewhere, it did not interfere with the core landscape space. The stone lions in front of the hall were suspected to be from Songjiang Chenghuang Temple during the Hongwu period of the Ming Dynasty, indicating a special sense of ritual and spatial temperament. During the period of the Republic of China, the thatch pavilion in the bamboos and stones secluded area in the north of the pond was converted into Letianxuan, which was also the result of the continuous expansion of architectural space.

## 3. Frequent changes of the focus

Since the early Qing Dynasty, the focal point of this garden's spatial scene seemed to have undergone three changes.

From the beginning of the Qing Dynasty to the middle of the Qing Dynasty, the focus of the whole garden was the pure and condensed Laoshuxuan in the south-east of pond.

In the Shantang era in the middle and late Qing Dynasty, this pavilion may have been rebuilt as Dahuting and became the focal point of the second generation. At that time, the old elm might no longer exist. It is speculated that Xiaohuting in the north-east of the pond was also built at this time, although it further enriched the gable facade of the residence, and can interact with the Taihu Stone Mudantai that was added on the slope of the pond and in the south of Simianting, after all, it blurred the focus of Dahuting, which was an unquestionable failure.

Perhaps in view of this, in the late Qing Dynasty, a hexagonal lake pavilion was built in the south-west of the pond to face the lake and became the center of the third generation scene, but it interfered with the image of the pond.

## 4. A pond of three meanings

The reconstruction since the middle of the Qing Dynasty had excessively increased the complexity of the surface of the pool and slightly divided the surface of the pool, giving it the illusion of scale like a lake. At the same time, the closing of the space on the pond gave it a slight vertical sense of Yuantan. One pond started to possess several meanings, and the purity of the scene was slightly lost, but it was also an inevitable phenomenon in the 'post' era.

生于扰。堂前原明洪武年间石狮疑来自松江府城隍庙，昭示着别样的仪式感与空间气质。民国年间池北竹石幽境中的茅亭被改建为乐天轩，亦是建筑空间持续膨胀的结果。

三、焦点叠变

自清初以来，此园之空间景象焦点似经历了三度变化。

清初至清中期，全园景象焦点为池东南清纯凝练的老树一轩。

清中后期的善堂时代，此轩可能被翻建为大湖亭，成为第二代景象焦点，其时老榆或已不存。推测池东北小湖亭亦建于此时，虽其进一步丰富了住宅的山墙立面，并能与四面厅南侧临池坡地上增设的湖石牡丹台形成互动；但毕竟模糊了大湖亭的焦点地位，系无可置疑的败笔。

或许有鉴于此，清末又于池西南增筑六角湖亭，挑临湖面，成为第三代景象中心，却又对池塘意象形成干扰。

四、一池三意

清中叶以来的改筑，过度增加了池周围界面的复杂程度，且微微划分了池面，令其渐有湖泊化般的尺度幻觉。同时池上空间的闭合，则令其微有纵向的渊潭化之感。一池而渐具多意，微失景象纯度，但亦是『后』时代的必然现象。

图十八　扬州徐园池塘型水面平面图，由朱宇晖步测绘制

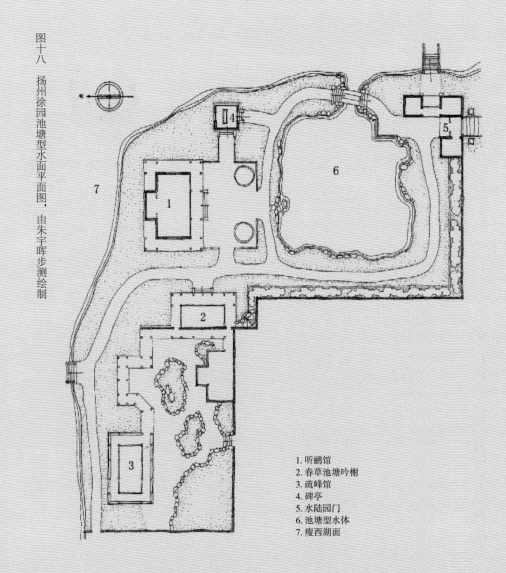

1. 听鹂馆
2. 春草池塘吟榭
3. 疏峰馆
4. 碑亭
5. 水陆园门
6. 池塘型水体
7. 瘦西湖面

Fig.18　Plan of Yangzhou Xu Garden and its pond, drawn and measured by steps by ZHU Yuhui

In contrast, the pond in Xu Garden on the Slender West Lake in Yangzhou, created in the early Republic of China, is purer and more typical.

## 5. Shanghai-style thatched cottage

Judging from the historical photos, Chishangcaotang in the garden was originally simple in appearance, with gentle wing angles and no roof ridge. Although it was huge, it can still continue the simple style of Simianting in the late Ming and early Qing Dynasties, without losing its identity as a 'thatched cottage'. Unfortunately, it was rebuilt by the Japanese invaders as a 'Japanese-style' during the Anti-Japanese War. In recent years, when it was restored, it seemed to emphasize its dominant position. The roof was raised and exaggerated. The 'thatched cottage' was changed to 'gorgeous hall', completely losing its original historical appearance. The buildings such as Yifang in the garden also give people the feeling of excessive decoration and scratching their heads. This has been a common problem in the renovation of Jiangnan gardens in recent years.

'Woshuxuan' in the garden that was built in the early stage of the Republic of China, has a round ridge roof currently. It is 'Shuiqiang Faqiang', light and elegant. The horizontal lines and the several old trees in front of it form a harmonious picture. The style is similar to that of the Linglongguan in Suzhou Zhuozheng Garden, and is actually a work with a more literati style.

In the late Qing Dynasty, 'Banshan banshui banshuchuang' pavilion had a steep and heavy roof, and a crane was erected on its top, like many late Qing architectures in Shanghai Yu Garden, having the tendency of 'Shanghai-style' in the early period. In the historical pictures, this pavilion was a warm pavilion with threshold windows under the eaves, which can avoid the disadvantages of top-heavy. The current situation of the other two pavilions near the pond in the garden are cool pavilions, which are not only unfavorable for the winter hustle and bustle, but also lack the virtuality and reality change in form.

## 五、海派草堂

从历史照片看，园中池上草堂原本外观简素，翼角平缓，不起屋脊，虽体量庞大，尚能延续明末清初四面厅之简远格调，不失『草堂』身份。惜抗战时被日寇改建为和样，近年再作复原时似为强调其主体地位，屋脊高起，发戗夸张，易『草堂』为『华堂』，全失历史原貌——园内疑舫等建筑亦给人过度装饰化、搔首弄姿的感受，这已是近年江南诸园修缮的通病了。

而园内初建于民国前期的卧树轩现状卷棚为顶，水戗发戗，清淡秀雅，以横卧的线条与轩前数株老树构成和谐画面，风格类似于苏州拙政园玲珑馆等建筑，实为尚具文人园气质的轩类作品。

晚清半山半水书窗亭屋顶陡峻沉重，宝顶高立一鹤，与上海豫园诸多晚清建筑同具有前期『海派』倾向。历史图片中此亭为檐下施槛窗之暖亭，可免头重脚轻之弊。园中临池另两亭现状也均为凉亭，既不利冬日盘桓，亦觉缺乏形态上的虚实变化。

相形之下，创作于民初的扬州瘦西湖徐园园池塘则更显纯粹典型。

图二十　晚清醉白池池上草堂旧影

图十九　原本檐脊质朴舒缓的池上草堂与四面厅于日据时期被改造为『和样』风格。左侧池上草堂前的木桥至一九六三年被改造为湖石拱桥，反倒破坏了池塘板桥的意象

Fig.19  The original Chishang Caotang and Simianting, with their low and humble ridges, were transformed into Japanese style during the World War II. The wooden bridge in front of Chishang Caotang on the left was converted into a Taihu Stone arch bridge in 1963, which in turn destroyed the image of the pond with slab bridge.

Fig.20  Old view of Chishang Caotang in Zuibaichi Garden in the late Qing Dynasty

# Same structure of hall and residence: Shantang originated from the GU family's residence

The east 'house' and west garden of Zuibaichi complex should originate at least from the GU Dashen era in the early Qing Dynasty.

*GU Dashen Ziding Nianpu Shougao* had recorded that 'on August 19th in 1663, I moved north with my family and farewelled with my parents in the west garden. On July 24th in 1664, my mother whose last name is XU die in the main bedroom of the west garden…' The 'West Garden' here should be Zuibaichi Garden in the south-west suburb of Songjiang City (the GU family's main residence was located in the prestigious area on the east side of Songjiang Government Office in the city), which was a complex combining the garden and the residence and had the 'main bedroom'. In the early years of Kangxi period, GU Dashen might have lived here in sorrow during the years of keeping his parents' filial piety, and later became the house of GU Sizhao in the early Qianlong years and Shantang in the middle and late Qing Dynasty. There is still a complete Qing Stone foundation under the granite foundation of Baochenglou of Shantang, which is suspected the foundation of the back building of GU Dashen's residence in the early Qing Dynasty, and it is also the 'main bed' where GU's mother lived in and died.

Now in Shantang, which is completely 'same structure' as the residence, three architectures still exist, namely entrance hall facing the east and near 'elm tree', lease hall facing the south and Baochenglou. They are all 'straight beams' which were randomly built in the middle and late Qing Dynasties. Among them, the lease hall is suspected to be the former site of the main hall of Gu Dashen's mansion, and its two compartments have been demolished during the 'park' renovation in recent years. There is still some extra space between the east side of Shantang and elm tree. It is speculated that there was an east axis in the old GU family's residence, otherwise the space seems to be insufficient for the family.

In the late Qing Dynasty, a granary courtyard was built on the south bank of the pond to show the functional attributes of Shantang. In recent years, small soft wooden works have been added to the renovation, making the entire courtyard a scene.

堂宅同构：源自顾宅的善堂建筑

醉白池建筑群的东「宅」西园格局应至少源自清初的顾大申时代。

《顾大申自订年谱手稿》载有「康熙二年「八月十九日，率家累北行，拜辞两大人于西园」；康熙三年「七月二十四日午时，先慈许宜人终于西园之正寝」……」——这里的「西园」应即松江府城西南郊的醉白池（顾氏正宅在府城内的松江府署东侧显赫地段），是拥有正寝的宅园一体建筑群。康熙初际，顾大申在为父母守孝家居的数年间可能即黯然盘桓在这里，后次第成为乾隆初年的顾思照宅与清中后期的善堂。今善堂宝成楼花岗石台基下尚存完整青石台基，疑即清初顾大申宅后楼故基，也即顾母日常起居并长逝于斯的「正寝」。

与住宅建筑完全同构的善堂今存临「榆树头」之东向门厅及南向征租厅与宝成楼三进，均为构架草草的清中晚期圆堂「直梁造」建筑。其中征租厅疑原即顾大申宅正厅旧址，其两厢于近年公园化修整中被拆除。善堂东侧与榆树头间尚有一定余地，推测旧日顾宅曾有东路建筑，否则空间似不敷家族使用。

晚清时于池塘南岸添筑粮仓合院一区，展现出善堂建筑的功能属性，近年改造中添加柔曲小木作，使整个院落景象化。

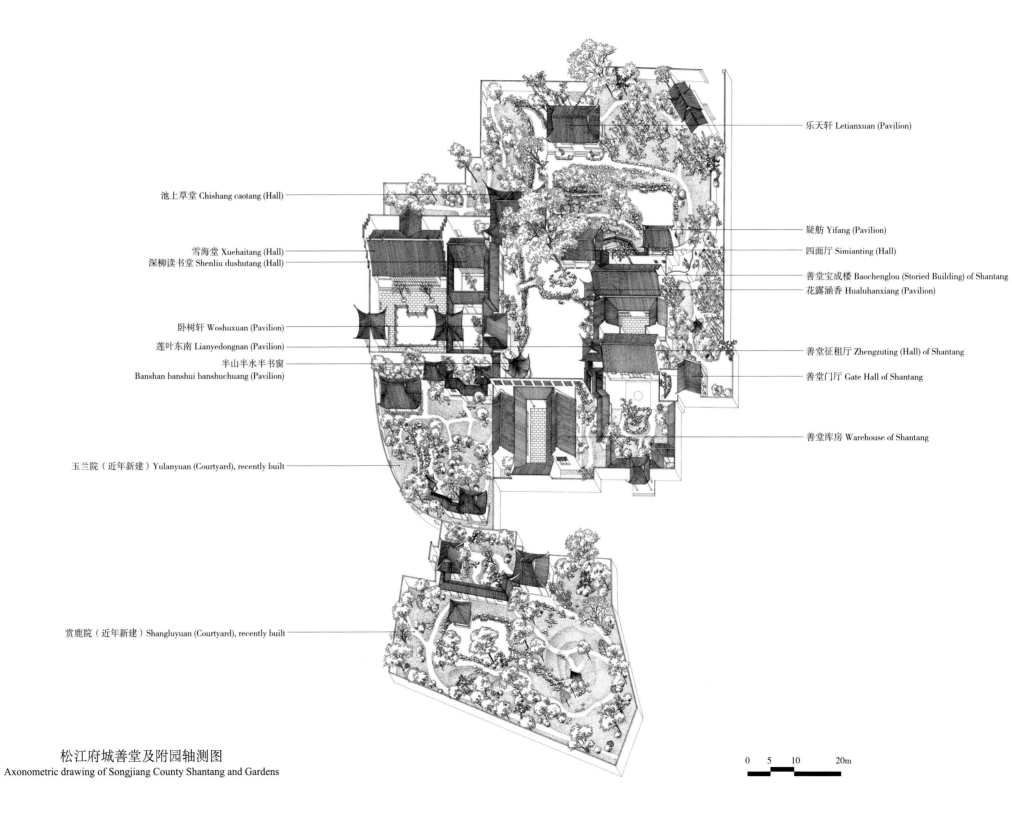

乐天轩 Letianxuan (Pavilion)

池上草堂 Chishang caotang (Hall)

疑舫 Yifang (Pavilion)

四面厅 Simianting (Hall)

雪海堂 Xuehaitang (Hall)
深柳读书堂 Shenliu dushutang (Hall)

善堂宝成楼 Baochenglou (Storied Building) of Shantang
花露涵香 Hualuhanxiang (Pavilion)

卧树轩 Woshuxuan (Pavilion)

莲叶东南 Lianyedongnan (Pavilion)

半山半水半书窗
Banshan banshui banshuchuang (Pavilion)

善堂征租厅 Zhengzuting (Hall) of Shantang

善堂门厅 Gate Hall of Shantang

善堂库房 Warehouse of Shantang

玉兰院（近年新建）Yulanyuan (Courtyard), recently built

赏鹿院（近年新建）Shangluyuan (Courtyard), recently built

松江府城善堂及附园轴测图
Axonometric drawing of Songjiang County Shantang and Gardens

0    5    10        20m

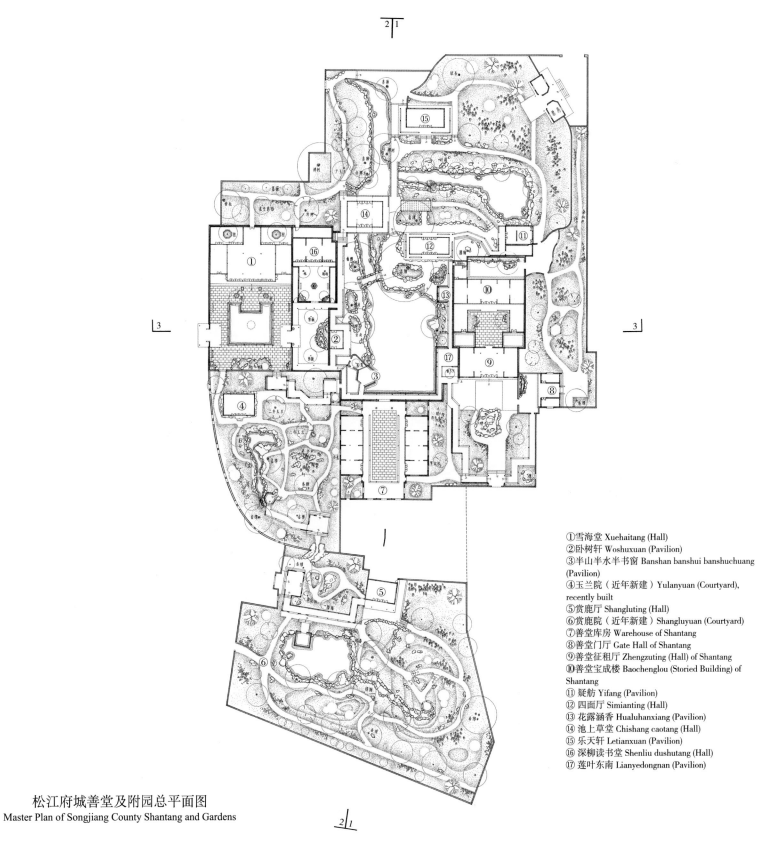

①雪海堂 Xuehaitang (Hall)
②卧树轩 Woshuxuan (Pavilion)
③半山半水半书窗 Banshan banshui banshuchuang (Pavilion)
④玉兰院（近年新建）Yulanyuan (Courtyard), recently built
⑤赏鹿厅 Shangluting (Hall)
⑥赏鹿院（近年新建）Shangluyuan (Courtyard)
⑦善堂库房 Warehouse of Shantang
⑧善堂门厅 Gate Hall of Shantang
⑨善堂征租厅 Zhengzuting (Hall) of Shantang
⑩善堂宝成楼 Baochenglou (Storied Building) of Shantang
⑪疑舫 Yifang (Pavilion)
⑫四面厅 Simianting (Hall)
⑬花露涵香 Hualuhanxiang (Pavilion)
⑭池上草堂 Chishang caotang (Hall)
⑮乐天轩 Letianxuan (Pavilion)
⑯深柳读书堂 Shenliu dushutang (Hall)
⑰莲叶东南 Lianyedongnan (Pavilion)

N
E

0  5  10     20m

松江府城善堂及附园总平面图
Master Plan of Songjiang County Shantang and Gardens

原善堂库房
Warehouse of
Shantang

赏鹿厅
Shangluting
(Hall)

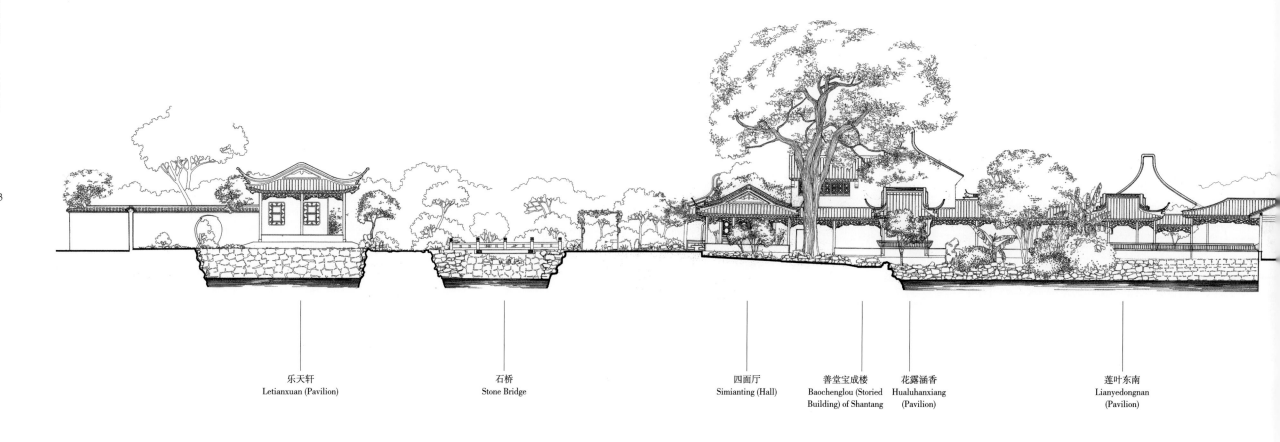

乐天轩
Letianxuan (Pavilion)

石桥
Stone Bridge

四面厅
Simianting (Hall)

善堂宝成楼
Baochenglou (Storied Building) of Shantang

花露涵香
Hualuhanxiang (Pavilion)

莲叶东南
Lianyedongnan (Pavilion)

0 1 2 5m

松江府城善堂及附园南北剖面图（东岸）
South-north section of Songjiang County Shantang and Gardens (the east bank)

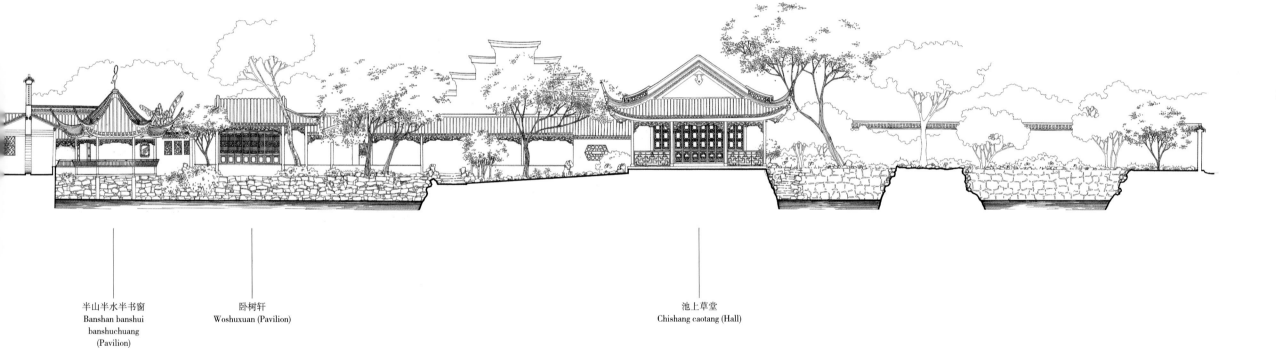

半山半水半书窗
Banshan banshui
banshuchuang
(Pavilion)

卧树轩
Woshuxuan (Pavilion)

池上草堂
Chishang caotang (Hall)

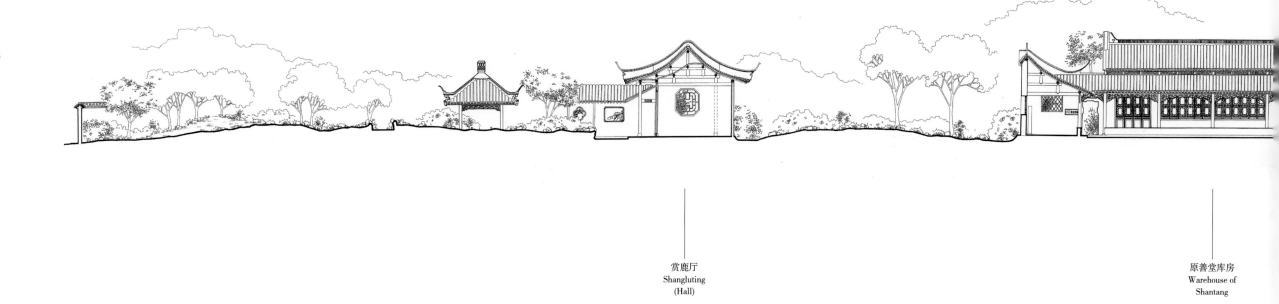

赏鹿厅
Shangluting
(Hall)

原善堂库房
Warehouse of
Shantang

松江府城善堂及附园南北剖面图（西岸）
South-north section of Songjiang County Shantang and Gardens (the west bank)

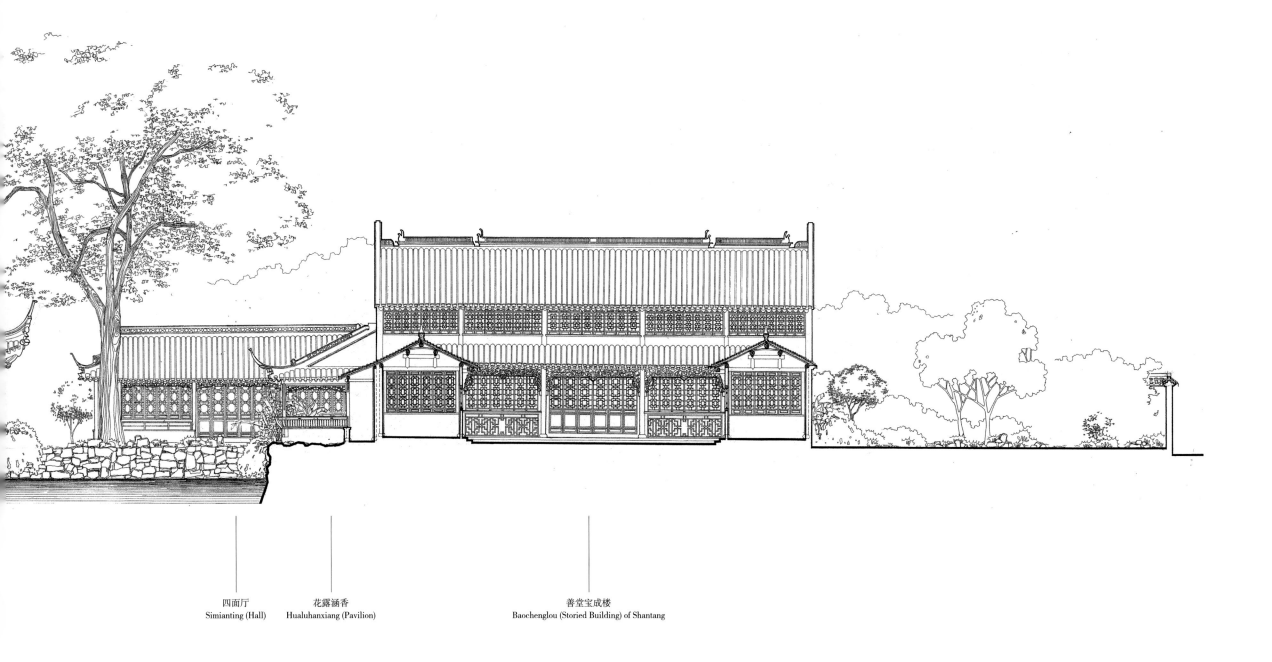

四面厅
Simianting (Hall)

花露涵香
Hualuhanxiang (Pavilion)

善堂宝成楼
Baochenglou (Storied Building) of Shantang

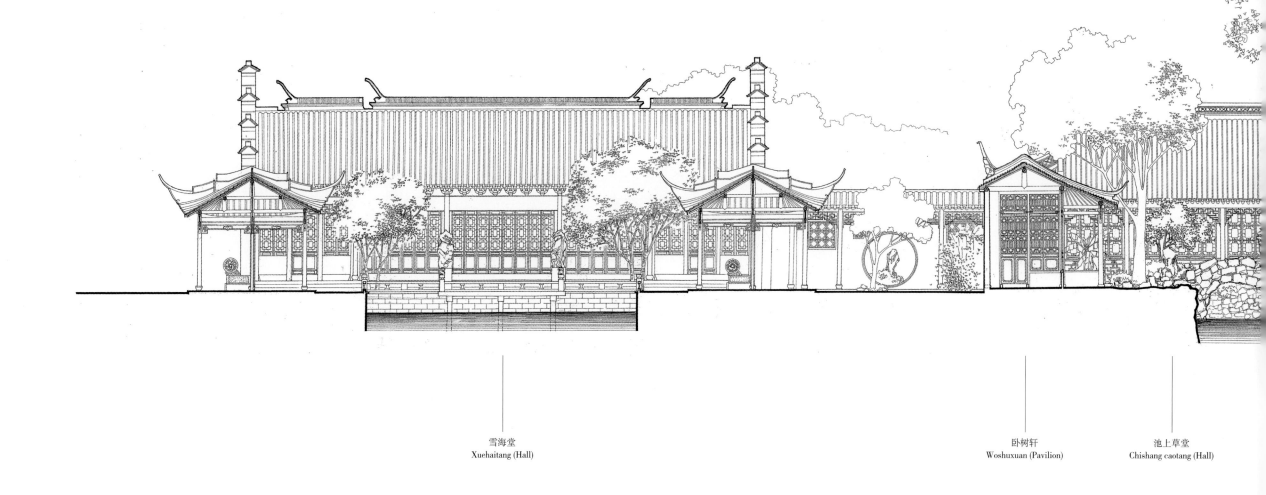

雪海堂
Xuehaitang (Hall)

卧树轩
Woshuxuan (Pavilion)

池上草堂
Chishang caotang (Hall)

松江府城善堂及附园东西剖面图（北岸）
East-west section of Songjiang County Shantang and Gardens (the north bank)

0  1  2  5m

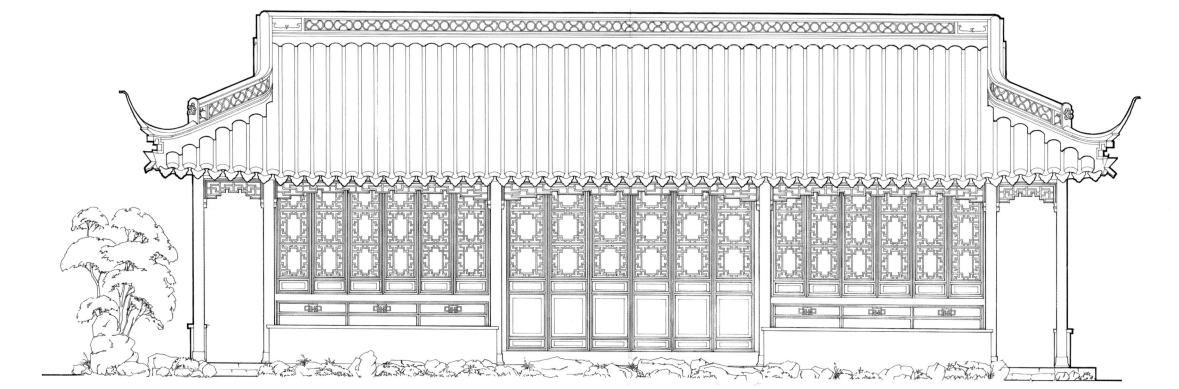

四面厅正立面图
Front elevation of Simianting

0　0.5　1　2m

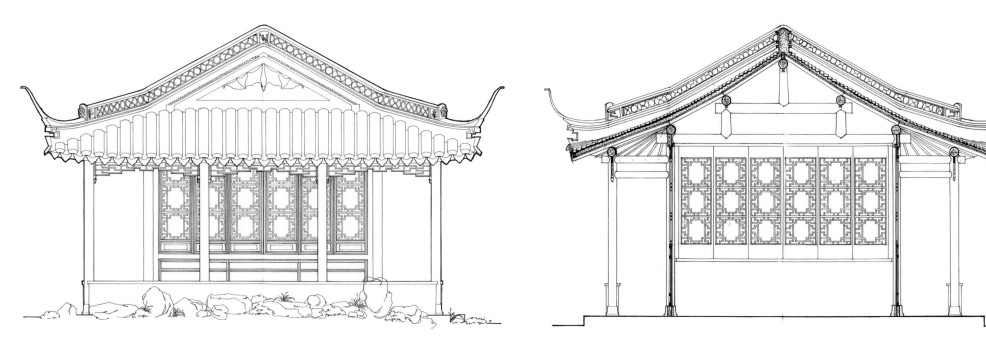

四面厅侧立面图
Side elevation of Simianting

四面厅横剖面图
Cross-section of Simianting

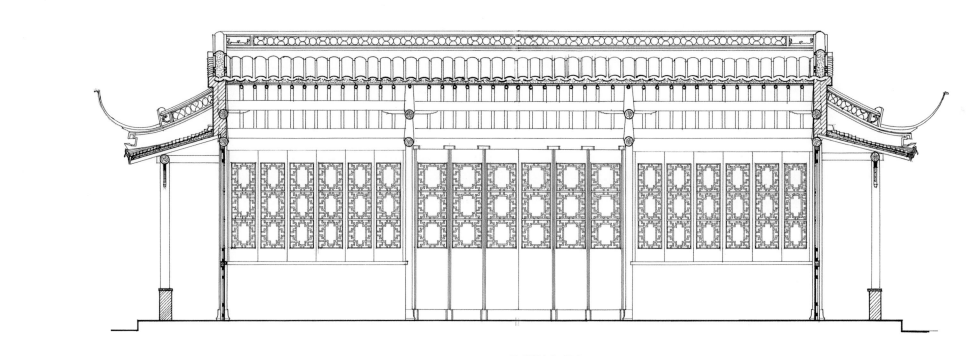

0    0.5    1         2m

四面厅纵剖面图
Longitudinal section of Simianting

花露涵香亭西立面图
West elevation of Hualuhanxiangting

0  0.25  0.5  1m

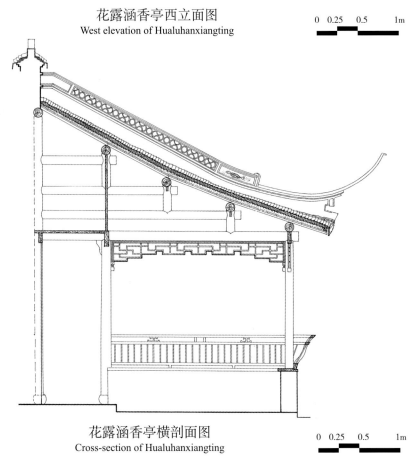

花露涵香亭横剖面图
Cross-section of Hualuhanxiangting

0  0.25  0.5  1m

花露涵香亭平面图
Plan of Hualuhanxiangting

0  0.25  0.5  1m

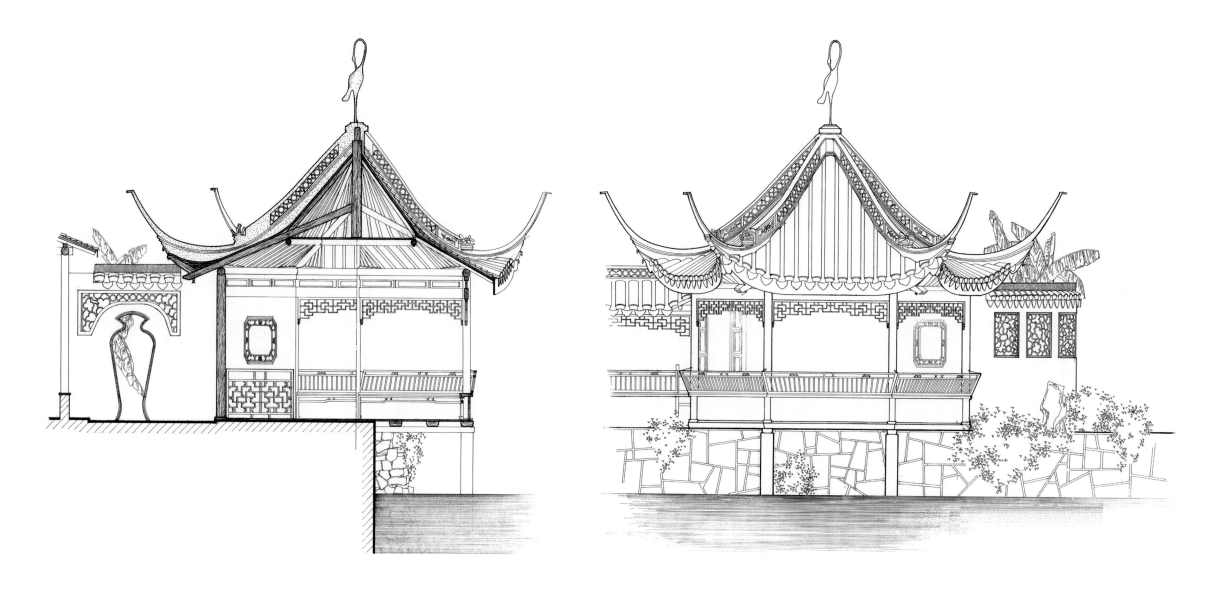

0　0.3　0.6　　1.5m

半山半水半书窗亭剖面图
Section of Banshanbanshuibanshuchuangting

半山半水半书窗亭立面图
Elevation of Banshanbanshuibanshuchuangting

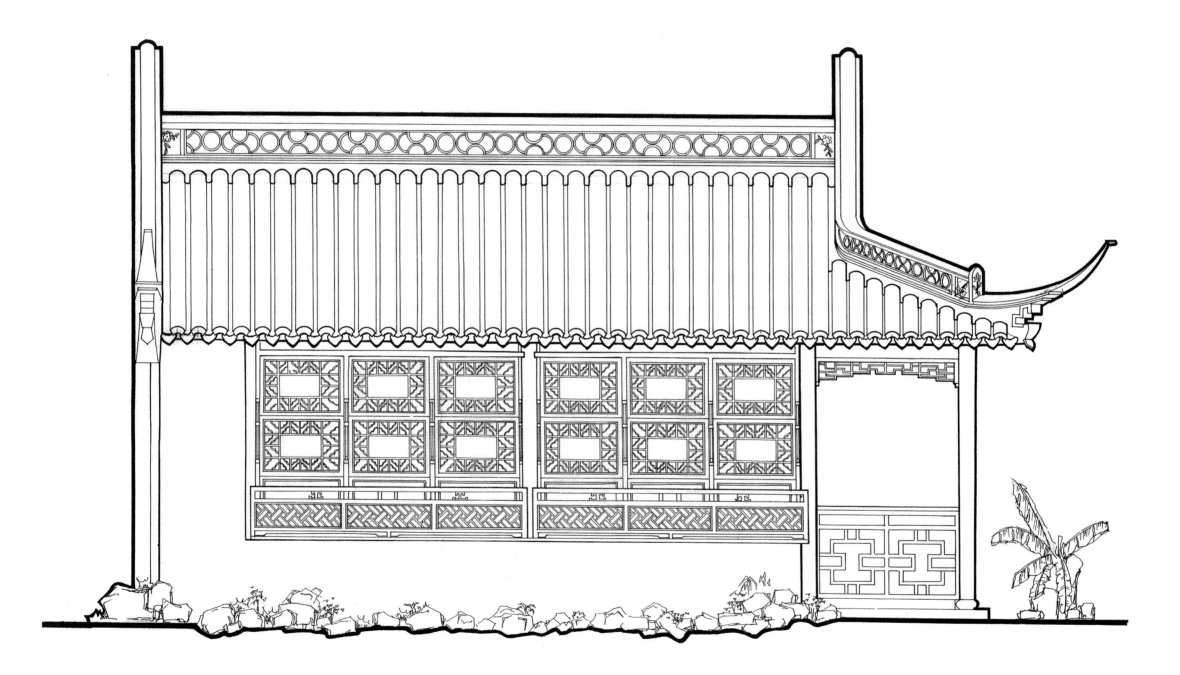

疑舫侧立面图
Front elevation of Yifang

0    0.5    1    2m

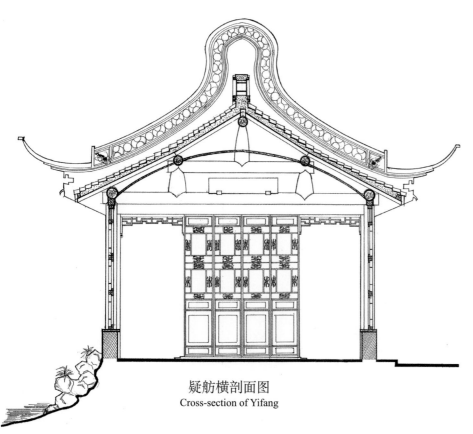

疑舫横剖面图
Cross-section of Yifang

疑舫正立面图
Side elevation of Yifang

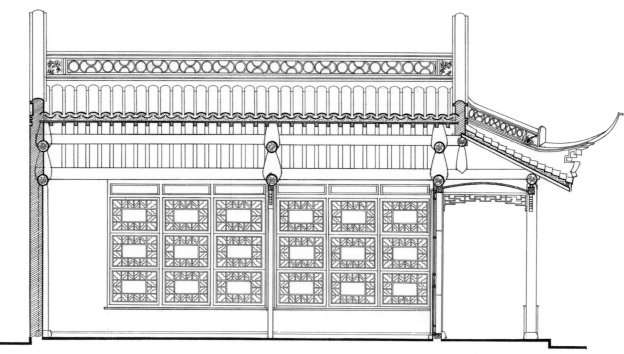

0 0.5 1 2m

疑舫纵剖面图
Longitudinal section of Yifang

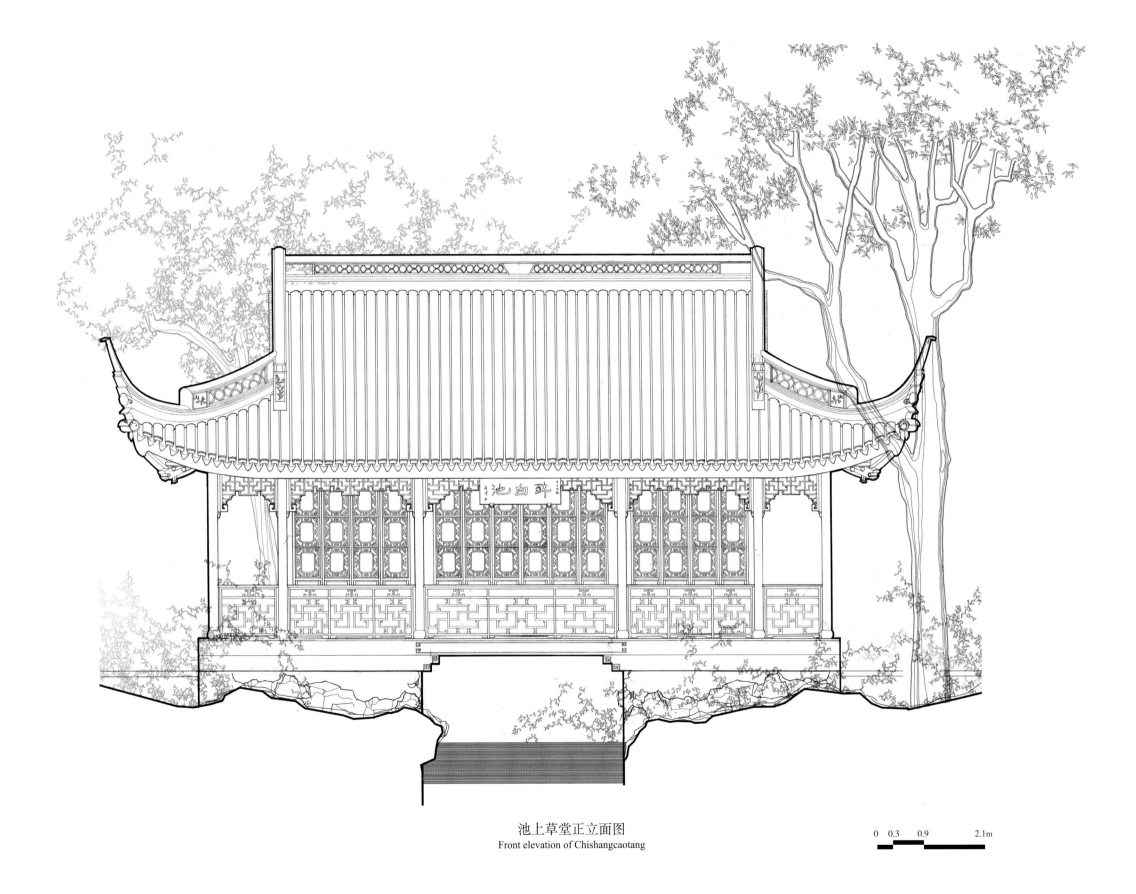

池上草堂正立面图
Front elevation of Chishangcaotang

0  0.3  0.9      2.1m

池上草堂侧立面图

Side elevation of Chishangcaotang

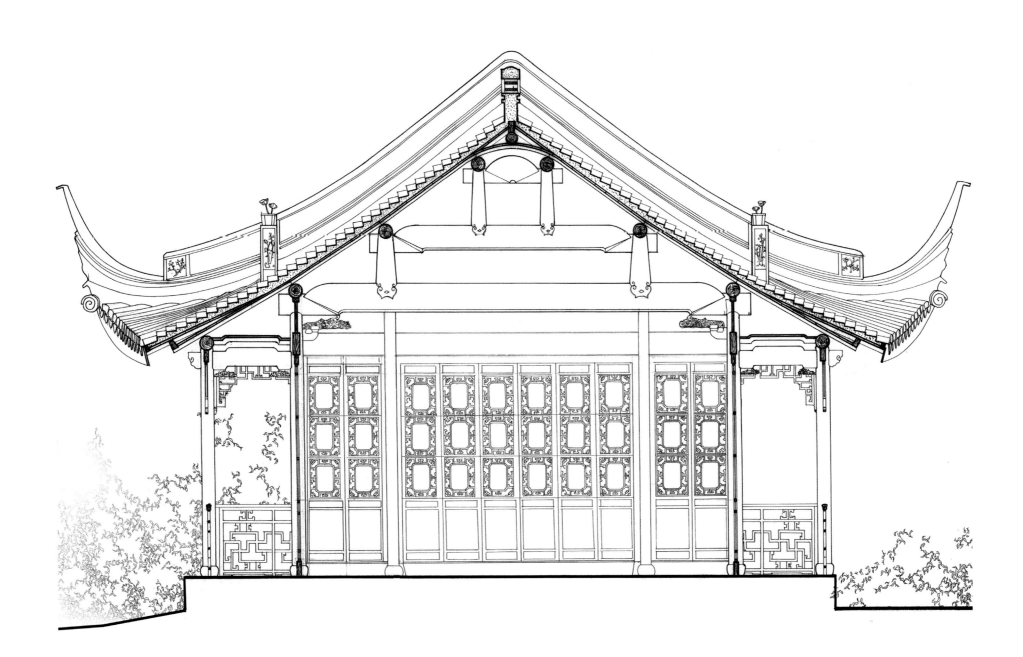

池上草堂横剖面图
Cross-section of Chishangcaotang

0　0.5　1　　2.5m

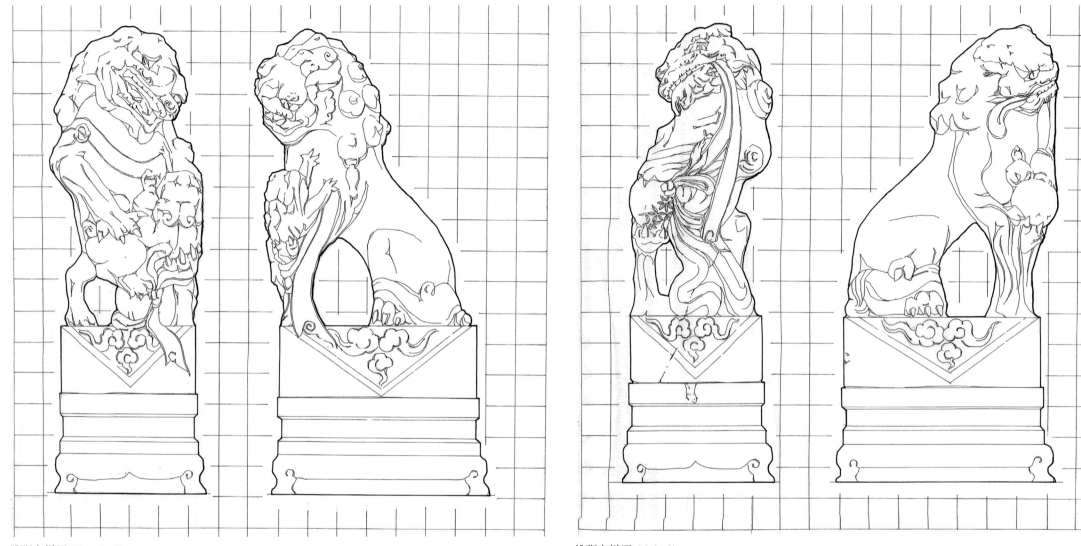

雌狮大样图 Female lion

注：今雪海堂前石狮，为一九八四年从明嘉靖隆庆年间松江籍内阁首辅徐阶的大夫人张氏墓地移来

雄狮大样图 Male lion

**雪海堂前明后期内阁首辅徐阶弟徐陟（1503-1583）墓石狮大样图**
The late Ming Dynasty stone lions of XU Zhi (1503-1583, the minister XU Jie's Brother)'s tomb in front of Xuehaitang

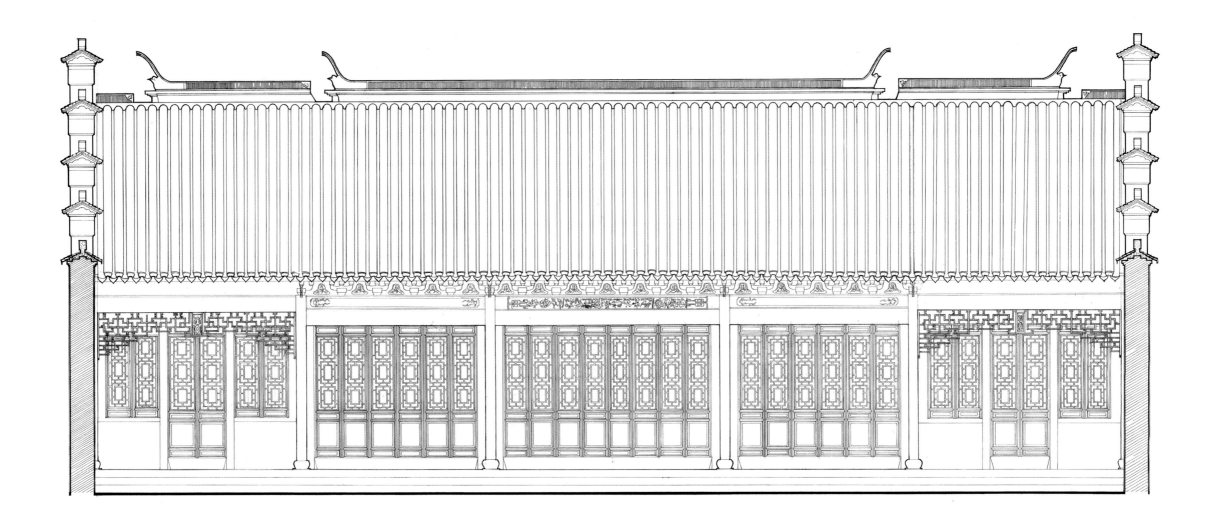

雪海堂正立面图
Front elevation of Xuehaitang

0      0.75      1.5m

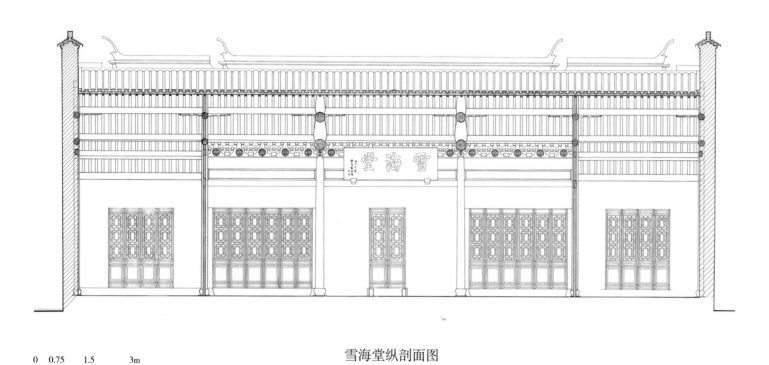

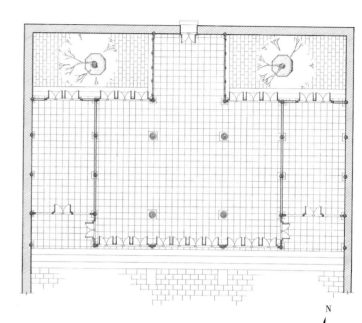

雪海堂纵剖面图
Section of Xuehaitang

0　0.75　1.5　　3m

雪海堂平面图
Plan of Xuehaitang

0　1.25　2.5　　5m

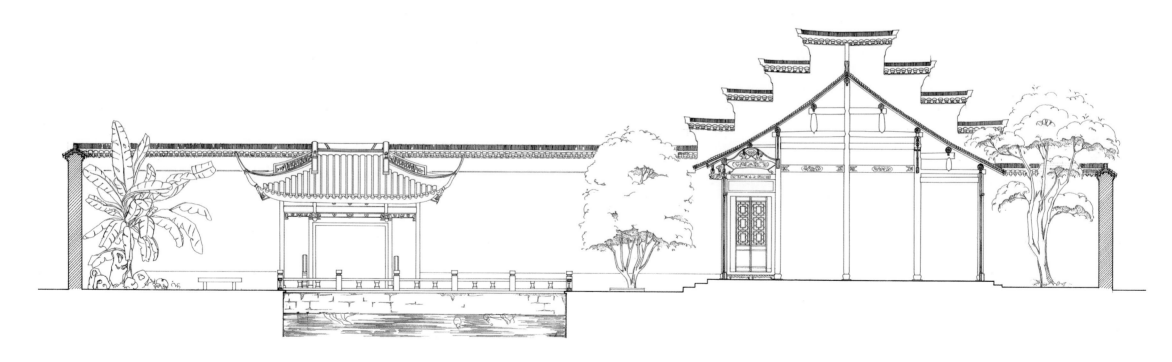

雪海堂院落南北剖面图
South-north section of Xuehaitang Courtyard

0　1.25　2.5　　5m

卧树轩东立面图
East elevation of Woshuxuan

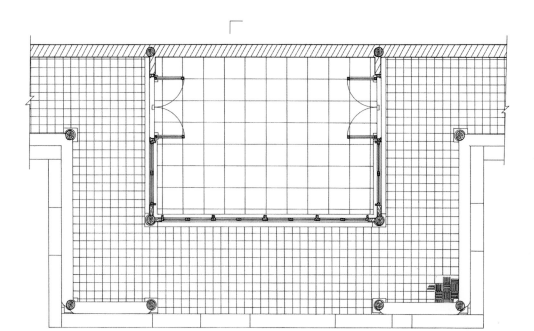

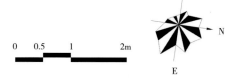

0    0.5    1    2m

卧树轩平面图
Plan of Woshuxuan

0    0.5    1    2m

卧树轩横剖面图
Cross-section of Woshuxuan

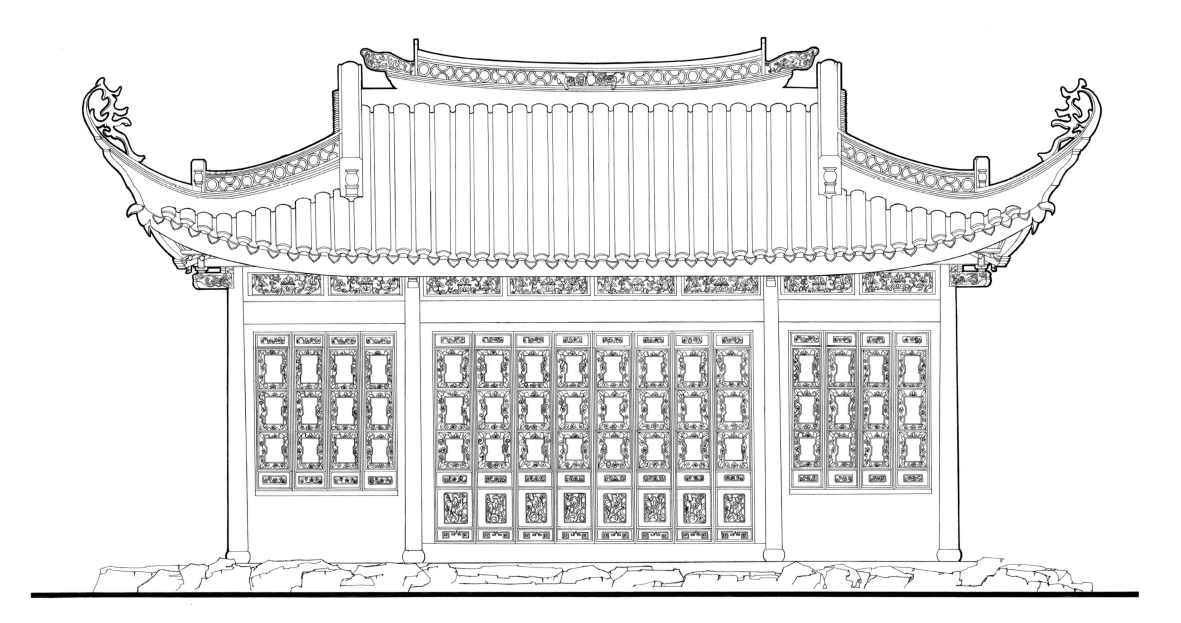

玉兰厅正立面图（一九八〇年代新建）
Front elevation of Yulanting (rebuilt in 1980s)

0　　0.5　　1　　　　2m

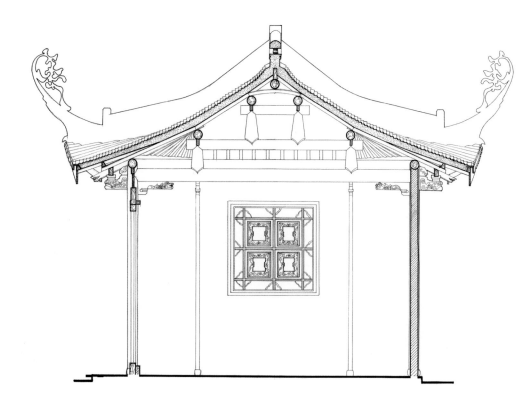

玉兰厅横剖面图（一九八〇年代新建）
Cross-section of Yulanting (rebuilt in 1980s)

0    0.5    1    2m

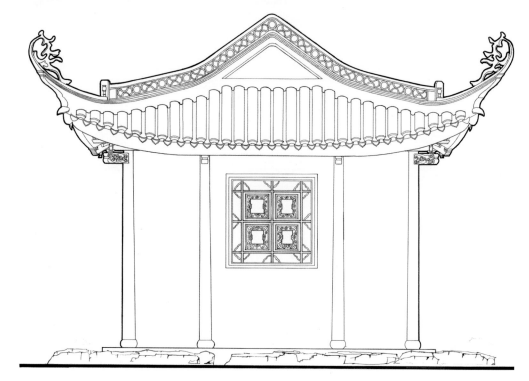

玉兰厅侧立面图（一九八〇年代新建）
Side elevation of Yulanting (rebuilt in 1980s)

0    0.5    1    2m

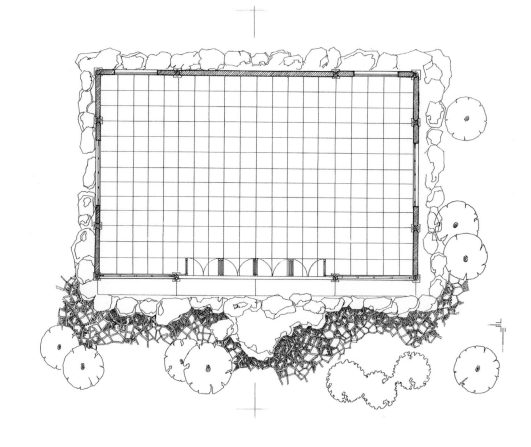

玉兰厅平面图（一九八〇年代新建）
Plan of Yulanting (rebuilt in 1980s)

0    0.5    1    2.5m

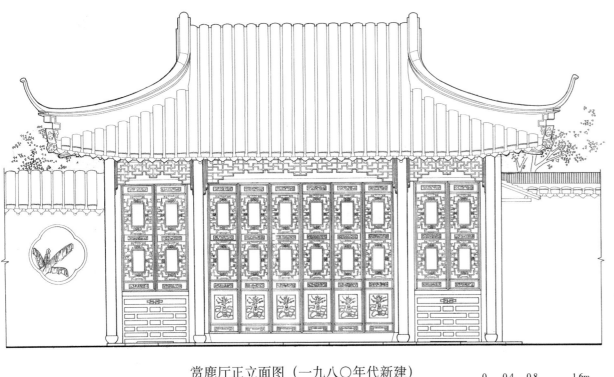

赏鹿厅正立面图（一九八〇年代新建）
Front elevation of Shangluting (rebuilt in 1980s)

0  0.4  0.8  1.6m

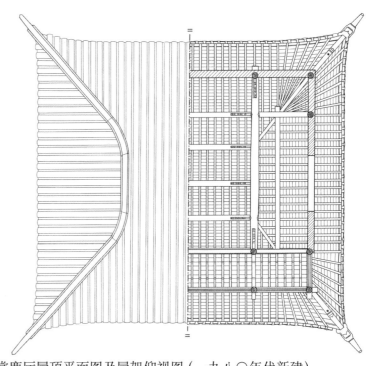

赏鹿厅屋顶平面图及屋架仰视图（一九八〇年代新建）
Roof plan and bottom view of Shangluting's beams (rebuilt in 1980s)

0  0.6  1.2  2.4m

赏鹿厅侧立面图（一九八〇年代新建）
Side elevation of Shangluting (rebuilt in 1980s)

0  0.4  0.8  1.6m

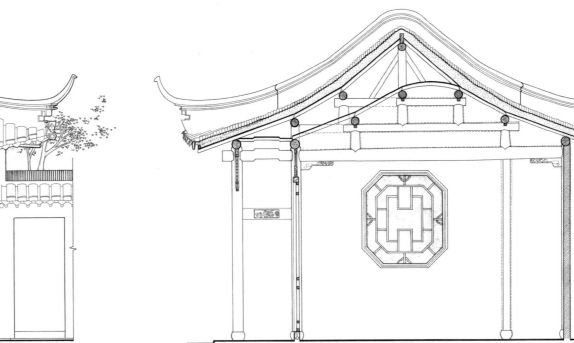

赏鹿厅横剖面图（一九八〇年代新建）
Cross-elevation of Shangluting (rebuilt in 1980s)

0  0.4  0.8  1.6m

中国古建筑测绘大系·祠庙建筑与园林建筑——上海庙园

345

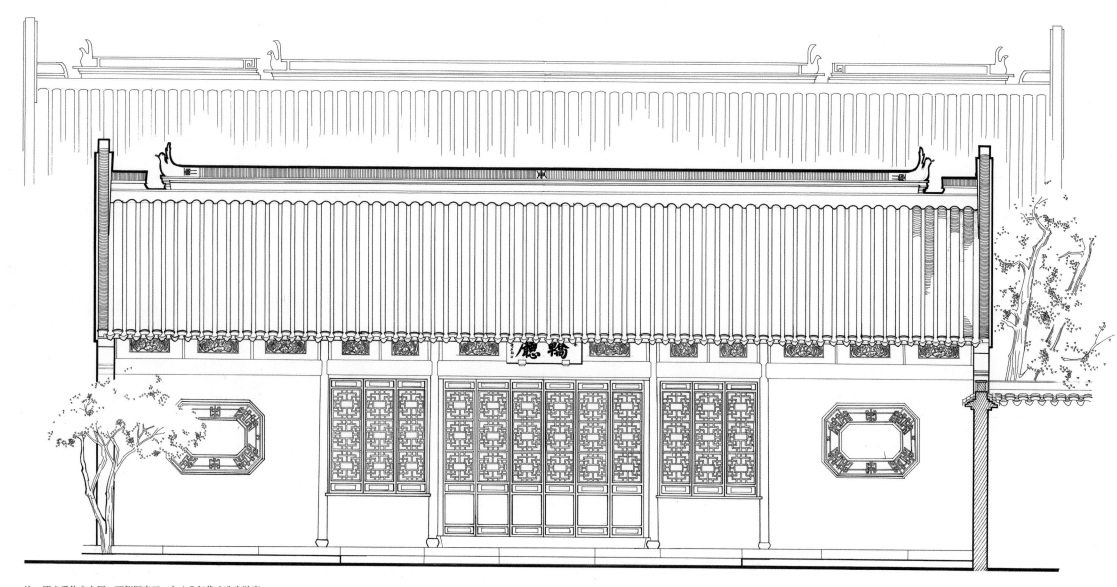

注：原应系住宅大厅，两侧厢房于一九八〇年代改造为游廊

0    0.25    0.5    1m

征租厅正立面图
Front elevation of Zhengzuting

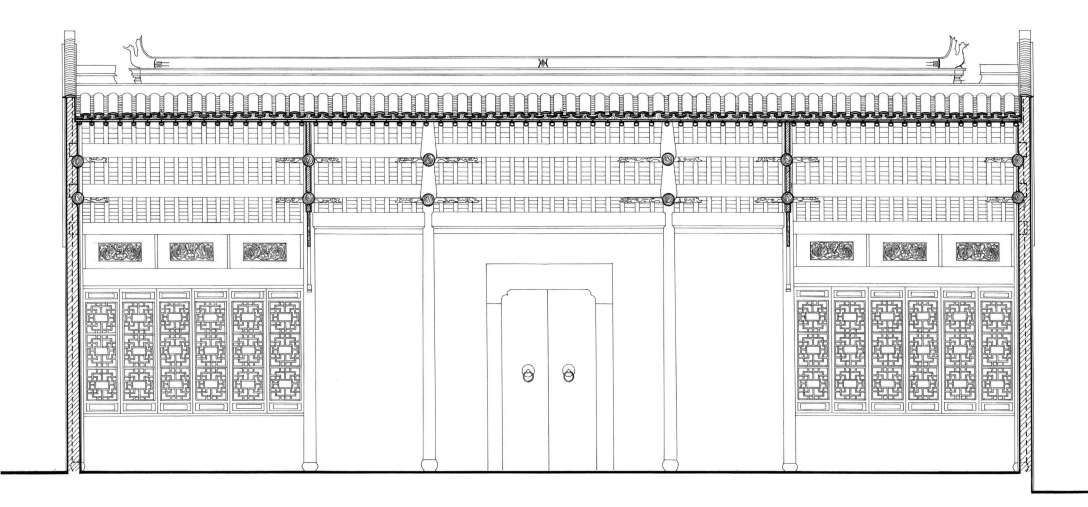

征租厅纵剖面图
Longitudinal section of Zhengzuting

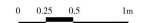

0　0.25　0.5　　　1m

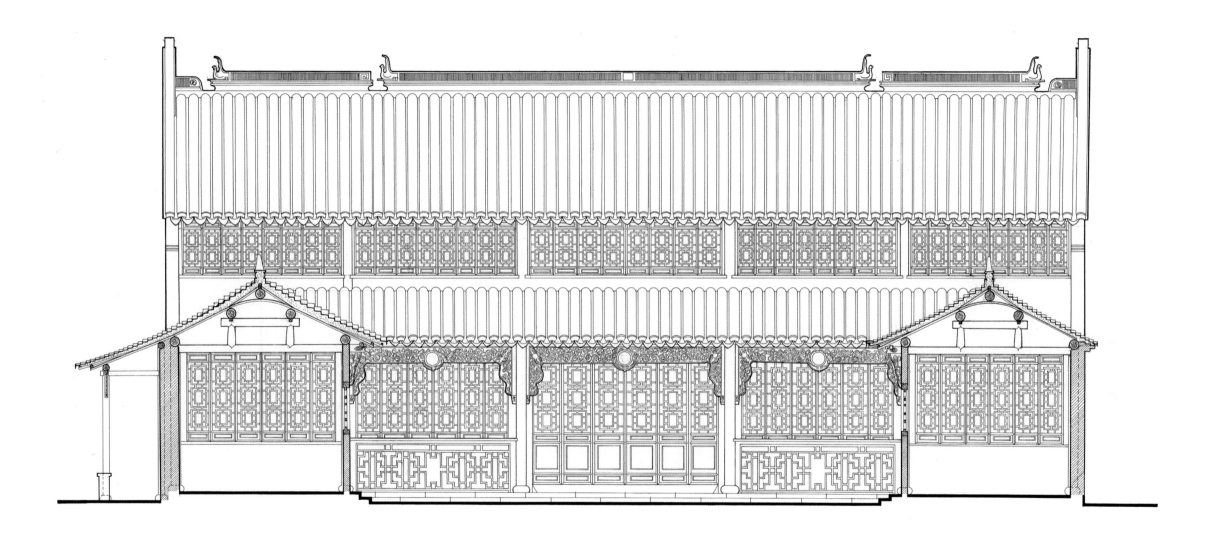

0　　　1　　2　　　　　　4m

宝成楼正立面图
Front elevation of Baochenglou

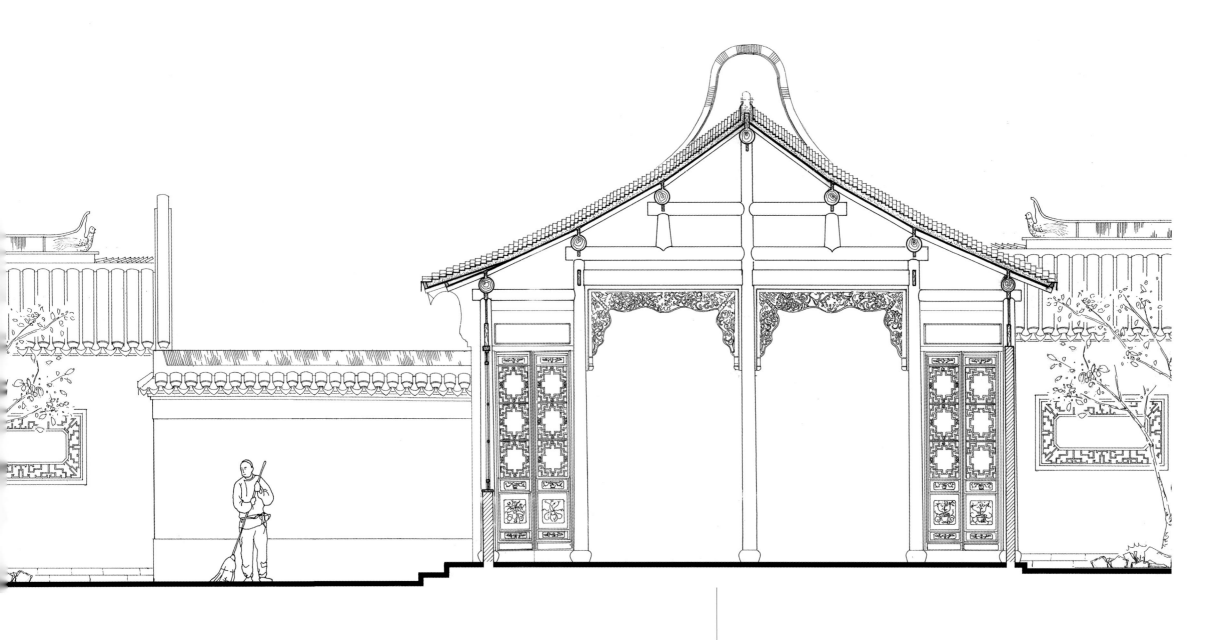

征租厅
Zhengzuting

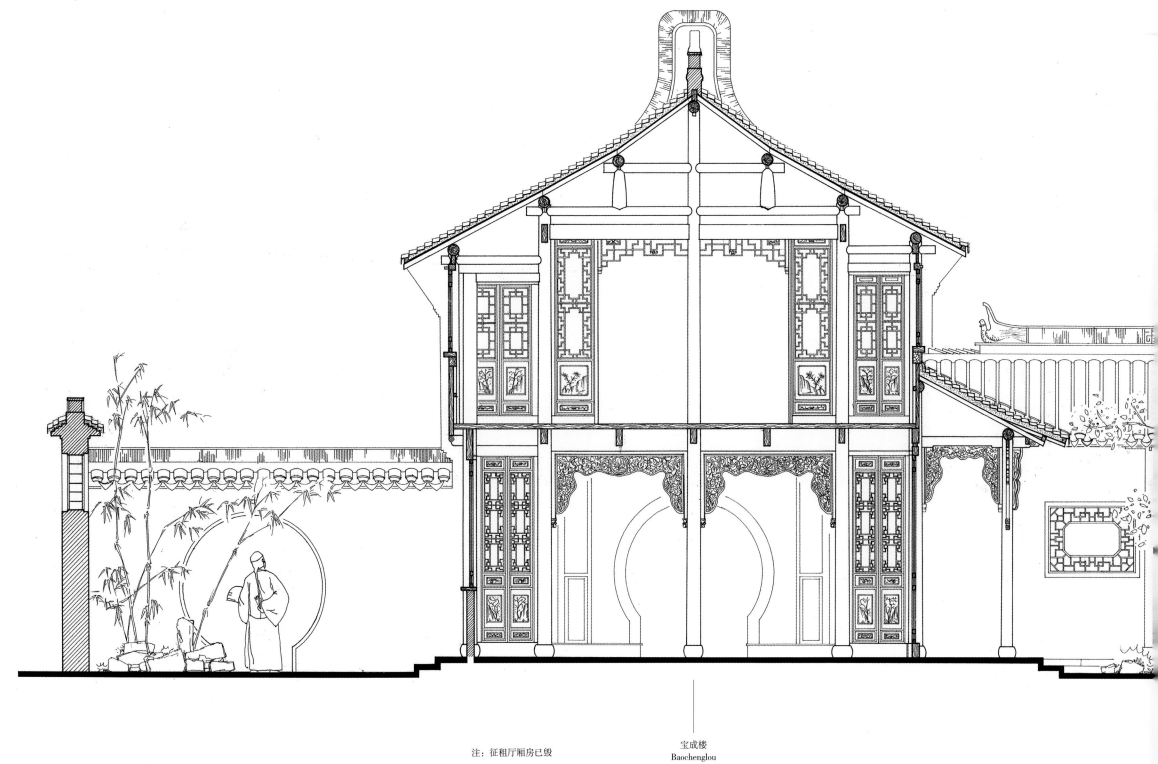

注：征租厅厢房已毁

宝成楼
Baochenglou

0    1    2         4m

松江府城善堂中轴剖面图
Section of Songjiang County Shantang's main axis

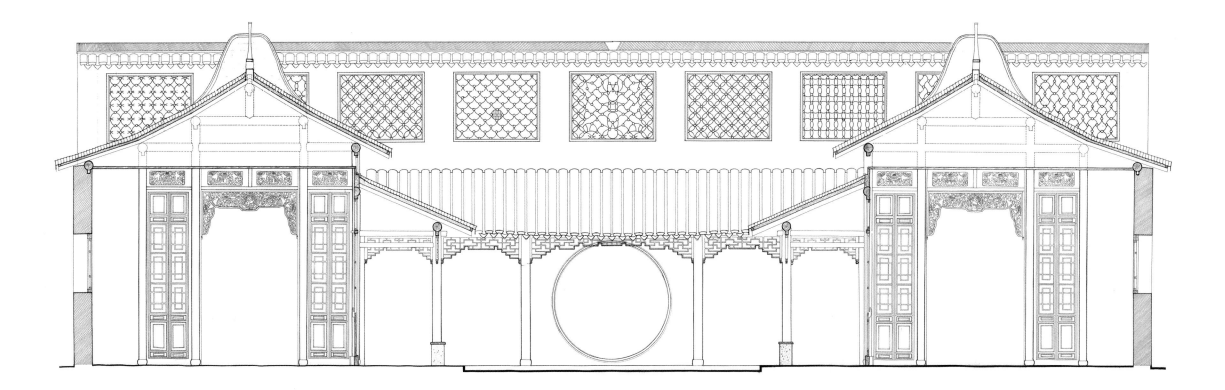

库房院落南北剖面图
South-north section of Storage Room's courtyard

0　0.5　1　2m

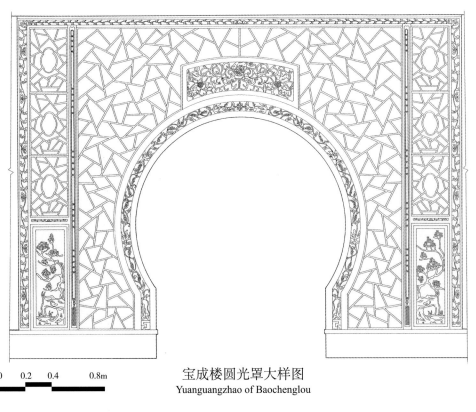

宝成楼圆光罩大样图
Yuanguangzhao of Baochenglou

0  0.2  0.4  0.8m

宝成楼裙板大样图
Qunban of Baochenglou

0  0.1  0.2  0.4m

门厅抱鼓石大样图
Baogushi of Menting

0  1  2  4m

玉兰厅梁头、梁垫大样图
Liangtou and Liangdian of Yulanting

0  0.1  0.2  0.4m

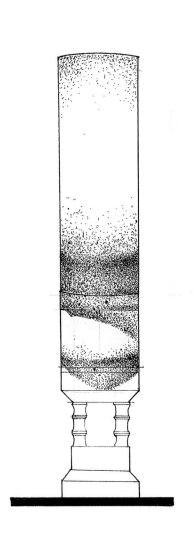

**Item of survey:** Shanghai County Guozhai and East Garden (Shuyin Building)

**Address:** No77, Tiandeng Lane, Huangpu District, Shanghai

**Age of construction:** 1763 (year 28 of Qianlong Emperor of Qing Dynasty)

**Site area:** 3.4 mu

**Competent organization:** Shanghai Cultural Heritage Management Committee; Guo
family

**Survey organization:** College of Architecture and Urban Planning, Tongji University;
Shanghai Cultural Heritage Management Committee

**Time of survey:** 2005

测绘项目：上海县郭氏宅及东园（书隐楼）

地　　址：黄浦区小南门天灯弄七十七号

始建年代：乾隆二十八年（一七六三年）

占地面积：三·四亩

主管单位：上海市文管会、郭氏家族

测绘单位：同济大学建筑与城规学院

　　　　　上海市文管会

测绘时间：二〇〇五年

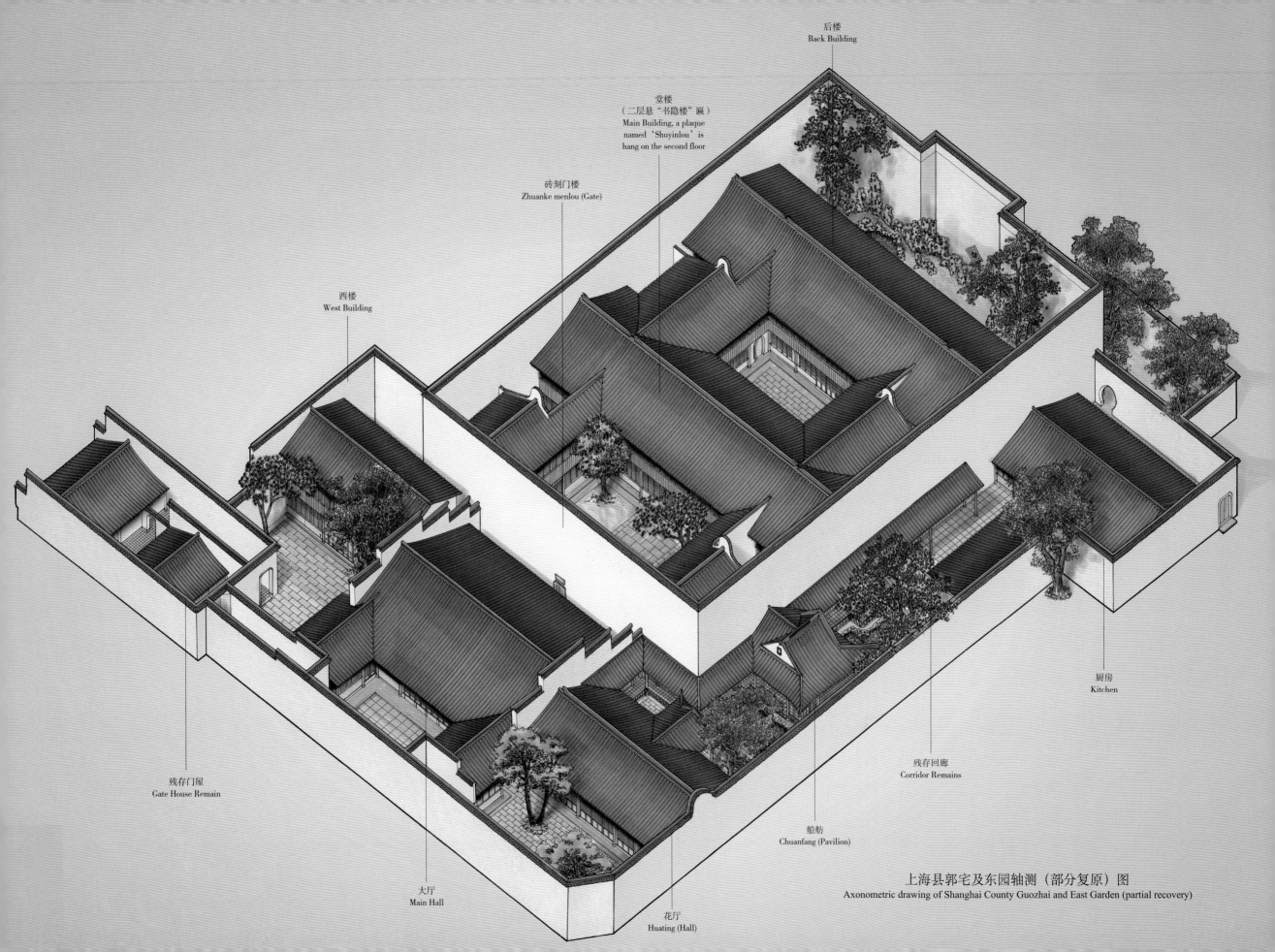

后楼
Back Building

堂楼
（二层悬"书隐楼"匾）
Main Building, a plaque
named 'Shuyinlou' is
hang on the second floor

砖刻门楼
Zhuanke menlou (Gate)

西楼
West Building

残存门屋
Gate House Remain

厨房
Kitchen

残存回廊
Corridor Remains

船舫
Chuanfang (Pavilion)

大厅
Main Hall

花厅
Huating (Hall)

上海县郭宅及东园轴测（部分复原）图
Axonometric drawing of Shanghai County Guozhai and East Garden (partial recovery)

# Introduction

The GUO Family's residence and garden complex (now called 'Shuyinlou', also commonly called 'Jiushijiu Jian Ban') is located in the south-east of Shanghai Old City, inlaid in the north of the southern section of Tiandeng Lane, the east of the northern section (formerly called Zhusutang Street), the west of Xundao Street and the south of Yinxian Lane, where has high-density mixed houses. Its residence locates in the north facing the south. The main hall (Yuruitang) of the mid-Qing Dynasty at the south of the main axis, three two storeys high buildings (including Shuyinlou and the back building) of the late Qing Dynasty in the north, two gates (were destroyed) and an affiliated house near Tiandeng Lane in the south-west corner, and Huating, Chuanfang and other affiliated houses in the east garden exist today. It covers an area of 2,272 square meters and a building area of 2,108 square meters. Because of the narrow and dense neighboring houses, it seems even more magnificent and impressive in scale.

The area looks eastward to the original Shanghai Daoshu, and faces several mansions of river transportation merchants. It is adjacent to Qiaojiabang YU family's house in the south, ZHU family's house on Yaojia Lane in Dadongmen and WANG family's house on Xiangua Street outside Dadongmen in the north-ast, and SHEN Family's house on Huayi Street in the south-east. The site was supposed to be part of Rishe Garden owned by Nanjing Taipusiqing CHEN Suoyun in the late Ming Dynasty. After repeated changes, the landscape space of the literati garden of the Tang family in the Ming Dynasty, the CHEN family in the late Ming Dynasty, and the LU family in the middle Qing Dynasty gradually resolved into 'ocean boats' giant GUO family's mansion, garden and pavilion including the surrounding houses in the city in the late Qing Dynasty. It is a precious specimen of wealth transfer, spatial evolution and value transformation in the old city.

# 导　言

郭氏宅园建筑群（今称『书隐楼』，民间俗呼『九十九间半』）位于上海老城东南，镶嵌于天灯弄南段以北、北折段（旧称『竹素堂街』）以东、巡道街以西、引线弄以南的高密度里弄杂屋内。其宅坐北面南，今存中轴南部之清中叶大厅（毓瑞堂）、北部之晚清走马楼厅（书隐楼厅及后楼）三进，以及西南角临天灯弄北段之门屋两进（遗址）、辅楼一座，和东侧庭园中的花厅、船舫及辅屋等，其占地二二七二平方米，建筑面积二一〇八平方米。因邻屋窄密，愈显堂皇恢宏，尺度摄人。

其地东瞻旧日的上海道署，并与几大沙船巨商宅邸遥遥相望——南临乔家浜郁氏宅，东北眺大东门内姚家弄朱氏宅、大东门外咸瓜街王氏宅，东南眺花衣街沈氏宅。其地原应为明末南京太仆寺卿陈所蕴日涉园的一部分，叠经变迁，由明代唐氏、明末陈氏、清中叶陆氏之文人园林景象空间，逐步解析为晚清『洋船』巨商郭氏豪宅园亭，及其周边市屋。成为老城厢财富转移、空间嬗变与价值转化的宝贵标本。

图一　乾隆帝在乾隆四十一年（一七七六年）「重华宫茶宴」后御赐陆锡熊的元末明初人杨基《淞南小隐图》

Fig.1  YANG Ji (late Yuan Dynasty and early Ming Dynasty)'s *Songnanxiaoyintu*, awarded by Qianlong emperor to LU Xixiong after the  tea banquet at the Chonghuagong in 1776

## Rishe Garden of the CHEN Family: 'High-tech' Literati Garden to Pile up Peaks and Display Stones

The city of Shanghai in the middle and late Ming Dynasty was still a space paradise where literati and officialdom lived together and indulged in. The PAN family (father and sons were all Jinshi) building complex in the north-east of the old city was connected across the street, including many luxury houses (including the present Shichuntang), famous gardens (today's Yu Garden), family ancestral hall, and family buddhist temple (today's Chenxiangge), and has been deeply imprinted in the spatial logic of the city till now. In the north-west, GU brothers' building complex, there were the famous Luxiang Garden and Wanzhushanfang, which nourished space derivatives such as GU Embroidery and juicy peaches, remaining in the spatial memory of the city. By the Qiaojiabang in the south of the city, there was the residential complex of the QIAO family (father and sons were Jinshi) in Chuansha and its villa Duhelou (the South Garden), whose spatial texture still faintly remains.

The south-east of the city which is just a few steps away, starting from Nanmeijia Lane (now Meijia Lane) in the west, Daoqiannan Street (now Xundao Street) in the east, East Meijia Lane and Yinxian Lane in the north, and Tiandeng Lane (or further south) in the south , was the urban space of more than 20 mu that was originally a river beach wetland between Qiaojiabang in the south of the city and Zhaojiabang in the city. Yupobang, traveling north from Qiaojiabang, may have extended here and expanded into a lake, and was operated as a Tang Garden. In the north of the garden, there is Tangjia Lane, named after TANG Yu, Jinshi in 1451, and TANG Xun, Jinshi in1457, so it may have been the place where the TANG family lived together. After the garden was abolished, it was purchased by CHEN Suoyun, who was also Jinshi in 1589 and took the position as Nanjing Taipusiqing in the late Ming Dynasty. The garden was built as Rishe Garden.

The central part of Rishe Garden faced the big lake with Zhusutang facing south, and the majestic main peak of Taihu Stone was on the south bank, which was still a facing the mountain across the water layout. There was a 'Yangyue' bridge leading to Donggaoting by the east lake of the main peak, which constitutes the focal point of this core landscape. There were also Yufuchiguan and Xiangxueling scene area in the south of the main peak. Wulaotang, Dianchunxuan, Cuiyunping, Yeshuchi scene area and Wulaofeng made of Taihu Stone and Yingde Stone Courtyard 'Xiaoyoudongtian' whose location was vague were in the north-west. 'Wanhushanfang' scene area made of Wukang Stone, Jinchuan

陈氏日涉园：分峰炫石的『高技派』文人园

明中晚期的上海城，仍是士大夫们聚族而居、纵情颐养的空间乐园，老城东北的『父子三进士』潘氏建筑群落连街跨坊，包含多路豪宅（包括今世春堂）、名园（今豫园）、家祠、家庵（今沉香阁），至今仍深刻烙印于城市的空间逻辑。西北的顾氏兄弟建筑群落有名传万口的露香园、万竹山房，流淌出顾绣与蜜桃这样的空间衍生品，留存于城市的空间记忆。城南乔家浜畔，则有川沙『父子进士』乔氏家族的住宅群落及其别业渡鹤楼（南园），其空间肌理至今依稀犹存。

而咫尺之遥的城东南，西起南梅家弄（今梅家弄）、东至道前南街（今巡道街）、北起东梅家弄与引线弄、南逾今天灯弄（或更南）的二十余亩城市空间，原系城南乔家浜与城中肇嘉浜间的江滩湿地。自乔家浜北行的郁婆浜可能曾延伸至此，并汇浜成湖，被经营为唐氏园。园北尚有因明景泰二年（一四五一年）进士唐瑜、天顺元年（一四五七年）进士唐珣得名的唐家弄，故其可能曾为唐氏聚居之地。此园废后，又为明末万历十七年（一五八九年）进士、南京太仆寺卿陈所蕴购得，营造为日涉园。

日涉园中部以竹素堂面南而临大湖，南岸为巍然的太湖石主峰，仍是隔水面山之局。主峰东麓湖畔，有『漾月』桥导向东皋亭，构成这一核心景域的景象焦点——主峰南麓另有浴凫池馆、香雪岭景域，西

图二 （明）林有麟《日涉园三十六景之日涉园图》（今存其十）中的全园核心景域，临池即主要厅堂竹素堂，疑在今郭宅毓瑞堂址，池南为过云峰，堂左则为飞云桥，堂后即灌烟阁，疑即今书隐楼赏楼与后楼址

Fig.2  The core area of Rishe Garden in LIN Youlin's Risheyuantu in 36 (10 remain) scenes of Rishe Garden, Ming Dynasty. Located on the waterfront is the main hall, Zhusutang, which is suspected locating at the site of Yuruitang in today's Guozhai. South of the pond is Guoyunfeng. On the left side of the hall is Feiyunqiao, and behind the hall is Zhuoyange, which is suspected locating at the site of Shuyinlou and Houlou today. LIN Youlin's *Risheyuantu*, Ming Dynasty

Stone and Fupi Stone was in the north-east. Eryatang scene area serving as entrance guidance was in the south-west. The owner of the garden was interested in stones and had a rich collection of stone. It was unavoidable that peaks were piled up to display stones in the garden, which was quite complicated and changeable, with a slight sense of dazzling. The scene had gone through the 'high-tech school' management of the famous gardening masters like ZHANG Nanyang, the 'Yunjian school' reproduction of the famous painters like SHEN Shichong and the text interaction of the famous scholars like LI Shaowen, which became famous for a while.

There were Zhixitang and Zhuoyange successively in the north of Zhusutang, and the three were likely to be on the same axis. In the east of Zhusutang, there was a water pavilion 'Xiuyuting' which was slightly like a boat, facing east on the long river from the north-east corner of the big lake, which was the second focal point besides Donggaoting. The image of this architecture group may have been passed down for more than 300 years.

北有五老堂、殿春轩、翠云屏、夜舒池景域和太湖石五老峰，以及位置模糊的英德石庭院『小有洞天』；东北为武康石、锦川石、斧劈石参差的『万笏山房』景域；西南为充当入口引导的尔雅堂景域——园主好石而藏石甚富，驱使此园分峰用石，不免纷繁峻峭，变化偏多，微有炫示之感。其景象历经造园名家张南阳们的『高技派』经营、画坛名家沈士充们的『云间派』再现与名士李绍文们的文本互动，名噪一时。

其中竹素堂以北依次有知希堂、画坛名家沈士充们的『云间派』再现与名士李绍文们的文本互动，名噪一时。

其中竹素堂以北依次有知希堂、濯烟阁，三者很可能在同一轴线上。竹素堂东侧有微似船舫的水阁——修禊亭，向东挑临于自大湖东北角北行的长河之上，为东皋亭之外的第二景象焦点——这一建筑组群意象可能历三百余年传承至今。

图三　（明）李绍箕《日涉园三十六景之过云峰图》（今存其十）中被大大夸张的全园太湖石主峰形象，近处小型建筑应即修褉亭，右侧数幢建筑应即竹素堂、知希堂、濯烟阁等建筑

Fig.3  The greatly exaggerated image of the Taihu Stone main peak of Rishe Garden in LI Shaoji's *Guoyunfengtu*, in 36 (10 remain) scenes of Rishe Garden, Ming Dynasty, in which the small building nearby should be Xiuxiting, and the buildings on the right should be Zhusutang, Zhixitang, Zhuoyange and other buildings

# Rishe Garden of the LU Family: The Course of Space Inheritance and Decomposition in the Middle Qing Dynasty

In the late Chongzhen period of the Ming Dynasty, not long after the completion of the Rishe Garden, the owner CHEN Suoyun passed away at an age of more than 80. The big garden had only been passed down for one generation, that is, it was transferred to a prominent family in Pudong, Yongning Zhixian LU Mingyun and his granduncle LU Shen, who was a famous minister and the owner of Lujiazui Houle Garden, which can be described as a famous garden and a family of famous officials. However, Yufuchiguan and Xiangxueling scene area in the south-west corner of this garden may have belonged to the Jinshi in 1737, Hunan Xunfu QIAO Guanglie (?—1765) in the early Qing Dynasty, and it was called Fengshu Garden, located in the north of his residence 'Zuiletang' in the north of Qiaojiabang.

The main part of the Rishe Garden was passed down to the fifth grandson of LU Mingyun, the Jinshi in 1761, and Duchayuan Fuduyushi LU Xixiong (1734—1792). In 1773, together with scholar Ji Xiaolan, he took up the position as the chief compiler of the Si Ku Quan Shu. Sixteen years later, As a famous resident in the county, he wrote a stele for the county's first landmark, the newly completed Chenghuang Temple West Garden (namely Yu Garden in the Ming Dynasty) Huxin Pavilion, which had survived to this day, and at that time Rishe Garden was also experiencing the last brilliance of its life. At that time, the entire garden's scene skeleton was not changed, but the main entrance of the space was changed from 'South Meijia Lane' on the west side to South Daoqian Street on the east side. The old house of CHEN Suoyun in the west of South Meijia Lane that had been deserted in the early Qing Dynasty was abandoned, while it was changed to face Dadongmen and Huangpu River Bund in Shanghai County, and was opposite to Shanghai Daotai Yamen, which would be erected in 1731, on the east side of South Daoqian Street, adjacent to the highest power space in the region of Shanghai and surrounding areas.

However, since 1782, LU Xixiong, together with Ji Yun and others, were demoted and compensated for the inexhaustible correction of the so-called mistakes in the *Si-ku Quan-Shu*'. Till ten years later, full of worries about the literary prison, he sadly died outside Shanhaiguan. The fate of the LU family may also experience the rapid decline due to this ominous book, which had caused the appearance and belonging of this urban landscape space to become superimposed and blurred.

陆氏日涉园：清中叶的空间传承与解析之路

明末的崇祯后期，日涉园建成未久，园主陈所蕴即以八十余岁高龄谢世，偌大名园只历一传，即辗转为浦东望族、永宁知县陆明允所有。其叔祖即陆家嘴后乐园主人、名臣陆深，可谓名园、名臣世家——但此园西南隅的浴凫池馆、香雪岭景域可能在清前期另归乾隆二年（一七三七）进士、名臣乔光烈（？—一七六五年），称『凤树园』，位于其乔家浜北的住宅『最乐堂』北。

日涉园主体部分则叠传至陆明允五世孙，清乾隆二十六年（一七六一年）进士、都察院副都御史陆锡熊（一七三四至一七九二年），他与名士纪晓岚联袂出任《四库全书》总纂官。十六年后作为一县人望，他为全县首要地标——新落成的城隍庙西园（即明代豫园）湖心亭书写了碑记，长存至今。而此时的日涉园也正经历着生命的最后辉煌——其时全园景象骨架未改，而空间主入口则已由西侧南梅家弄改易至东侧的道前南街，背弃了清初即已荒废的南梅家弄西的陈所蕴老宅，而面向了上海县城大东门与黄浦江滩，并与道前南街东侧、即将于清雍正九年（一七三一年）矗立起的上海道台衙门望衡对宇，邻接了上海及周边地区的最高权力空间。

然而乾隆四十七年（一七八二年）起，陆锡熊即与纪昀等人一起，因《四库全书》中改不尽的所谓错失，被降职罚赔，直至十年后，满怀对文字狱的忧惧，黯然去世于关外。陆氏家族的命运，可能也因这部不祥之书急速由盛转衰，导致这片城市景象空间的面貌与归属变得叠合迷离。

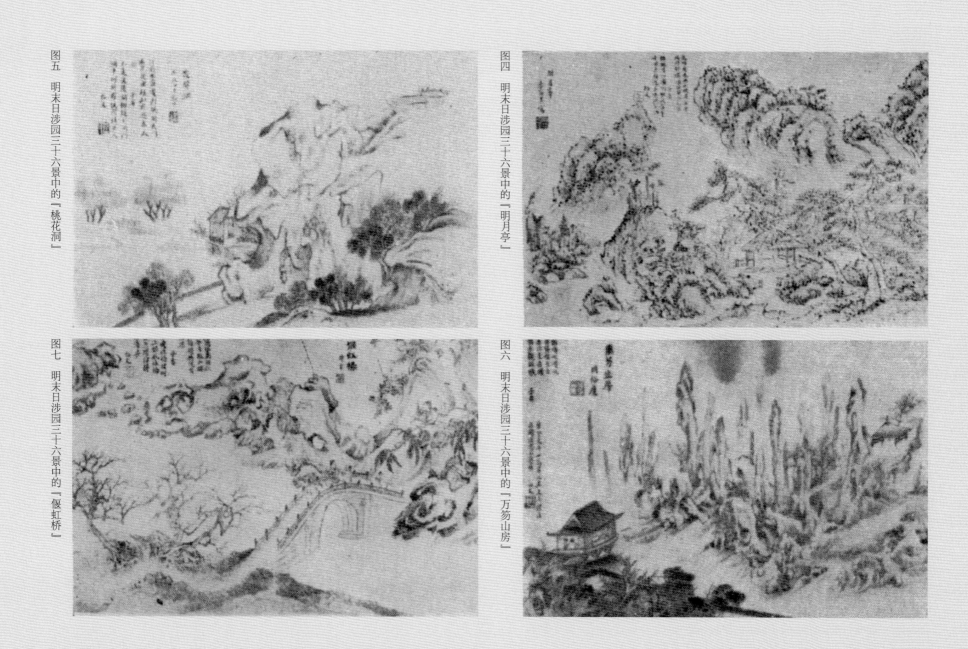

图五　明末日涉园三十六景中的『桃花洞』

图四　明末日涉园三十六景中的『明月亭』

图七　明末日涉园三十六景中的『偃虹桥』

图六　明末日涉园三十六景中的『万笏山房』

Fig.4　Mingyueting, one of the ten scenes of Rishe Garden in the late Ming Dynasty
Fig.5　Taohuadong, one of the ten scenes of Rishe Garden in the late Ming Dynasty
Fig.6　Wanhushanfang, one of the ten scenes of Rishe Garden in the late Ming Dynasty
Fig.7　Yanhongqiao, one of the ten scenes of Rishe Garden in the late Ming Dynasty

Only 22 years later, in the image of 1814's *Shanghaixianzhi*, Zhongdafafang Street in the north of the garden had been extended to the south, and the space was divided into two parts (it could only be a whole at first), which was obviously the result of the family's property decomposition and even the change of the surname, and the need for another access road to enter and exit, showing that the original structure of the scene in the garden had been fatally split. Three years later, it was said in *Songjiangfuzhi* that the garden had been hanging for two hundred years, and the architectures had changed a lot, gradually becoming a ruined foundation, but the LU family had been guarded for the whole life, and the remains still existed, proving the astonishing speed of the garden's rundown and decomposition. The *Xianchengneiwaijiexiangtu* in 1871's *Shanghaixianzhi*, the newly extended street had been named after the main hall in the park, call 'Zhusutang'.

The last of the LU family's hard work to maintain the garden was 'Wulaotang' scene area in the north-west of the family sacred site, which kept Wulaofeng and Songnan Xiaoyin relics of the ancestors. The scene was rich and independent, and continued the bloodline of the literati era. Therefore, *Shanghaixianzhi* in 1871 called 'Wulaotang, where the LU family lived, still existed today'. In the Republic of China, on the west side of Zhusutang Street, 'relying on Meijia Lane, there were several old houses, where the grandchildren of the LU family lived', which was also more in line with the location of Wulaotang.

The first thing that the LU family sold should be the eastern part of the whole garden, including the later GUO family's 'Shuyinlou'. There were speculations that in 1819, ZHAO Wenzhe, ZHAO Bingchong and ZHAO Rong family in the same county that reached the peak of 'brothers serving as Hanlin' and 'three generation working at Junjichu' may have shared this house space with the LU family, or this area. In the early period of the Republic of China, when *Shanghaigujitan* referred to Shuyilou, it said that 'There was a giant house on Zhusutang Street, used to be the residence of ZHAO Mingzhao, a gentry in the city. There were as many as five or six halls, while the back door was on Yinxian Lane, and became GUO *Shen* residence today. It was probably the same ZHAO family.

仅仅二十二年后的清中期嘉庆十九年（一八一四年）《上海县志》图中，园北中大夫坊街已南延，将该空间中分为二（其原先只能是一个整体），显然是家族析产甚至易主他姓、需要另辟便道进出的结果，足见此园原有景象结构已发生致命割裂。约三年后的《松江府志》则称：『园垂二百年，台榭多变更，渐作颓垣断础，而陆氏至今世守，遗迹犹有存者』，旁证了其惊人的破败解析速度。晚清同治十年（一八七一年）《上海县志》县城内外街巷图中，该新延街道已被冠以园中最主要厅堂『竹素堂』之名。

其中陆氏家族苦苦维系至最后的，是存留着先祖深遗石五老峰和淞南小隐这一家族圣地的西北五老堂景域。其景象丰富而又独立成篇，赓续着士大夫时代的一线血脉。所以晚清同治上海县志称『今存五老堂，陆氏居之』。而民国时竹素堂街西侧『靠梅家弄还有旧屋几椽，为陆家裔孙居住』，也较符合五老堂位置。

而陆氏首先出让的，应是包括后来的郭氏书隐楼在内的全园东部。有观点猜测，在嘉庆二十四年（一八一九年）达到『兄弟翰林』『三世军机』巅峰的同县赵文哲、赵秉冲、赵荣家族，可能一度与陆氏分享此宅园空间，或即此区。民国初的《上海古迹谈》言及今书隐楼时，称『竹素堂街街有巨宅一，昔为邑绅赵名照所居，大厅共有五六之多，其后门在引线弄，今为郭绅住宅』。此赵即彼赵的可能性很大。

图九　明末日涉园三十六景中的『浴凫池馆』

图八　明末日涉园三十六景中的『漾月桥』，右为东皋亭，远处为修禊亭

图十一　明末日涉园三十六景中的『灌烟阁』

图十　明末日涉园三十六景中的『蒸霞径』

Fig.8　'Yangyueqiao', one of the ten scenes of Rishe Garden in the late Ming Dynasty, on the right is the Donggao Pavilion, with Xiuqi Pavilion in the distance

Fig.9　'Yufuchiguan', one of the ten scenes of Rishe Garden in the late Ming Dynasty

Fig.10　'Zhengxiajing', one of the ten scenes of Rishe Garden in the late Ming Dynasty

Fig.11　'Zhuoyange', one of the ten scenes of Rishe Garden in the late Ming Dynasty

In the late Qianlong period, during the last glorious moment of Rishe Garden, the owner LU Xixiong once participated in the emperor Qianlong's annual New Year small-scale 'top-level' tea feast held in his youthful birthplace, Chonghuagong (the former west part of Qianqinggong) for his repeated work in the compilation of 'Si-ku Quan-Shu'. The participating princes, scholars, and Hanlins were no more than 28 people. As usual, they joined the poems with the emperor Qianlong on the spot, and the banquet was bestowed with antiques, paintings and calligraphy. LU Binghu, the father of LU Xixiong, was called 'Songnan Laoren', and LU Xixiong was once given the picture of *Songnan Xiaoyin* by YANG Ji, one of the 'Four Sons of Wuzhong' in the late Yuan and early Ming Dynasty, and the imperial title had seven quatrains, which can be imagined. Before that, LU Binghu had rebuilt Dianchunxuan in the garden into 'Chuanjing Shuwu'. This time it was renamed 'Songnan Xiaoyin', and the painting was honored to commemorate this rare occasion that the whole city talked about.

From 'Chuanjing Shuwu' to 'Songnan Xiaoyin', the spatial expression of bring 'books' to 'live in seclusion' had been determined since then.

在乾隆后期，日涉园的最后华光时刻，园主陆锡熊曾屡因总纂《四库全书》之功，参与乾隆帝每年新春于自己的年少发祥之地——重华宫（原乾清宫西二所）举行的小范围顶级茶宴，参与的诸王、大学士、翰林最多不过二十八人，照例即席与乾隆帝赋诗联句，宴罢则御赐古玩书画等——陆锡熊之父陆秉笏别号为『淞南老人』，而陆锡熊即曾被赐元末明初『吴中四子』之一杨基的《淞南小隐》图，并御题有七言绝句，其感泣可想而知。此前陆秉笏曾将园中殿春轩改建为传经书屋。这次又将其改名为『淞南小隐』，敬奉此图，以纪念这一令举城津津乐道的难得际遇。

从『书』屋到小『隐』，携『书』而『隐』的空间表述就此确定。

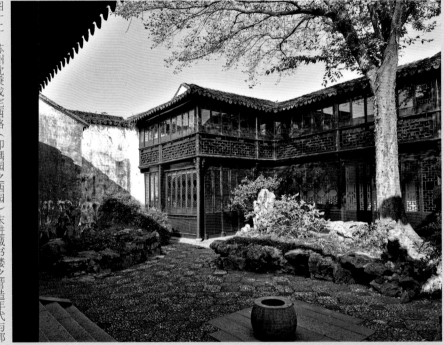

图十二 苏州沈秉成宅西路（即耦园之西园）末进藏书楼之营造年代与郭宅堂楼相近，其界面、檐口之流转整饬亦无二致。只是沈宅书楼以必要的单调沉郁，将深院中的老树丛花映衬得格外明艳动人，不同于郭宅之阔大沉雄

图十三 晚清时的上海县城隍庙湖心亭

图十四 湖心亭景域全景

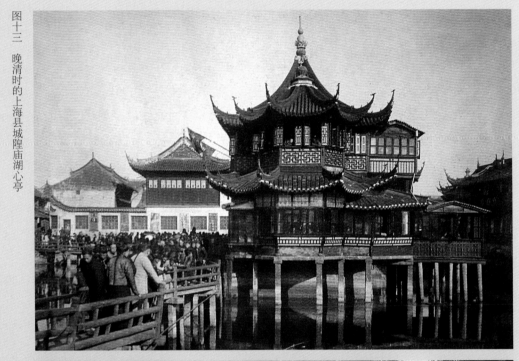

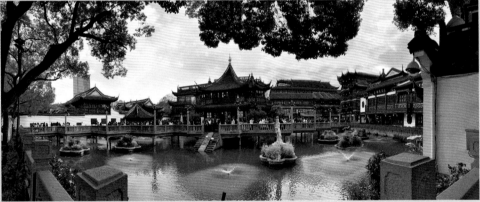

Fig.12 The library building, the last house on the west axis of Shen Bingcheng's mansion (namely the west garden of Ou Garden) in Suzhou, was built in similar time with that of Guo family's mansion, and their facades and eaves are also similar. Shen family's library building, with its necessary simplicity and somberness, highlights the beauty of the trees and flowers in the deep garden, which is different from the magnificence pursued by Guo family.

Fig.13 Huxin Pavilion of Shanghai County Chenghuang Temple in the late Qing Dy nasty

Fig.14 A panoramic view of Huxin Pavilion area

## Shuyinlou of the GUO Family: Space Interception and Expansion in the Late Qing Dynasty

During Jiaqing and Daoguang periods of the Qing Dynasty, with the prosperity of marine trade between the north and the south brought by the 'opening of the seas' in Kangxi period, the role of Beijing-Hangzhou Grand Canal and the Liu River waterway started to degenerate, while Shanghai was the intersection of domestic coastal trade and the inland river trade of the Yangtze River and Tai Lake, welcoming the tide of marine trade in the heavy confinement. Sandboats (square-head flat-bottomed wooden sailboats) merchants specializing in offshore, especially Beiyang trade were extremely prosperous, and the GUO family, the giants of the ocean boats (large sharp-bottomed wooden sailboats), which were mainly engaged in Nanyang trade, also quietly emerged.

Approximately in 1755, the GUO family, from Longxi County (the county seat is also the city), Zhangzhou, Fujian Province, came to Shanghai's Yanghang Street (later called Yangshuo Road, no longer existing today) outside Xiaodongmen where merchants from Fujian and Guangdong gathered together and specialized in Nanyang goods, and opened the ship transportation company 'GUO Wanfeng'. Till 1809, the family had started a clan and established a family ancestral hall in Pudong. In 1855, the GUO family had been already the only ocean ship transportation company among the eight largest ship companies in Shanghai. In addition, the GUO family opened up Ruitaisi tea company, Fengtai wood company, Changfeng bank, Wanyi bank, and other business companies on Yanghang Street, and widely sold real estate, stretching across a large area of the Huangpu River. So-called 'full of the Huangpu River bank' showed the wide expansion of its space. Around Tongzhi period to the early years of Guangxu period in late Qing Dynasty, the GUO family followed the LU family and the ZHAO family to the east of the original Rishe Garden, namely the part of 'Shuyinlou' east of the 'Zhusutangqian' street.

During Tongzhi period of the Qing Dynasty, the second industrial revolution was in the ascendant, and the traditional sailing merchants who still used wind-driven wooden sailboats were gradually declining, and Wanfeng ship transportation compamy was officially diverted tens of thousands of taels of silver. Due to internal and external difficulties, the GUO family had to sell the Jinliyuan Wharf at the newly opened estuary of the Huangpu River to Ship Business Soliciting Bureau on April 3rd of 1882, and completely shifted the main business from ship transportation to the original silver trade. His family also separated into three respective ones. The eldest GUO Changzuo the owner

# 郭氏书隐楼：晚清的空间截取与膨胀

清嘉庆、道光年间，随着康熙一度「开海」带来的南北洋海上贸易的勃兴，京杭大运河与太仓浏河航道的退化，上海作为国内沿海贸易与长江、太湖内河贸易的交汇点，于重重禁锢中迎来海洋贸易的高潮。专营近海，尤其北洋贸易的沙船（方头平底木帆船）商帮极度兴盛，而主营远洋贸易的洋船（大型尖底木帆船）业巨商郭氏家族也悄然崛起。

郭氏自清乾隆二十年（一七五五年）前后，自福建漳州府龙溪县（县城即府城）来上海小东门外、闽粤商人萃集、专营南洋货品的洋行街（后称「阳朔路」，今不存）开设「郭万丰」船号，至嘉庆十四年（一八〇九年）开宗立脉，在浦东建造家族祠堂。咸丰五年（一八五五年）时，郭氏已是上海八大航商中的唯一洋船商。此外，郭氏尚在洋行街开设有瑞泰丝茶号、丰泰木行、长风银号、万益钱庄等商号，并广置房产，绵延于大片黄浦江岸，所谓「外滩兜转里滩通」，足见其空间扩张之广。大约在晚清同治至光绪初年，郭氏继陆氏、赵氏之后染指原日涉园东部，也即竹素堂前街以东的今书隐楼部分至今。

清同治年间，第二次工业革命方兴未艾，仍使用风力木帆船的传统航商日渐式微，而万丰船号又被官方挪借白银十数万两——内外交困之下，郭氏不得不于光绪八年（一八八二年）四月初三日将黄浦江新开河口的金利源码头出售给轮船招商局，由船运主业彻底转向原本的银楼副业。其家族亦约于此年前后分三房析居。长房壶瑞堂郭长祚一脉坚守洋行街至民国初年，始迁居吴家弄；三房诒瑞堂郭长钰一脉

图十六 《四库全书》书影

图十五 沈阳故宫文溯阁曾用于存放《四库全书》

图十七 纪昀（纪晓岚）像

Fig.15 Wensuge in Shenyang Gugong, which was used to keep *Si-ku Quan-Shu*
Fig.16 The front covers of *Si-ku Quan-Shu*
Fig.17 The portrait of JI Yun（JI Xiaolan）

of Huruitang adhered to Yangxing Street until the beginning of the Republic of China, and then moved to Wujia Lane. The youngest GUO Changyu the owner of Yiruitang moved to QIAO Guanglie's former residence in the original QIAO family community in Qiaojiabang, almost near to the north of Shuyinlou. The second eldest GUO Changdi moved to the previously purchased Shuyinlou residence and garden complex, and a large-scale renovation and expansion were carried out immediately, giving it different spatial characteristics and artistic interest. About five years later in the mid-autumn of 1887, PAN Zuyin, a well-known minister and Taizi Taibao At the end of the Qing Dynasty, inscribed 'Yuruitang' (Tanghao of ) plaque for the main hall of GUO the family, which may be a sign of the completion of this large-scale renovation and expansion.

Until the social transformation in 1949, the GUO family's industry had been operating normally, ensuring the continuity of the garden. However, the rest of the large space in Rishe Garden was gradually 'formatted' by the high density of lane buildings, and was sadly eliminated by the long river of urban space changes. The relics of the garden stones collected by CHEN Suoyun for half a lifetime and carefully piled up by ZHANG Nanyang were purchased by the newly built Hatong Garden in the late Qing Dynasty and restructured according to the logic of the Shanghai-style garden, while a similar destiny shift was completed.

迁至乔家浜原乔氏聚居群落中的乔光烈故宅，几乎北接书隐楼；二房郭长地一脉则迁至此前购置的书隐楼宅园建筑群，并应即进行了大规模的改扩建，赋予其不同的空间特征与艺术趣味——约五年后的光绪丁亥年（一八八七年）仲秋，清末中枢名臣、太子太保潘祖荫为郭宅大厅题写了「毓瑞堂」（郭氏二房堂号）匾额，或许正是这次大规模改扩建完成的标志。

直到一九四九年社会转型前，郭氏银楼业均正常经营，确保了这方宅园空间的承续。而日涉园的其余大片空间则逐渐被高密度里弄建筑所「格式化」，黯然消泯于城市空间变迁的长河。园中经陈所蕴搜罗半生、张南阳精心堆掇的遗石被清末新建的哈同花园购去，按海派园林逻辑重新叠构，也完成了类似的命运转移。

①上海县城隍庙与湖心亭（原
　为明代潘氏豫园）
②郭氏金利源码头
③里洋行街郭氏商号群与祖宅
④郭氏长房居所壶瑞堂
⑤郭氏二房居所毓瑞堂
⑥郭氏三房居所诒瑞堂
⑦沙船巨商郁氏居所宜稼堂
⑧沙船巨商朱氏宅邸
⑨沙船巨商王氏宅邸
⑩沙船巨商沈氏宅邸

图十八　上海老城厢郭氏码头、商号、宅邸分布图

上海一邑为通商要口其间中外交涉公私尤聚敛
尤必详绘以固庶商实性来潜以地势者有所平
间世谁祖界里每易其名玫右图间有石符
之谍主所难免阅者谅之

Fig.18  The map of the Guo Family's wharves, merchant houses and residences in the Old Town of Shanghai

## From Zhuoyange to Zoumalou: The Gentrification and Fujian-Zhejiang Style of the Literati Garden Space

The current GUO family garden can be roughly summarized by the front hall with the back buildings, and the west houses with an east garden.

Yuruitang with 'Bianzuo' beams at the front of the middle axis is three rooms wide, and on both sides there are extremely narrow and small rooms that seem to have been used to connect 'Binong'. Its depth can be regarded as six frames, and the beam style is 'Neisijie with front and back Danbu (the front and back should have Ketouxuan originally)', similar to the Song style 'Sichuanfu and front and back zhaqian with four columns'.

This hall has very gentle Juzhe, and the Qing Stone drum shape column bases with Baofujin pattern on four sides under the columns only show the upper part, As the outdoor ground rises, the platform base is almost at the same level with the ground. The above points are closer to the architectural features of the Ming Dynasty. However, its flat-shaped moon beam feels a bit stiff and tends to be decorative, which resembles the style of the late Qianlong period of the Qing Dynasty, with mixed techniques. However, its Bianzuo moon beams feel a bit stiff and tend to be decorative, which resemble the style of the late Qianlong period of the Qing Dynasty, with mixed techniques. Based on comprehensive judgment, this hall may be after the mid-Qianlong period of the Qing Dynasty, and it was rebuilt on the basis of the original structure of the Ming Dynasty, and the houses on both sides were added.

The back of the middle road, which is also on the north side of this hall, are two two-story buildings with majestic volume, steep Juzhe, rich decoration, and vigorous style. Each building is five rooms wide and six frames deep, and the beam style is 'Neisijie with front and back Danbu', surrounded by the wall to block fire that is more than ten meters high, and may be built in the late Qing Dynasty from the perspective of their style. Among them, the front building has Bianzuo moon beams, which are soft and delicate and rich in decoration, the back building is equipped with 'Yuantang' straight beams, and the entire interior of the second floor is covered with a thin plate shaped like a boat canopy to top the ceiling.

The east axis of the residence is a small, long and narrow 'waterscape' garden from north to south. Due to the excessive slumping, its layout can only be vaguely understood.

# 从濯烟阁到走马楼：文人园空间的绅商化与闽浙化

今存郭氏宅园大致可用前厅后楼，西宅东园加以概括。

中路前部之『扁作』大厅毓瑞堂面阔三间，两侧尚有极狭小而似曾用以连接『避弄』的梢间。其进深可视为六架，梁架贴式为内四界前后单步（前后单步位置原尚设有磕头轩），类似宋式之『四椽栿前后劄牵用四柱』。

此厅举折平缓，柱下为四面搭包袱锦文青石鼓形柱础但仅露出上部。随着室外地面的层累升高，台基几欲与地面相平——以上几点较接近明代建筑特征。但其扁作月梁稍觉生硬而偏于装饰化，似清乾隆中期稍后之风，且手法错杂。综合判断，此厅颇似清乾隆中期稍后，于明代原构基础上翻建而成，并添建两侧厢屋。

中路后部，也即此厅北侧，为体量雄巨、举折陡峻、装饰富丽、风格浑成之二层二进走马楼，每进均面阔五间，进深六架，梁架贴式为『内四界前后单步』，周遭则有高逾十米之森然封火墙围绕，从风格看，显系晚清一次建成——其中前进楼厅施『扁作』月梁，柔曲纤弱而富装饰性；后楼则施『圆堂』直梁，二层室内整体施形似船篷的薄板回顶天花。

宅园东路为南北狭长的小型水景庭园，因颓圮过多，其布局意匠只能依稀识之。

图十九 郭氏宅园花厅清前期梁架，由徐瑞彤绘制

Fig.19 The early Qing Dynasty beam of Huating of the GUO family's garden, drawn by XU Ruitong

Its main hall (commonly called Huating) runs across the south of the whole garden, which is the south-east corner of the whole residence. This hall has a simple and elegant structure with a moon beam, and seem that it was renovated by the LU family in the middle and early Qianlong period, with the highest artistic value. It adjoins the east wall of the hall to the west, as if the hall extends eastward, while it in the south faces the stage across the courtyard and in the north faces the courtyard with Baosha. With one hall and three sides, the logical connection and transformation of the middle axis ceremony space, the east axis tour space and the south-east observation space are completed.

On the west side of the garden, there is a veranda facing east, and Huating and the main hall are connected to the south, forming a curved and converging space for entering the garden. It lean on the large east wall of the middle axis to the west, like white drawing paper, and it lean against the wall on the way to the north, the single-eave gable and hip roof Chuanfang protruding eastward, adding a north-south spatial levels and forming the focal point of the entire garden. The Bianzuo moon beams of the boats are blunt and gorgeous, and similar to the verandas and Zoumalou buildings, should be of the new constructions in the late Qing Dynasty. Inferred from the status quo, the garden seems to have a long and narrow 'Haopu-shaped' water surface running through the north and south, and the verandas and Chuanfang are facing the water surface to the east. The owner of the garden recalled that in the old days this water was accessible by boat, so it is most likely to be the northern extension of Yubobang River in the south of the garden, and a remnant of the old Rishe Garden's water system.

Contrasting with *Risheyuantu* and CHEN Suoyun's article at the beginning of the garden, and the precious traces of the landscape in the *Songjiangfuzhi* of Jiaqing period in the middle Qing Dynasty, the relationship of Yuruitang, Zoumalou buildings and Chuanfang in the east still existing today resembles the reappearance of Zhusutang, Zhixitang, Zhuoyange and Xiuxianting on the east side in the late Ming Dynasty. The three-room Yuruitang is suspiciously still in the place of the old Rizhusutang, while Zhixitang and the lightly stretched Zhuoyange in the back of the Taihu Stone mountain, which had appeared in *Risheyuantu* for at least three times, were transformed into the gigantic Zoumalou buildings surrounded by high walls. The newly build Zoumalou embodied the gentry and merchant temperament that is very different from the literati garden or residence.

The depth and height of the main building and the wing building of Zoumalou building are relatively close, different from the ordinary Jiangnan style, and closer to the atmosphere

其主要厅堂（俗呼之『花厅』）横贯于全园南部，也即全宅东南角。此厅扁作月梁架构简洁优美，似乾隆中前期陆氏翻建，艺术价值最高。其向西邻接大厅东墙，仿佛大厅向东延出；向南隔庭院而对戏台；向北出抱厦而对庭园——以一厅三面，完成了中路仪式空间、东路游赏空间与东南观演空间的逻辑关联与转换。

园西侧则为一线面东的游廊，南接花厅与大厅，形成宛曲而收敛的入园空间铺垫；西倚大片高峻的中路东墙，仿佛洁白画纸；倚墙北行途中，向东探出单檐歇山顶船舫，增添了南北向的空间层次，并构成全园景象焦点。船舫之扁作月梁生硬而富丽，与游廊及走马楼建筑群类似，均应属晚清新构——由现状推断，此园旧日似有狭长形水面贯穿南北，游廊与船舫均向东挑临水上，园主回忆旧日此水可泛舟而入，则其极可能为园南郁婆浜的北延段，并为旧日日涉园水面的残留。

对照建园之初的《日涉园图》及陈所蕴园记，以及清中期嘉庆《松江府志》对园景的宝贵追溯，今存毓瑞堂大厅与走马楼及东侧船舫的位置关系，仍仿佛明末竹素堂、知希堂、濯烟阁及东侧修褉亭的再现。颇疑面阔三间之毓瑞堂仍在旧日竹素堂位置，而后部太湖石山夹峙中的知希堂和濯烟阁则变身为高墙四锢的庞然走马楼，濯烟阁轻盈舒展，至少三度进入《日涉园图》；新构的晚清走马楼则体现了与文人园林或住居迥异的绅商气质。

该走马楼之正楼与厢楼进深、高度均较为接近，不同于普通吴地风格，而更接近于闽浙围屋气息。

图二十一　书隐楼匾额

图二十　书隐楼「古训是式」砖刻门楼

Fig.20　Guxunshishi Menlou of Shuyinlou
Fig.21　The plaque of Shuyinlou

of 'Weiwu' in Fujian and Zhejiang province. The treatment of the north tip room on the first floor of the first building's wing building is full-face brick carving, which is a typical Fujian practice, but the stone is changed into bricks, and the craft style is similar to that of Jiangnan, which can be regarded as Zhangzhou's style but Jiannan's technique. On the second floor of this building's middle room, a plaque of 'Shu Yin Lou' inscribed by 'SHEN Chu (1729-1799) from Pinghu', the vice president of 'Si-ku Quan-Shu' in 1762, was hung, like a collecting meaning of 'Chuanjing Shuwu' to 'Songnan Xiaoyin'. Perhaps when the LU family changed the Chuanjing Shuwu into Songnan Xiaoyin, he also changed the Zhuoyange into Shuyinlou, with a plaque written by colleague SHEN Chu. In the end, this plaque, together with the Zhusutang and Zhuoyange building complex, was transferred to the ZHAO and GUO family. Of course, this plaque may have been obtained from elsewhere.

Looking at the GUO family's residence and garden complex, it not only has the grandeur of the middle axis and the secluded the east axis, but also implicts the style changes from gentry to merchants, from Jiangnan to Fujian, and may even include the three generations from the late Ming Dynasty to the middle Qing Dynasty to late Qing Dynasty's building remains. Such as Huating of the LU family's Rishe Garden in the early Qing Dynasty (which is suspected that was relocated from the original location of Zhixitang in the north of the main hall to the present site) the mian hall renovated in the middle of the Qing Dynasty, Zoumalou buildings in the residence and Chuanfang and verandas in the garden in the late Qing Dynasty and even Zhusutang's basement of the CHEN family's Rishe Garden in the late Ming Dynasty, they outlined ups and downs, hidden deeply, in the dense market, and they jointly completed the space inheritance and image reconstruction of a generation of literati gardens.

其前进厢楼一层北梢间之处理为整面砖细雕刻，更是典型福建做法，只是易石为砖，工艺风格亦近吴地，可称为『漳州样、苏州工』——此楼前进二层明间曾悬乾隆二十八年（一七六二年）榜眼、《四库全书》副总裁、平湖沈初（一七二九至一七九九年）题写的『书隐楼』匾额，似集『书』屋与小『隐』之意而成，或许当年陆氏改传经书屋为淞南小隐的同时，也曾改濯烟阁为书隐楼，由同僚沈氏题写匾额。最终此匾连同竹素堂、濯烟阁建筑群整体出让至赵、郭二氏。当然此匾亦有微小可能得自他处。

综观郭氏宅园建筑群，不只兼得了中路之宏大与东路之幽邃，也潜藏了士族与商家、吴地与闽域间的风格变化，甚至可能包含着从明末到清中、清末的三代建筑遗迹。如疑似自大厅之北知希堂原址迁建至今址）的厅（疑似清前期陆氏日涉园时代的花厅，疑似清中期翻建之大厅，疑似晚清郭氏新建的走马楼建筑群与庭园中的船舫、游廊，甚至疑似明末陈氏日涉园时代的竹素堂基——它们起伏勾勒，深藏若虚，于绵密市肆间，共同完成着对一代文人名园的空间传承与意象重构。

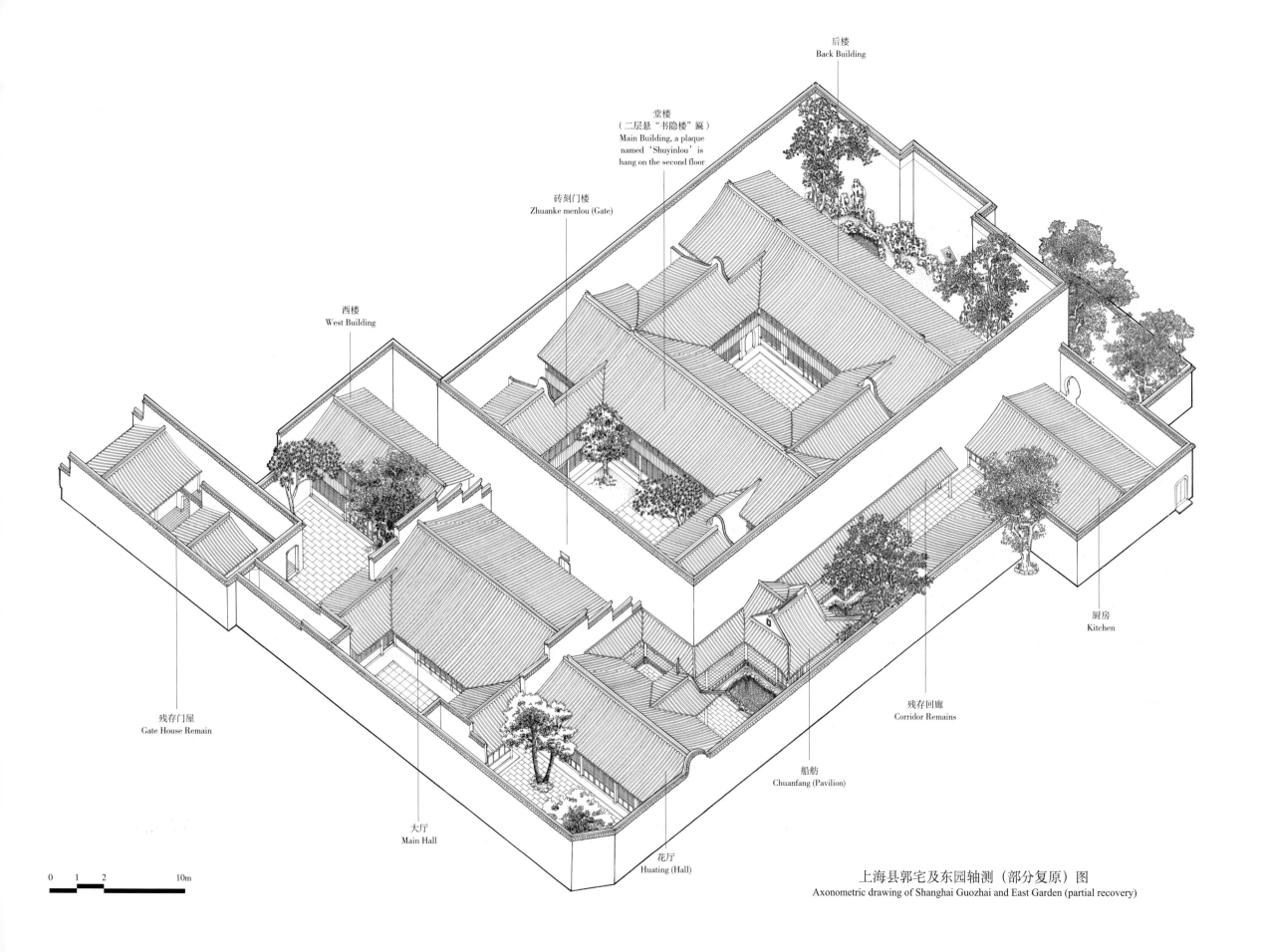

后楼
Back Building

堂楼
（二层悬"书隐楼"匾）
Main Building, a plaque
named 'Shuyinlou' is
hang on the second floor

砖刻门楼
Zhuanke menlou (Gate)

西楼
West Building

残存门屋
Gate House Remain

大厅
Main Hall

花厅
Huating (Hall)

船舫
Chuanfang (Pavilion)

残存回廊
Corridor Remains

厨房
Kitchen

0  1  2      10m

上海县郭宅及东园轴测（部分复原）图
Axonometric drawing of Shanghai Guozhai and East Garden (partial recovery)

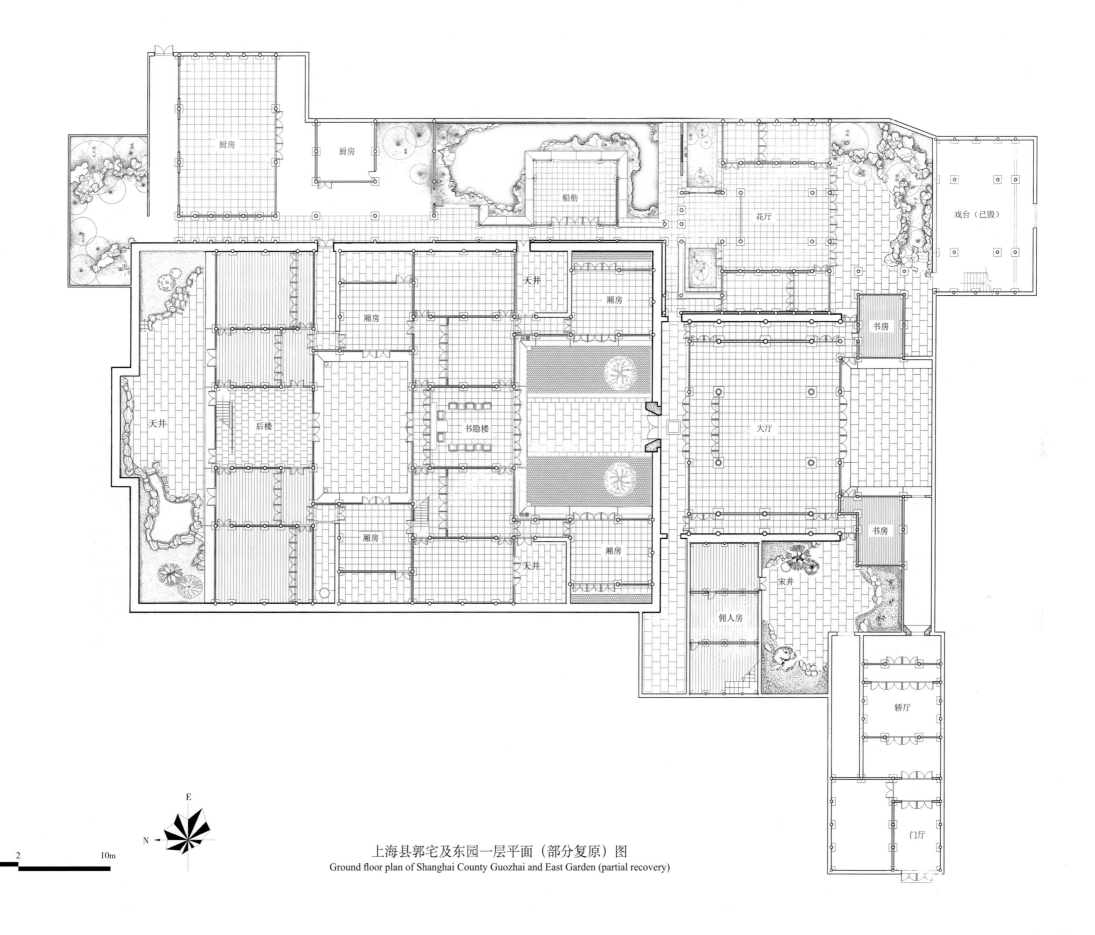

厨房

厨房

船舫

花厅

戏台（已毁）

天井

厢房

书房

厢房

天井

后楼

书隐楼

大厅

厢房

厢房

书房

天井

宋井

佣人房

轿厅

门厅

E

N

0　1　2　　　　　　10m

上海县郭宅及东园一层平面（部分复原）图
Ground floor plan of Shanghai County Guozhai and East Garden (partial recovery)

上海县郭宅及东园二层平面（部分复原）图
Second floor plan of Shanghai County Guozhai and East Garden (partial recovery)

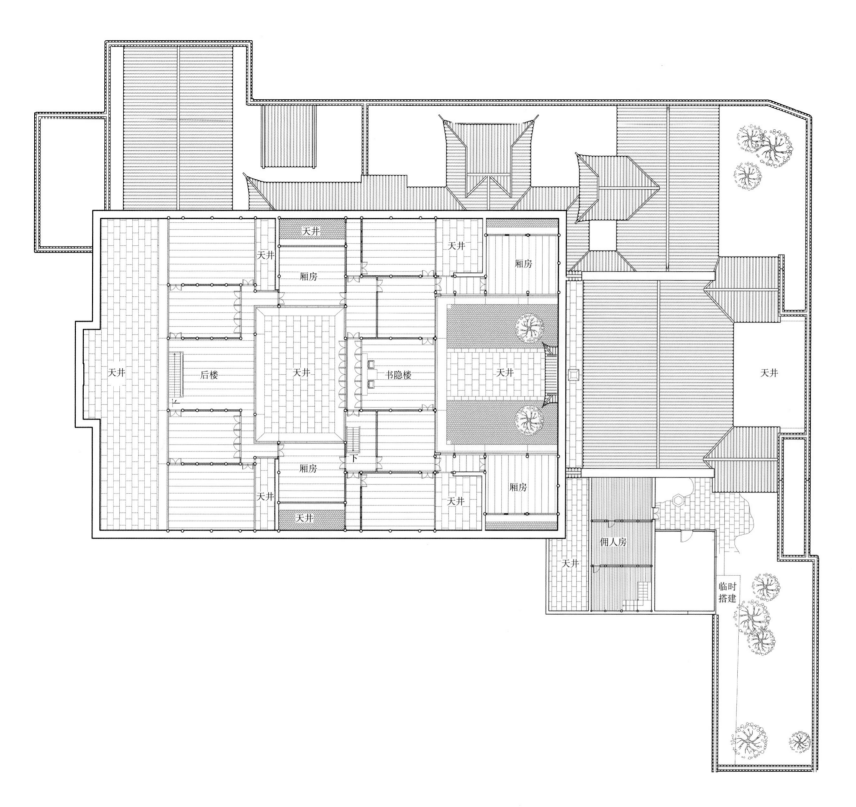

天井

天井

厢房

天井

厢房

后楼

天井

书隐楼

天井

天井

下

厢房

天井

天井

天井

厢房

佣人房

天井

临时搭建

E

N

0 1 2 10m

上海县郭宅及东园一层平面（现状）图
Ground floor plan of Shanghai County Guozhai and East Garden (present situation)

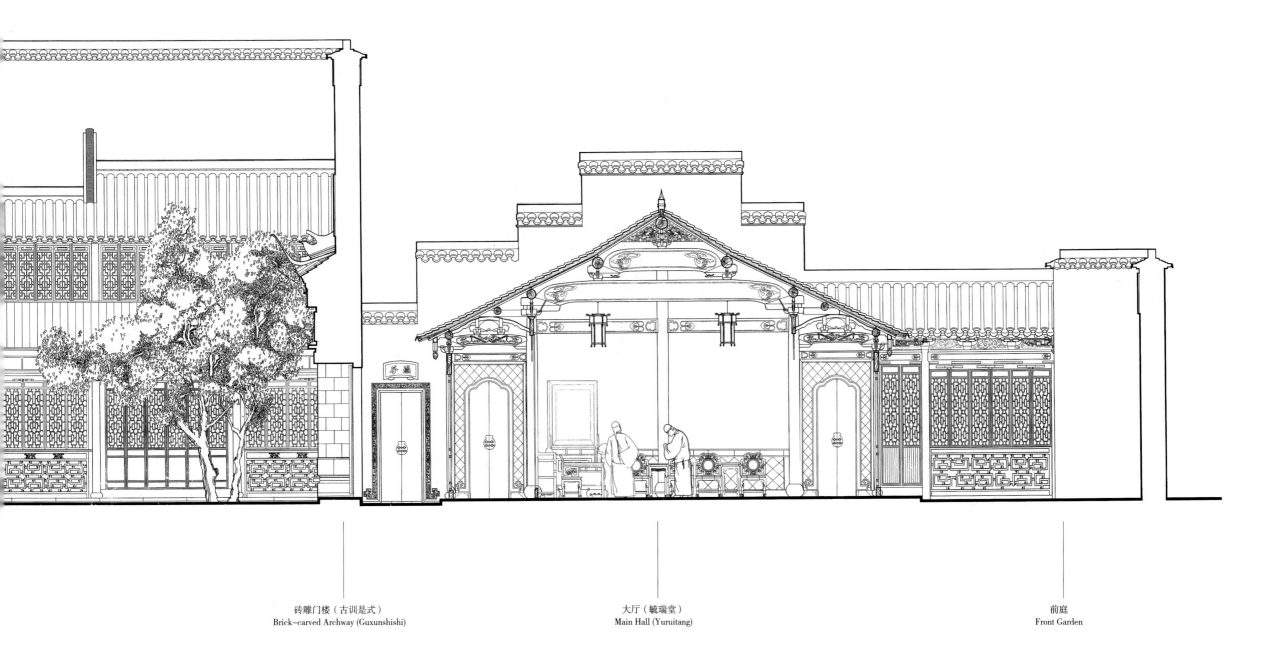

砖雕门楼（古训是式）
Brick-carved Archway (Guxunshishi)

大厅（毓瑞堂）
Main Hall (Yuruitang)

前庭
Front Garden

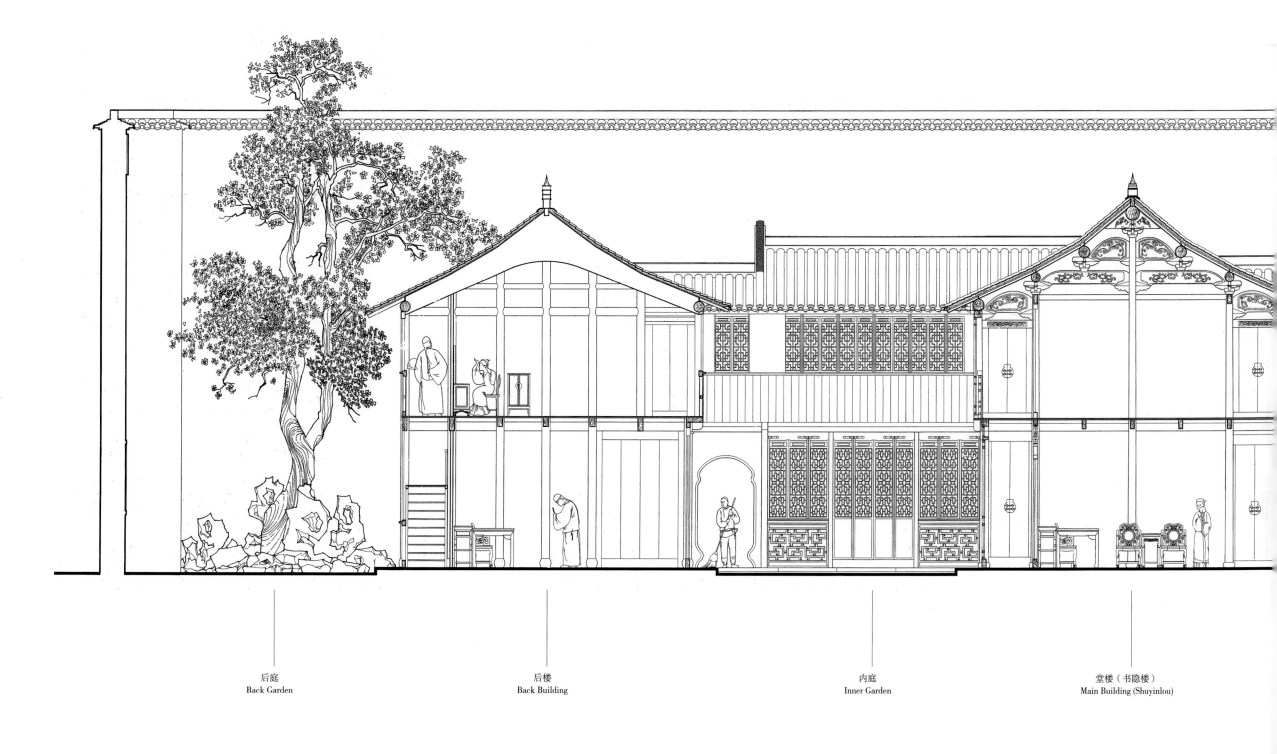

后庭
Back Garden

后楼
Back Building

内庭
Inner Garden

堂楼（书隐楼）
Main Building (Shuyinlou)

0   0.75   1.5        3m

上海县郭宅及东园正落中轴剖面图
Section of Shanghai Guozhai and East Garden's middle lane

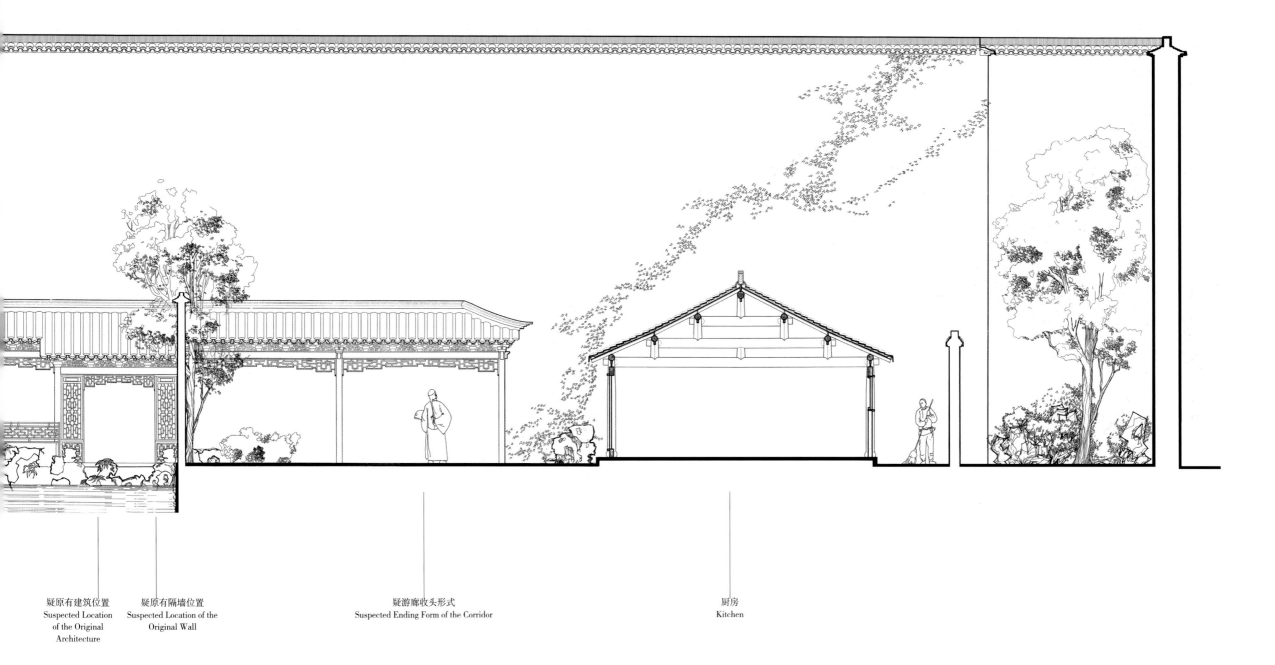

疑原有建筑位置
Suspected Location
of the Original
Architecture

疑原有隔墙位置
Suspected Location of the
Original Wall

疑游廊收头形式
Suspected Ending Form of the Corridor

厨房
Kitchen

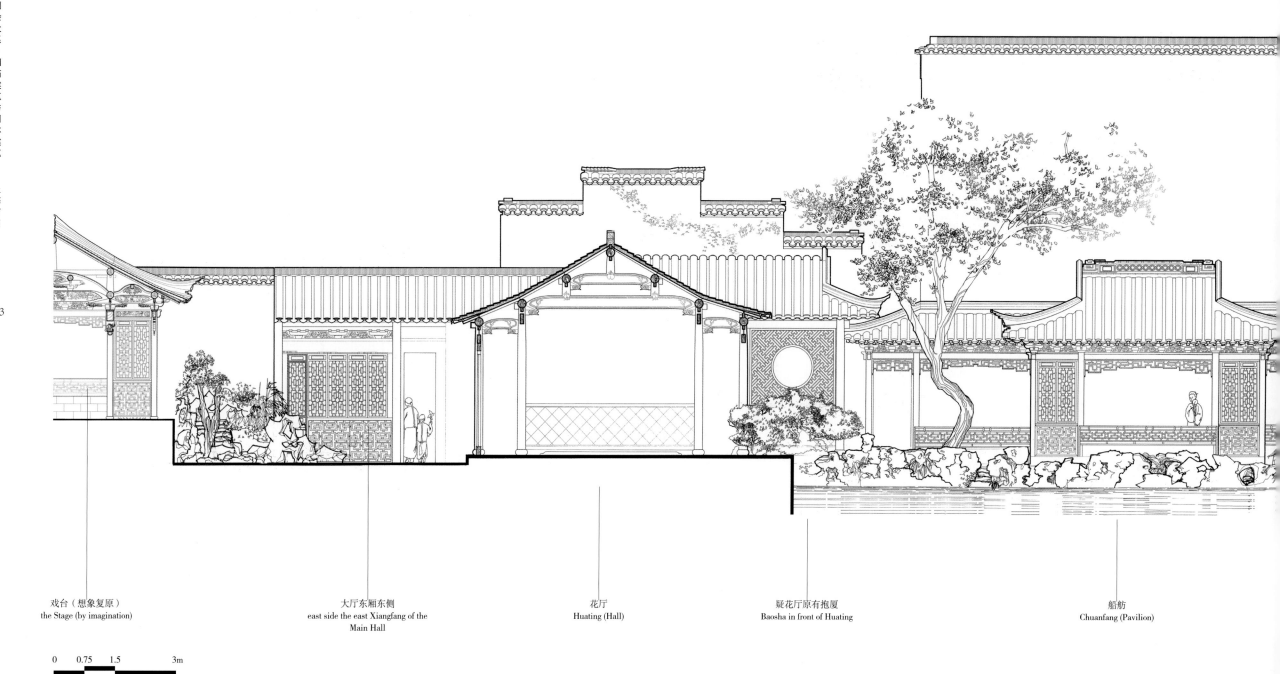

戏台（想象复原）
the Stage (by imagination)

大厅东厢东侧
east side the east Xiangfang of the
Main Hall

花厅
Huating (Hall)

疑花厅原有抱厦
Baosha in front of Huating

船舫
Chuanfang (Pavilion)

0    0.75    1.5        3m

上海县郭宅及东园东落中轴剖面图
Section of Shanghai Guozhai and East Garden's east lane

中国古建筑测绘大系·祠庙建筑与园林建筑——上海庙园

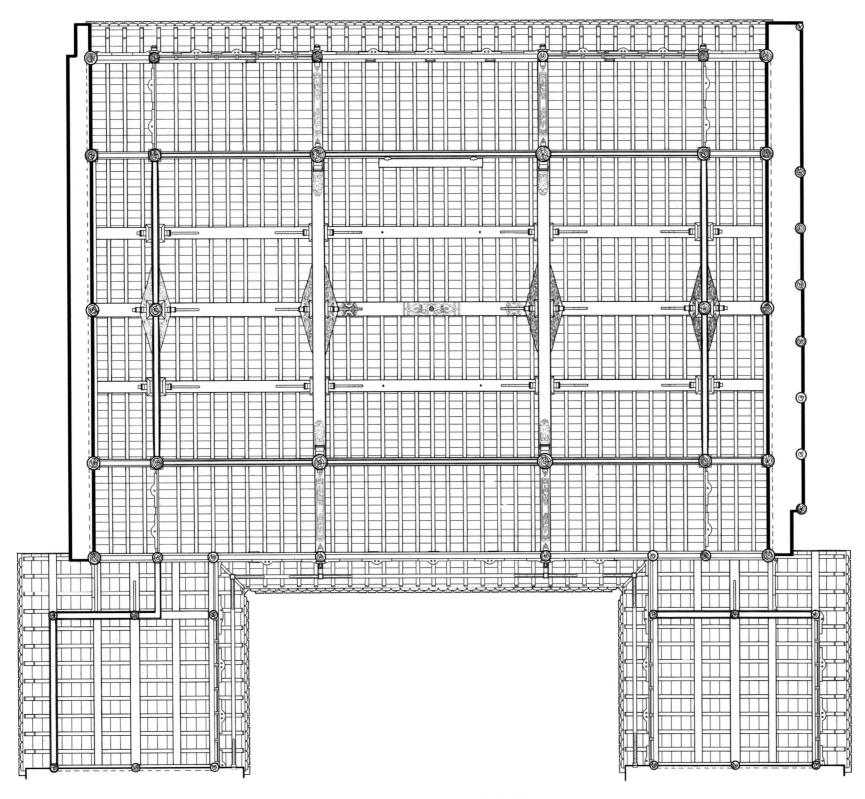

0 0.5 1 2.5m

郭氏宅邸大厅及厢房屋架仰视图
Bottom view of the main hall and Xiangfang's beams of Guozhai

屋架
Roof Structure

二层
Second Floor

一层
First Floor

郭氏宅邸轴测解析图
Axonometric and exploded drawing of Guozhai

3m

1.5

0.75

0

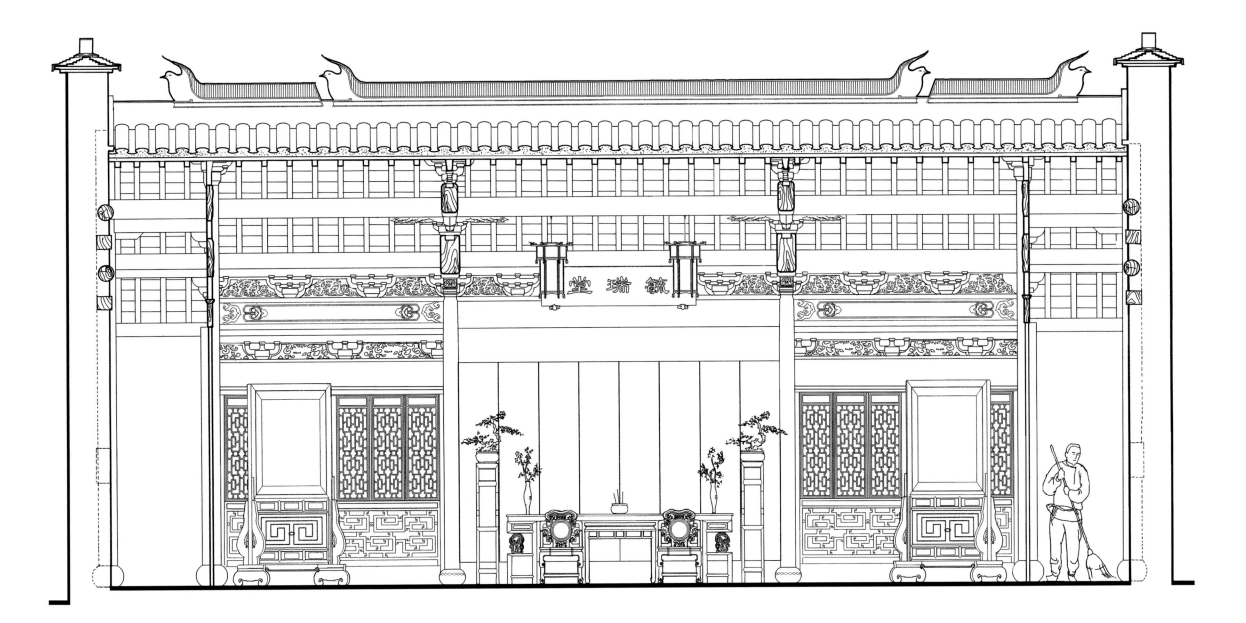

锦瑞堂

大厅纵剖面图
Longitudinal section of the main hall

0    0.5    1    2m

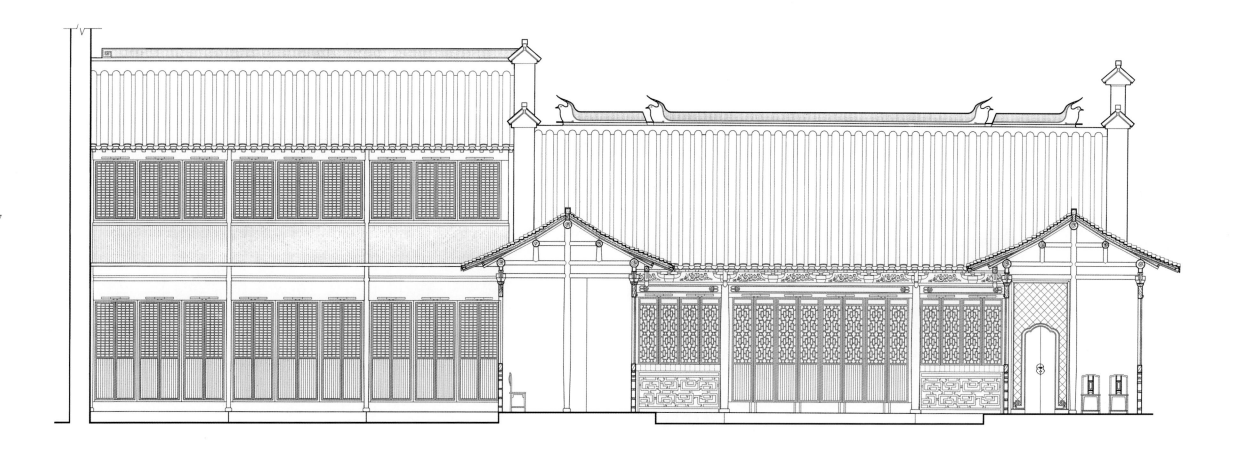

大厅及厢房剖立面图
Section and elevation of the main hall and Xiangfang

0    0.5    1         2m

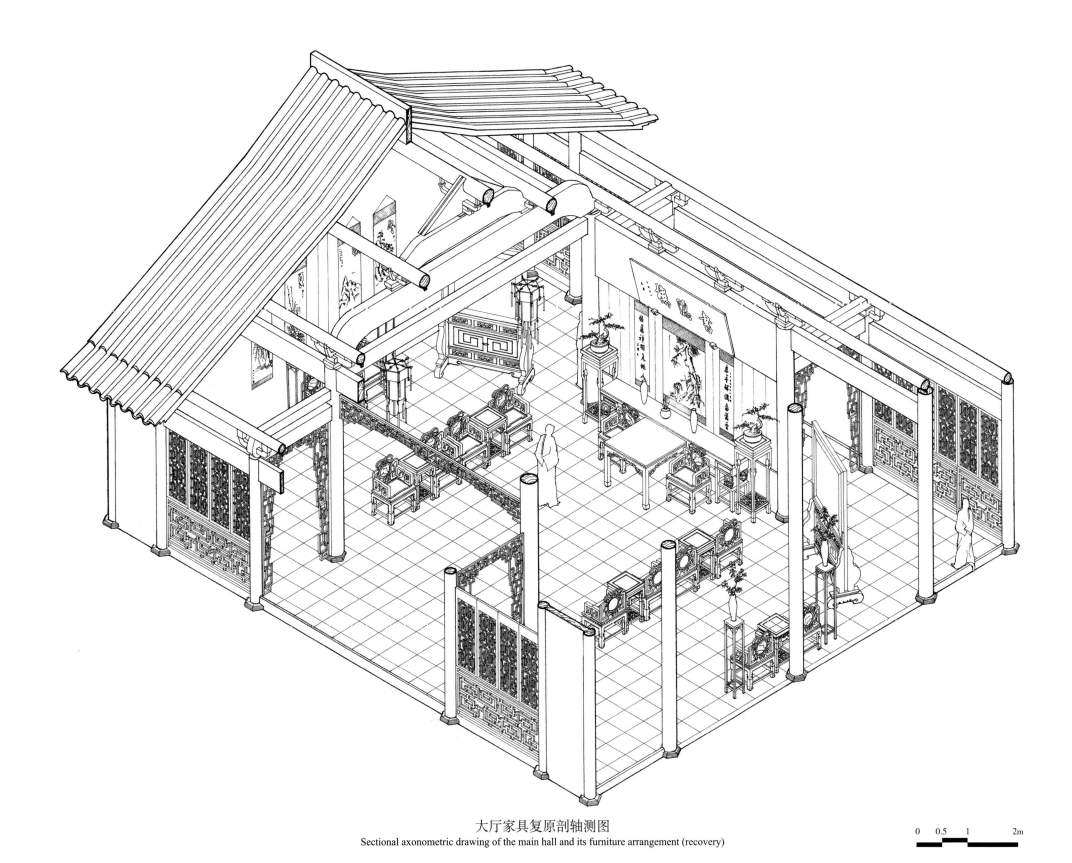

大厅家具复原剖轴测图

Sectional axonometric drawing of the main hall and its furniture arrangement (recovery)

0　0.5　1　　2m

堂楼前"古训是式"砖刻门楼正立面图
Front elevation of Guxunshishi Menlou in front of Tanglou

堂楼前"古训是式"砖刻门楼侧立面图
Side elevation of Guxunshishi Menlou in front of Tanglou

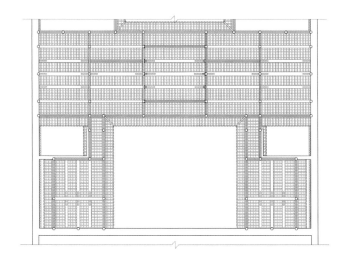

堂楼（书隐楼）屋架仰视图
Bottom view of Tanglou (Shuyinlou)'s beams

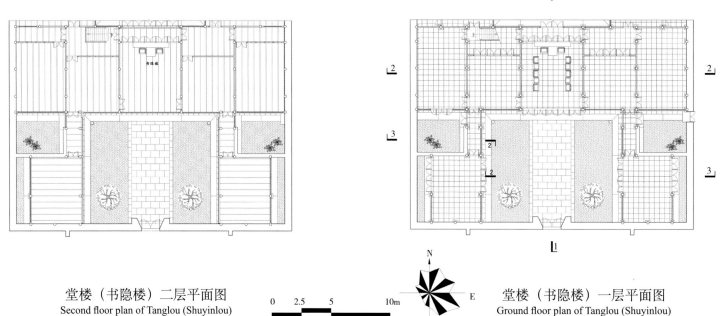

堂楼（书隐楼）二层平面图
Second floor plan of Tanglou (Shuyinlou)

0  2.5  5  10m

N  E

堂楼（书隐楼）一层平面图
Ground floor plan of Tanglou (Shuyinlou)

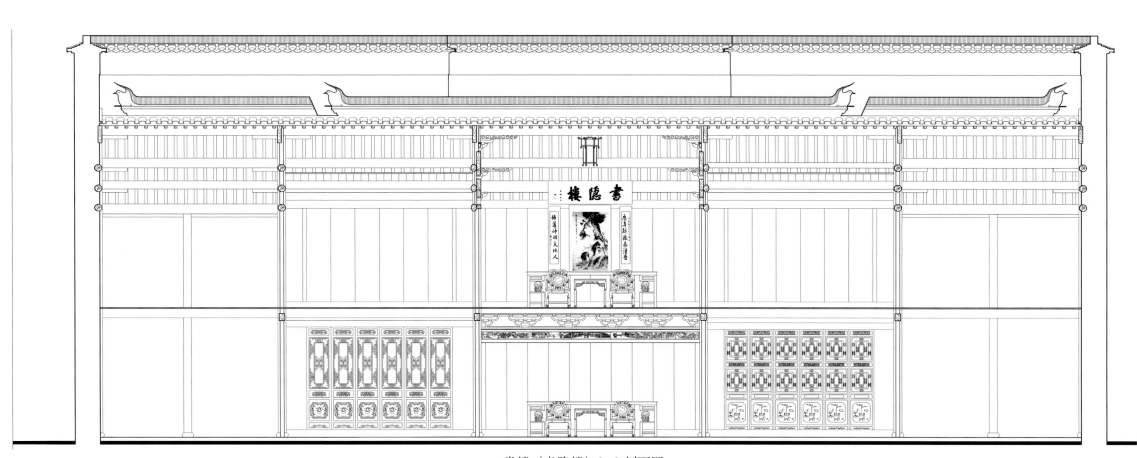

堂楼（书隐楼）2-2 剖面图
Section 2-2 of Tanglou (Shuyinlou)

0  0.3  0.18  0.6m

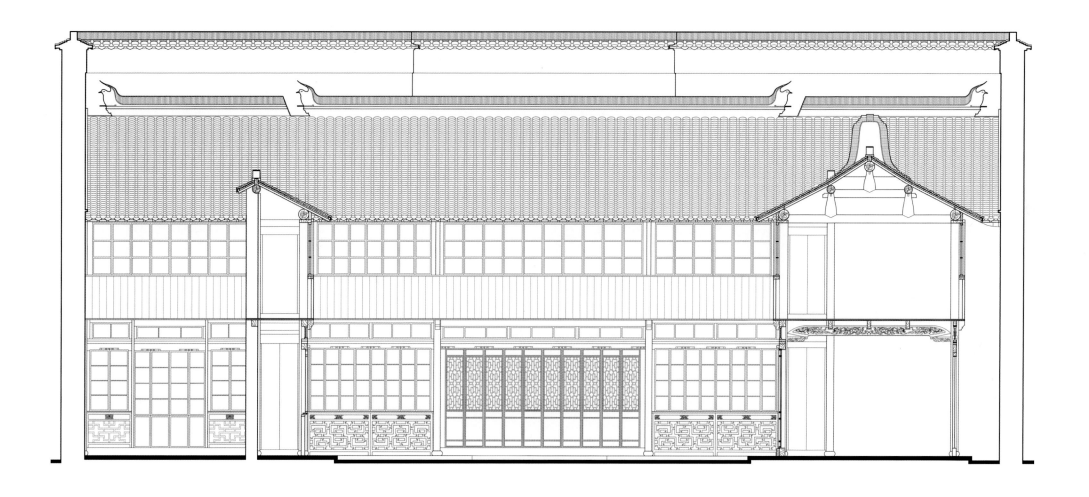

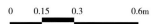

0    0.15    0.3        0.6m

堂楼（书隐楼）3-3 剖面图

Section 3-3 of Tanglou (Shuyinlou)

堂楼（书隐楼）西翼 "八仙游山" 砖雕仿木屏门大样图

Baxianyoushan brick-imitating-wood four crosspieces door of Tanglou (Shuyinlou)'s west side

0.3m 0.12 0.06 0

堂楼（书隐楼）东翼 "三星贺寿" 砖雕仿木屏门大样图

Sanxingheshou brick-imitating-wood four crosspieces door of Tanglou (Shuyinlou)'s east side

0    0.06   0.12                    0.3m

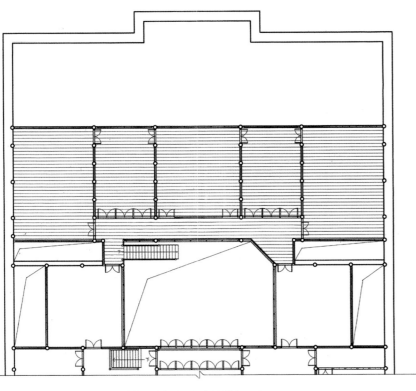

后楼二层平面（现状）图
Second floor plan of Houlou (present situation)

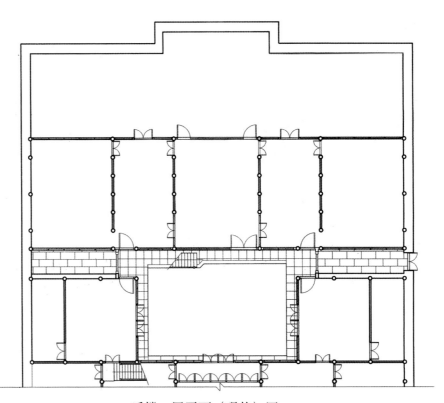

后楼一层平面（现状）图
Ground floor plan of Houlou (present situation)

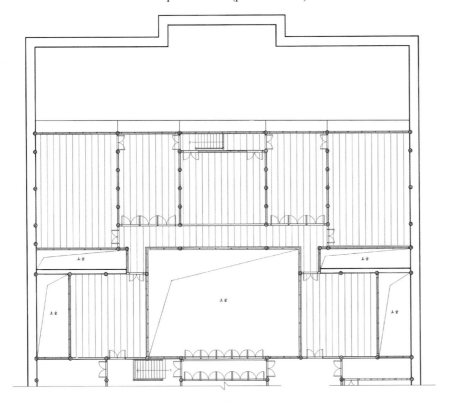

后楼二层平面（复原）图
Second floor plan of Houlou (recovery)

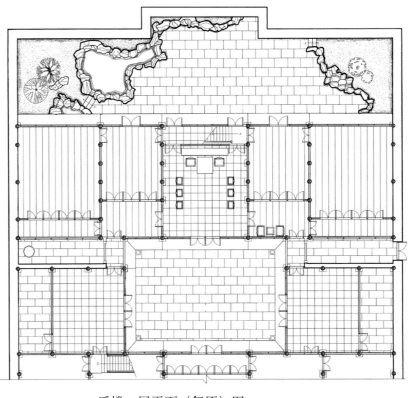

后楼一层平面（复原）图
Ground floor plan of Houlou (recovery)

0  2  4          10m

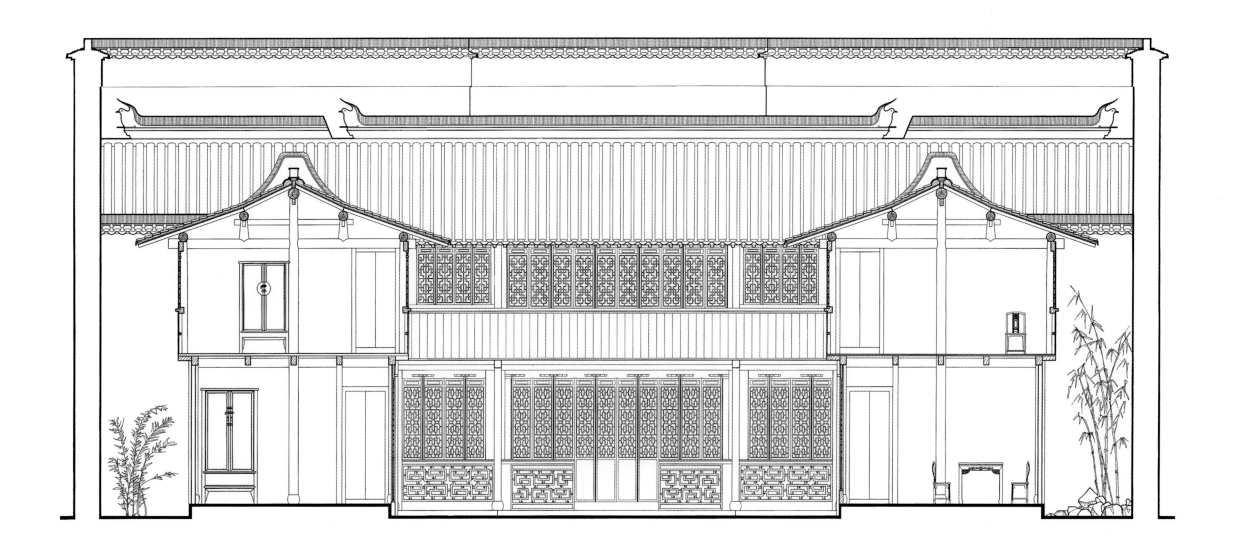

后楼及厢房剖立面图
Section and elevation of Houlou and Xiangfang

0  1  2  5m

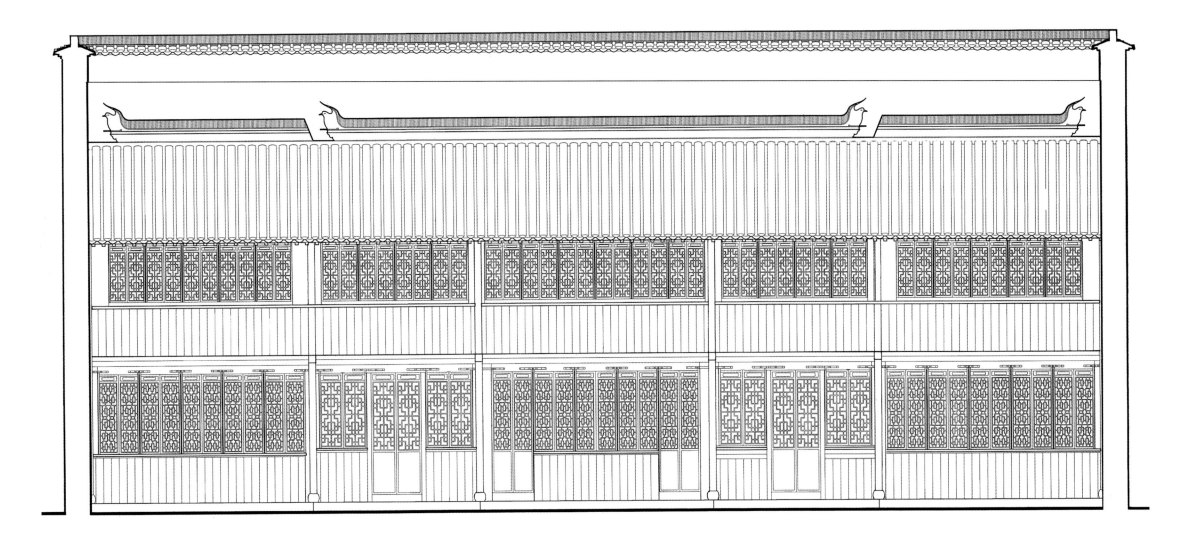

后楼北立面图

North elevation of Houlou

0 1 2 5m

397

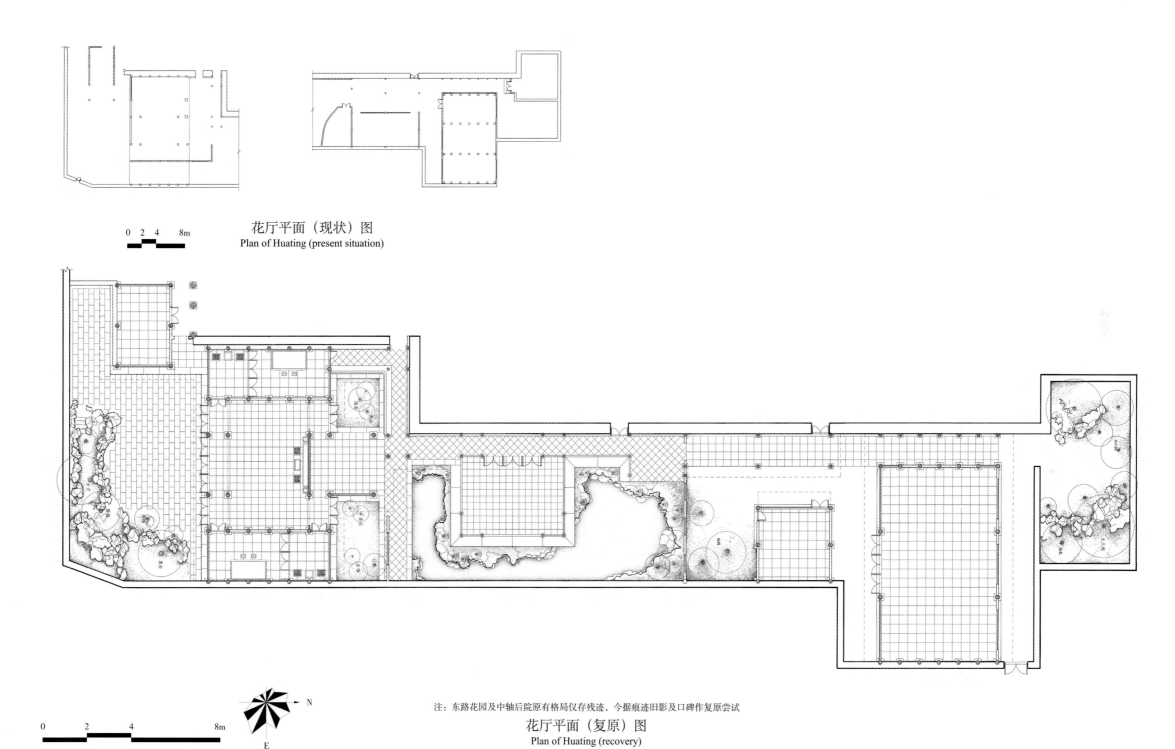

0 2 4 8m

花厅平面（现状）图
Plan of Huating (present situation)

注：东路花园及中轴后院原有格局仅存残迹，今据痕迹旧影及口碑作复原尝试

花厅平面（复原）图
Plan of Huating (recovery)

0 2 4 8m

N

E

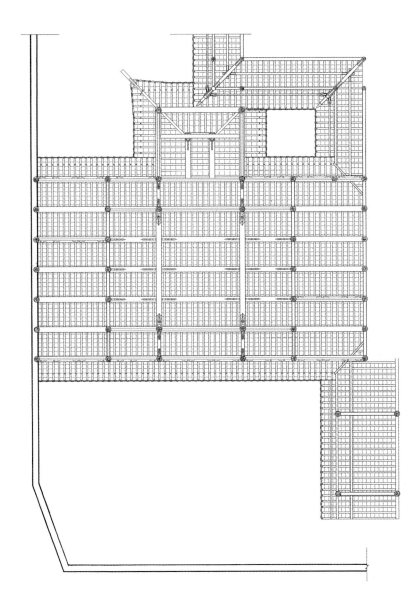

花厅梁架仰视图
Bottom view of Huating's beams

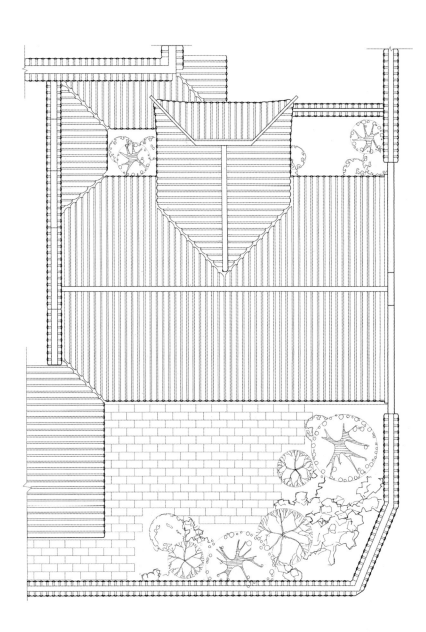

注：疑原有抱厦今不存，故据痕迹及口碑作复原尝试

花厅屋顶平面图
Roof plan of Huating

0　1　2　　4m

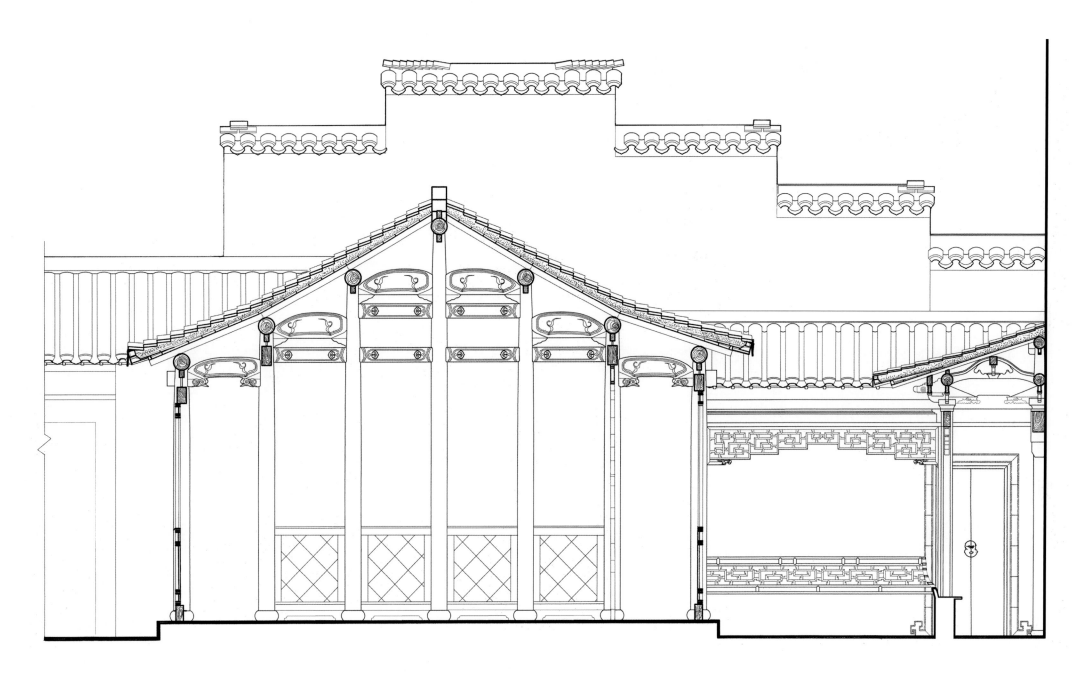

0    0.5    1    2m

花厅边帖横剖面图
Cross-section of Huating's Biantie

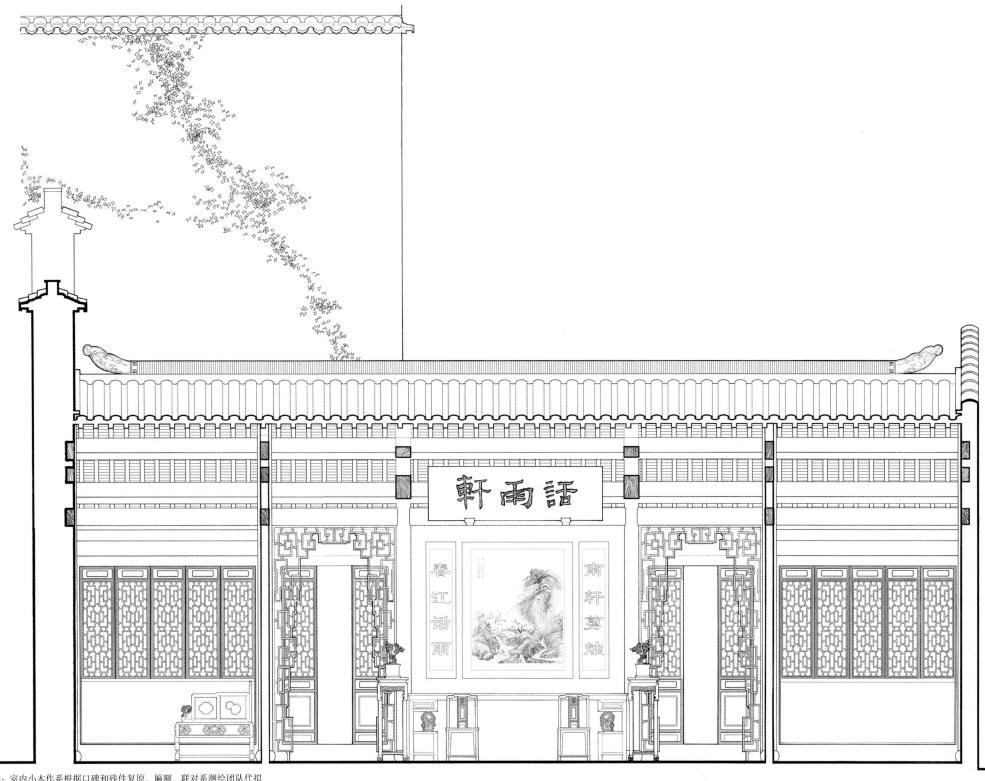

注：室内小木作系根据口碑和残件复原，匾额、联对系测绘团队代拟

花厅纵剖面图
Longitudinal section of Huating

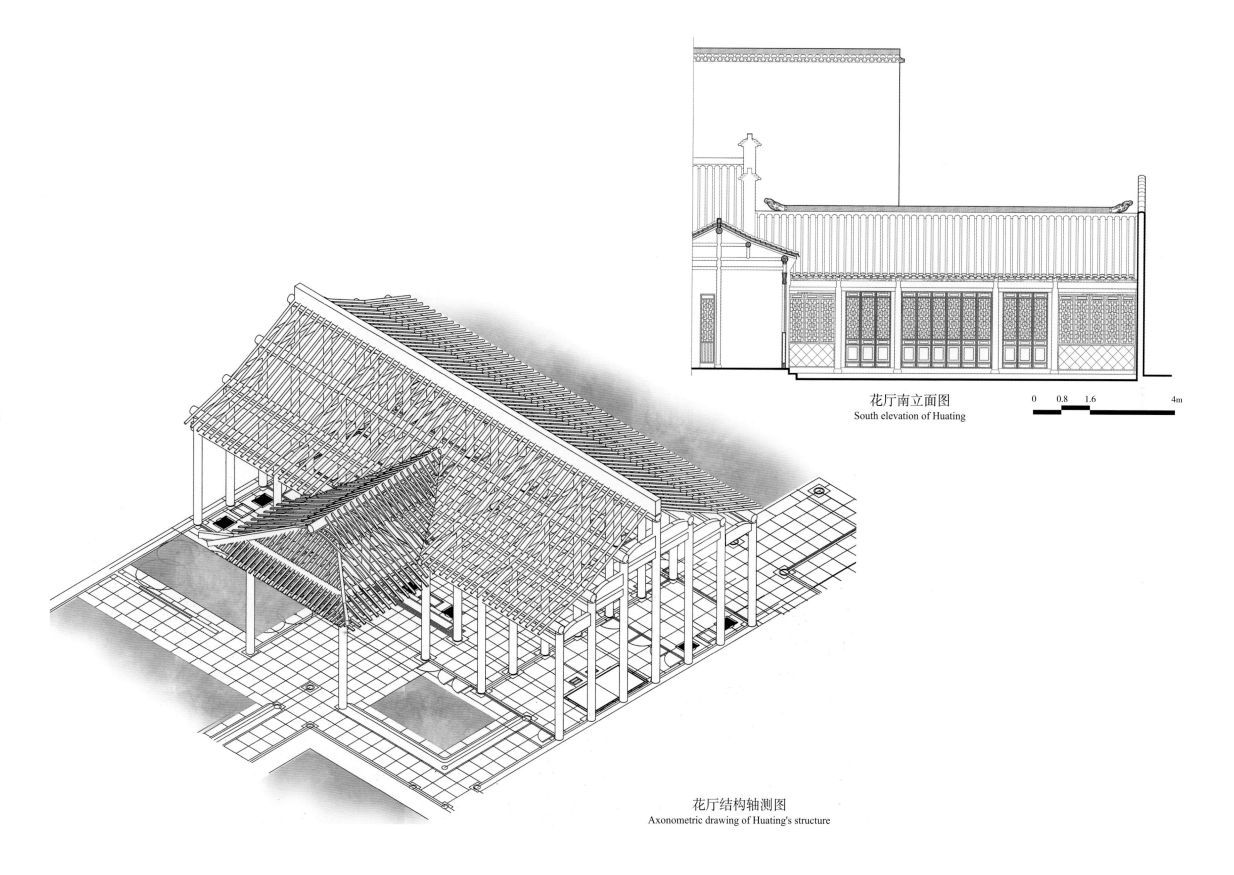

花厅南立面图
South elevation of Huating

0    0.8    1.6          4m

花厅结构轴测图
Axonometric drawing of Huating's structure

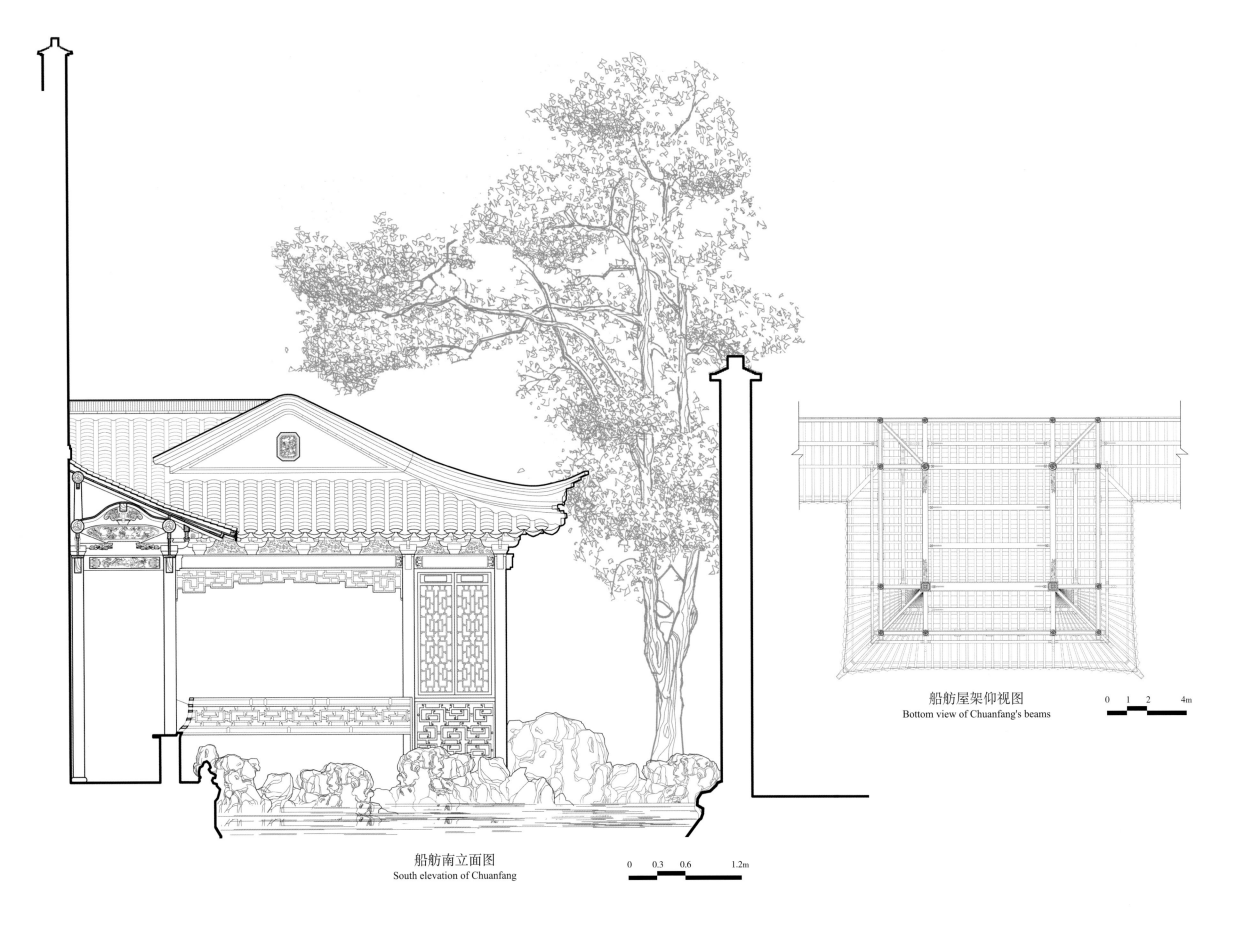

船舫屋架仰视图
Bottom view of Chuanfang's beams

0　1　2　　　4m

船舫南立面图
South elevation of Chuanfang

0　0.3　0.6　　1.2m

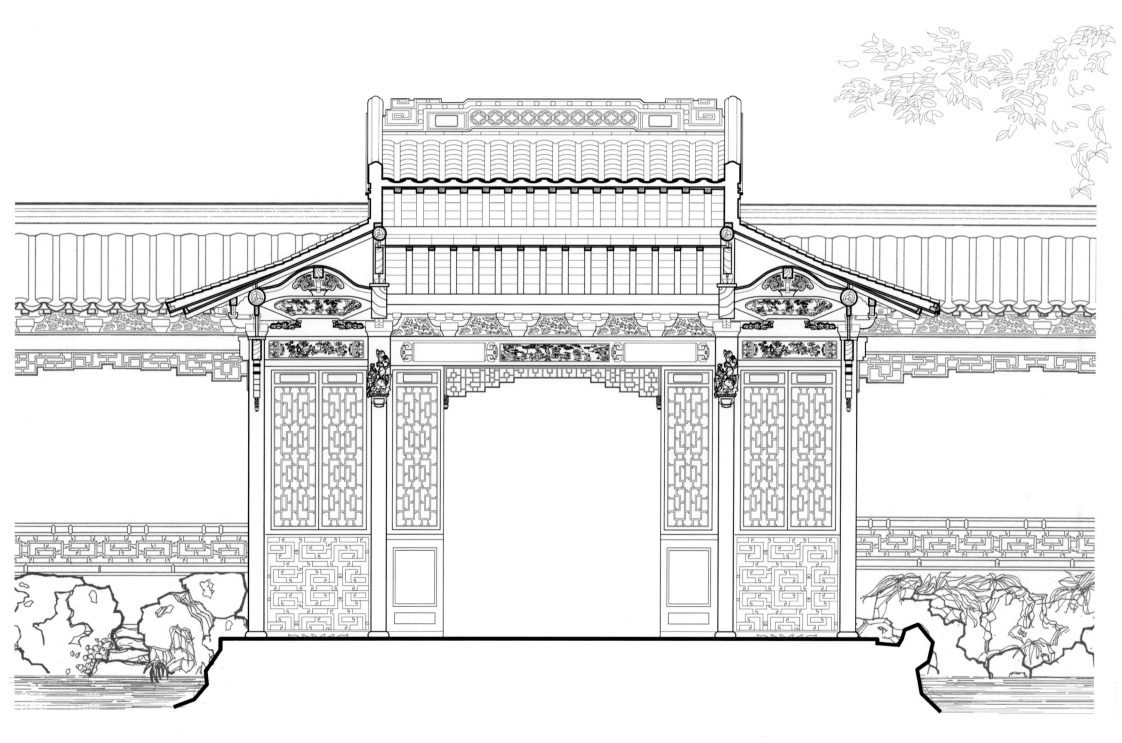

船舫纵剖面图
Longitudinal section of Chuanfang

船舫横剖面图
Cross-section of Chuanfang

0　0.3　0.6　　1.2m

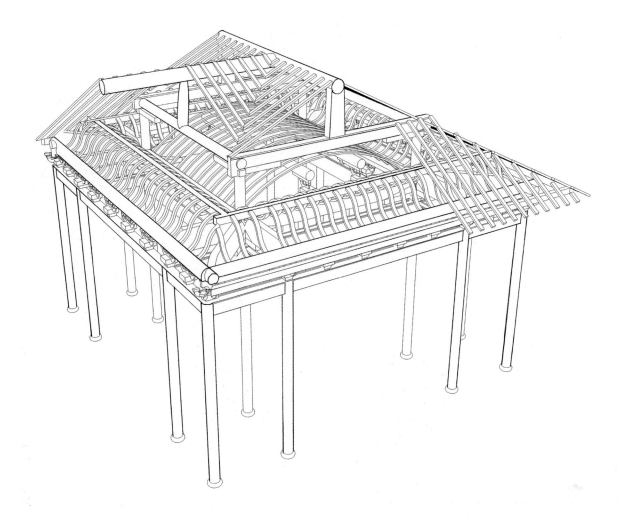

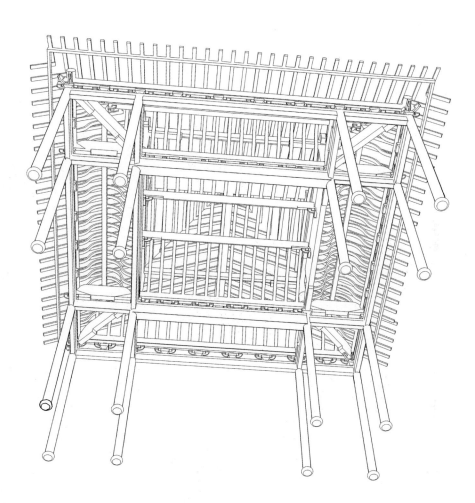

船舫结构透视图（一）
Perspective view of Chuanfang's structure 1

船舫结构透视图（二）
Perspective view of Chuanfang's structure 2

**Item of survey:** Songjiang County Yi Garden and Nearby Houses

**Address:** No1172, West Songhui Road, Songjiang District, Shanghai

**Age of construction:** the late Ming Dynasty

**Site area:** 4.95 mu

**Competent organization:** Shanghai No.4 Social Welfare Institution

**Survey organization:** College of Architecture and Urban Planning, Tongji University

**Time of survey:** 2006

测绘项目：松江府城颐园及邻近宅第

地　　址：上海市松江区松汇西路一一七二号

始建年代：始建于明末

占地面积：四·九五亩

主管单位：上海市第四福利院

测绘单位：同济大学建筑与城规学院

测绘时间：二〇〇六年

Songjiang County Yi Garden
and Nearby Houses

松江府城颐园及邻近宅第

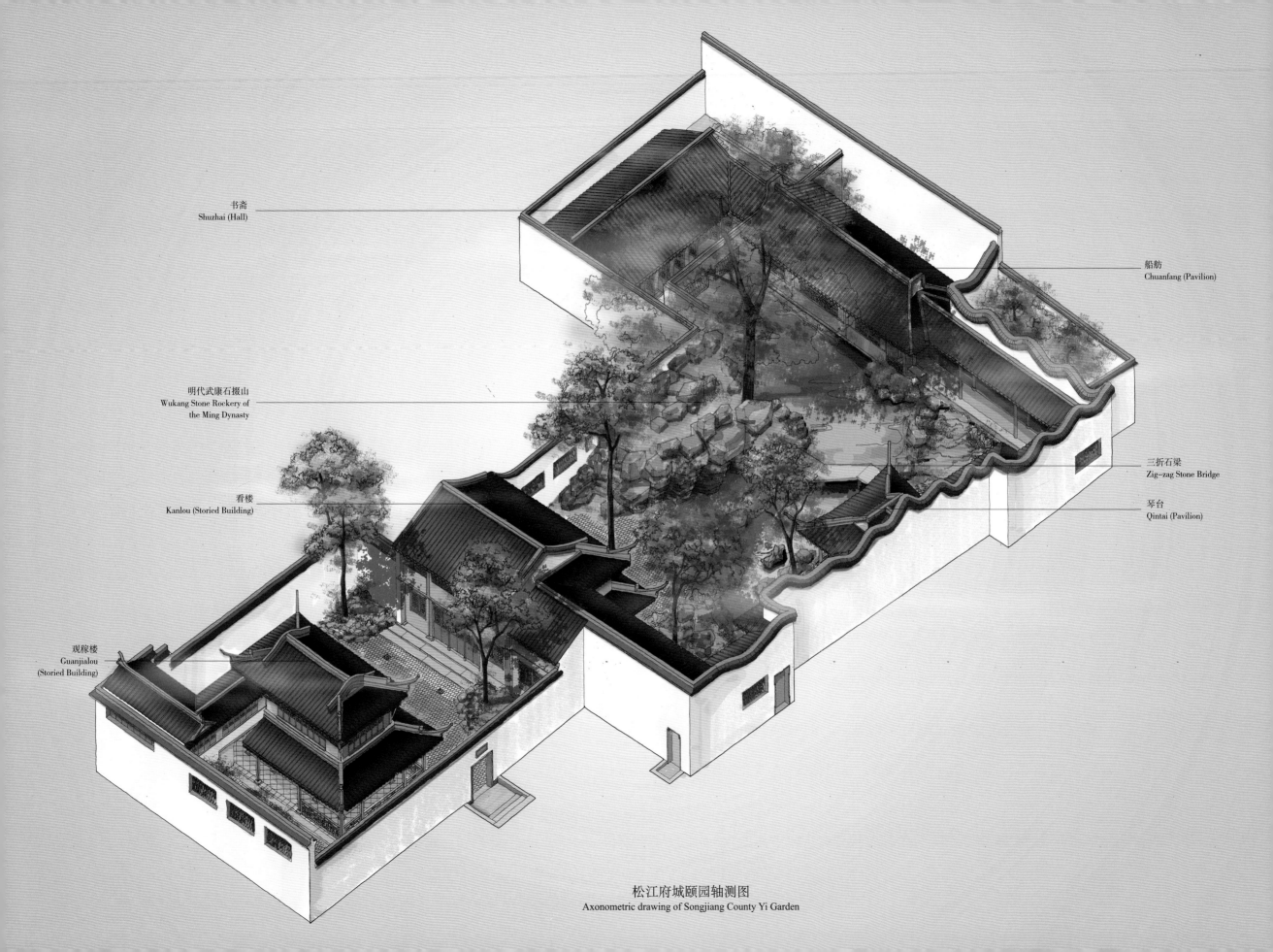

书斋
Shuzhai (Hall)

明代武康石掇山
Wukang Stone Rockery of
the Ming Dynasty

看楼
Kanlou (Storied Building)

观稼楼
Guanjialou
(Storied Building)

船舫
Chuanfang (Pavilion)

三折石梁
Zig-zag Stone Bridge

琴台
Qintai (Pavilion)

松江府城颐园轴测图
Axonometric drawing of Songjiang County Yi Garden

# Introduction

Yi Garden and its neighboring houses are located in the west of Xiunan Bridge, south of Xiunan Street, and east of Chenjia Lane, outside the west gate of the original Songjiang Prefecture, which is today's area inside and outside the Fourth Welfare Institute of Shanghai. Its landscape only occupies 2 mu, and it should be a small garden near the residence in the late Ming Dynasty. On the east and north sides, the residential parts that were originally adjacent to Xiunan Street in the north and near the city river (Xiuzhoutang River) had been separated and superimposed by later generations, and only the part of the central axis is likely to still retain the basic appearance of the same period as Yi Garden.

As the highlight who almost led the prosperity of Songjiang Prefecture of Jiangnan area in the late Ming Dynasty, Yi Garden was unfortunately returned from the prosperous time to a remote place in the cold alley, but fortunately became a garden that nearly retains the style, the structure and the landscape and detail of the Ming Dynasty in China. The opera building (Guanjialou) of the late Ming Dynasty, the Wukang Stone mountain and pool, and the main hall and back building which are suspected to belong to the residence, can be comparable to the gardens and houses of Suzhou Yipu Garden in the same period, and are particularly subtle and typical.

In the former prosperous crevices of the city, the garden was compatible with the extremely condensed vertical and horizontal landscape changes such as cliffs, tunnels and pools, lakes, as well as the strong reflection of the cornices in the same line of the buildings. It is a precious demonstration of the peak and change period of Chinese gardening art represented by Jiangnan in the late Ming and early Qing Dynasty. Its style of the era is also comparable to Changzhou Jin Garden, Wuxi Jichang Garden, Jiading Qiuxiapu Garden, Suzhou Yipu Garden, and the central part of Suzhou Zhuozheng Garden. As a typical mountain landscape garden in the late Ming Dynasty, this garden is comparable to Suzhou Chang Garden in the late Qing Dynasty, which covers the similar area, as the two famous small gardens with mountain and water landscape.

The layout of Suzhou Huanxiu Shanzhuang created by GE Yuliang, a famous master in the middle Qing Dynasty, resembled the symmetrical enlarged version of this garden.

导　言

颐园及其邻宅位于原松江府城西门外秀南桥西、秀南街南，陈家弄东侧，今上海市第四福利院内外。其景象空间占地仅两亩许，应为晚明时盈盈傍宅一园；；而其东侧与北侧、原本北临秀南街而近市河（秀州塘）的住宅部分已历经后世割裂叠加，惟中轴部分很可能仍存留与颐园同时代的基本面貌。

作为晚明时近乎引领江南的松江一府鼎盛时期的吉光片羽，该宅园不幸由繁华一度归于冷巷僻处，却幸而为中华存留一明风、明构乃至明代空间景域与细节均较为完整的江南宅园。其晚明戏楼（观稼楼）、武康石平崖与濠濮及疑似原住宅大厅、后楼等构筑，堪比同时期苏州艺圃之园宅并古，而尤为精微而典型。

该宅园于曾经的城市繁华隙地中兼容了峭壁、隧函与濠濮、湖泊等极尽凝练的纵横景象变化，和一线建筑檐口的劲利映带，是明末清初以江南为代表的中华造园艺术发展至巅峰期与求变期的珍贵实证。其时代风格亦足与常州近园、无锡寄畅园、嘉定秋霞圃、苏州艺圃、苏州拙政园中部等同列。作为典型的晚明山景园，此园还堪与面积相仿之晚清水景园——苏州畅园并称为小园山水双璧。

清中叶名家戈裕良之苏州环秀山庄布局，则酷似此园之对称放大版。

图一　上海松江府城与仓城之间的秀州塘（古市河），远方即松江府城，左侧即今中山路，右侧即秀南街，颐园即在右前方远处树丛中

Fig.1  Xiuzhoutang River (the original city river) between Songjiang Prefecture and Cangcheng in Shanghai, with Songjiang Prefecture in the distance, today's Zhongshan Road on the left, and Xiunan Street on the right, and Yi Garden in the trees in the distance on the right side

## From Business Paddles on the River to the Small Building Called Guanjia, the Return of Space Aphasia

In the middle and late Ming Dynasty, the two prefectures of Jiangnan, Suzhou and Songjiang once stood side by side like two peaks, leading the wave of economy, culture and gardening art for more than 100 years. In the later period, the tide started to move eastward to the sea. Although it was once suspended because of the war set off by the soldiers of the Qing Dynasty, it was finally inherited by the Shanghai County after the opening of the port. Outside the west gate of Songjiang Prefecture, the prosperous urban area that broke through the city wall and between the city river and headed westward of the Shuicicang city, was the testimony of this historical wave. Sails gathered on the river and merchants converged and once gathered and transported one-sixteenth of the country's water transportation grain. Junzhi Street (now Zhongshan Road) on the north bank and Xiunan Street on the south bank were facing each other. The residential houses were sprawling, and the gardens and pavilions faces each other, owning the convenience of land and water transportation, as well as the charm of the market and the countryside. Since hundreds of years' changing, the remaining residential complex are still regulated and magnificent, spanning the Ming and Qing Dynasties, while the waterfront building interface is also magnificent, occupying the first place in the south of the Yangtze River.

Yi Garden and its residences are located deep on the south side of Xiunan Street on the south bank of the ancient city river.

# 从长河听橹到小楼观稼——空间失语者的回归

明中晚期的江南，苏州、松江二府曾如双峰并峙，联驾并驱，引领了百余年的经济、文化和造园艺术浪潮，后期更有潮头东移向海之势。虽一度中止于清兵的南下，却最终被开埠后的上海县承继至今——

松江府城西门外，突破城墙禁锢、夹市河而向水次仓城西行的数里繁华城市带，就是这一历史浪潮的见证。

河上帆樯云集、商贾辐辏，一度聚合、输送着帝国十六分之一的漕粮，北岸郡治大街（今中山路）一线与南岸秀南街一线则夹河而峙，宅邸绵亘，园亭相望，既得水陆交通之便，又兼收市肆与郊野之趣——

百年陵替至今，所存留住宅建筑群仍规制宏巨，跨越明清，水岸建筑界面亦恢宏浩荡，甲于江南。

颐园及其宅邸就深居于古市河南岸之秀南街南侧。

图二 上海松江林景旸、林有麟宅明代走马楼厅

Fig.2  The Ming Dynasty Zoumalou of Lin Jingyang and Lin Youlin's mansion in Songjiang District, Shanghai

或许因其规模精微，主人异类，又乏名人题咏，颐园曾长年湮没于海上造园史的长河，失语于多种重要志书，崖岸深池，仅供闭门独对。从今存建筑及掇山看，其始建至迟在晚明，观稼一楼更曾历经碳十四之精确测年，；却或许直至晚清时归于浙江归安县令、金石书画家许威（号『铁山』），改称颐园时，此园才正式转入文人士大夫之手，其园史变得清晰可溯。民国时此园更转售给名动一时的文人团体——南社之活跃人物、金山县张堰高吹万次子高君藩，一时成为柳亚子等南社巨子和沪上文人社团的雅集会文之地。

百年冰封，一朝倾动。晚明空间的流风余韵，或许至此终与文人不羁的辞采遇合，由滔滔市河上的繁忙商橹，转入文人视野的小楼观稼。

Perhaps because of its small scale, heterogeneous owner, and lack of celebrity chants, Yi Garden has been obscured for many years in the history of gardening, aphasia in many important chronicles, and the cliffs and deep pools are only for closed doors. Judging from the existing buildings and the mountain, it was first built in the late Ming Dynasty, and Guanjialou has been accurately dated by carbon 14 However, perhaps it was not until the late Qing Dynasty that it was returned to XU Wei (whose pseudonym is 'Tieshan' ), who was Zhejiang Guian Xianling and a painter and calligrapher, and renamed as Yi Garden, the garden was officially transferred to the hands of literati, and its garden history became clear and traceable. In the Republic of China, the garden was resold to GAO Junfan, an active figure of the famous literati group 'Nanshe', and the second son of GAO Chuiwan in Zhangyan, Jinshan County, and became a place for the magisters of the Nanshe giants such as LIU Yazi and the Shanghai cultural associations.

It has been frozen for 100 years, and once moved. The style and rhyme of the space in the late Ming Dynasty may eventually meet with the literati's poetry chapters, from the busy business paddles on the city river to the small building with the view of the literati.

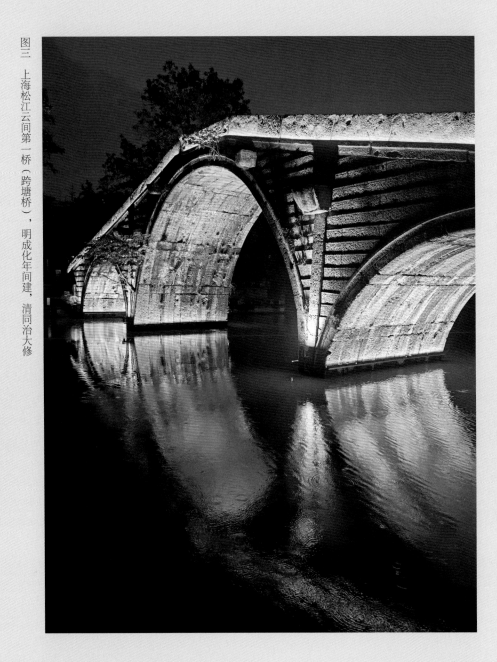

图三 上海松江云间第一桥（跨塘桥），明成化年间建，清同治大修

Fig.3 Yunjiandiyiqiao (Kuatangqiao) in Songjiang, Shanghai, which was built in Chenghua period in the Ming Dynasty and renovated in Tongzhi period in the Qing Dynasty

## From Upright View to Side View, the Disenchant-ment of the Mountain Landscape Layout

As far as the current appearance of the garden is concerned, the main hall (building) is the second-storey 'Kanlou' with single eave flush gable roof in the north of Stage. The south is for viewing performances, while the north is for viewing mountains, and the east is connected to the old residential halls through the corridors, thus dividing the long and narrow space of the whole garden into the main landscape space on the north side and the small viewing space on the south side. Its spatial role resembles Huating on the east side of the GUO Family's residence and garden complex (Shuyinlou) in Shanghai County.

However, the landscape space on the north side is dominated by mountain views. Kanlou and Chuanfang respectively stand in the south and the north. 'Qintai', the core pavilion, locates in the east, facing the main mountain in the west to look up at its dignity and the 'high and far' form. Kanlou then from the south side, gaze at the 'flat and far' trend. As if the predecessor's poem, 'The winding of the Xiang River allows the sails to turn continuously with the water, and down the river, you can see the different perspectives of Mount Heng.', that different perspectives have different characteristics.

This main hall's 'side view' layout is different from the layout of the 'upright view' mountain island since BAI Juyi's 'Lushan Caotang' in the Tang Dynasty highlighting the remaining authoritarian color. It was earlier than the similar layout of Jichang Garden in Wuxi of the early Qing Dynasty, the eastern part of Suzhou Ou Garden suspected of the early Qing Dynasty, and the Xiaopangu Garden in Yangzhou of the middle Qing Dynasty.

# 从正仰到侧眺——山景布局的祛魅者

就此园今存面貌而言，实以戏台北侧的单檐硬山顶二层『看楼』为主要厅堂（楼厅），南则观演，北则观山，向东则通过楼廊与旧日的住宅厅楼相连，并将南北狭长的全园空间划分为北侧的主体山水空间与南侧的小型观演空间——其空间角色与上海县郭宅（书隐楼）东侧庭园中的花厅酷似。

而其北侧之山水景象空间实以山景为主，以看楼与船舫分居南北，以核心亭榭『琴台』居东，正对西侧主山，以仰其峻拔高远之形，看楼则自南侧眺其透迤平远之势。仿佛前人歌咏之『帆随湘转，望衡九面』，不同的视角各具景象特征。

这一主要厅堂之侧眺布局与唐人自居易『庐山草堂』以来的正仰山岛而残留威权色彩的布局不同，较清初无锡寄畅园、疑似清初之苏州耦园东部、清中叶扬州小盘谷之类似布局则更早。

图五　上海松江府城隍庙放生池前云纹勾栏（现位于松江方塔园内）

图四　北宋兴圣教寺方塔（现位于松江方塔园内）

Fig.4　Fangta of Xingshengjiao Temple of the Northern Song Dynasty (now loceted in Fangta Garden in Shanghai Songjiang District)

Fig.5　Cloud pattern *goulan* in front of Fangsheng Pond of Shanghai Songjiang Chenghuang Temple(now located in Fangta Garden in Shanghai Songjiang District)

## From Facing the Mountain across the Water to One River with Two Banks, Progress and Intensification of Water Landscape Layout

Below the high mountains, there must be Yuantan (deep pool) and Haopu (moat), and the water landscape in the mountain view garden must always be vertical.

The water landscape of this garden is a line of deep sinking Haopu gushing out from the mouth of the southeast veranda. When it flows north to the middle of the garden, 'Qintai' suddenly appears on the soil slope on the east bank, and faces the Wukang Stone 'Flat Cliff' that seems to break into the west wall, forming a core dialogue relationship full of tension as well as the focus of 'seeing' and 'being seen'. At the same time, the long and narrow water space between north and south is divided into east and west, forming a similar 'one river with two banks' landscape.

In the north of 'Flat Cliff', the original vertical Haopu is horizontally displayed as a lake. At the end of its north bank, there is the horizontal Chuanfang, which faces the north wall of the Wukang stone 'Flat Cliff' on the south bank across the lake, forming a classic layout of 'facing the mountain across the water'. The endless meaning of the lake seems to overflow to the north-west. In terms of the purity of the scene, the lake seems to be better treated as a 'vertical' Yuantan. And it is also a pity to fail to shape the lingering veins in the small north-west courtyard across the wall.

The transformation from a relatively stable and homogeneous layout of 'facing the mountain across the water' to a more conflicting and expressive layout of 'one river with two banks' is a major symbol of the disenchantment of water landscape in the history of gardening. And in this garden, the vertical Haopu develops into Pinghu (calm lake), combining the two states, and the landscape of the water and the mountain is almost equal.

从隔水面山到一河两岸——水景布局的进步与集约

高山之下，必有渊潭、濠濮，山景园中的水景总须以纵向为主。

此园水景系自东南游廊下之水口涌出一线深陷的濠濮，向北蛇行至全园中部时，东岸土坡忽现『琴台』，与仿佛破西墙而入的武康石『平崖』隔濠相对，近在眉睫，形成饱含张力的核心对话关系，与看与被看的焦点，同时对狭长的南北水上空间作东西向的层次划分，形成近似一河两岸的景象。

『平崖』以北，原本纵向的濠濮则横展为湖泊，其北岸尽头有横舫一抹，隔湖与南岸之武康石『平崖』北壁相对，形成经典的隔水面山景象。湖泊之不尽余意则仿佛向西北溢出——如就景象的纯度而言，此部湖泊似仍以处理为纵向的渊潭为佳；而未能于隔墙之西北小院中塑造点滴余脉，亦为恨事。

由相对平稳而匀质的隔水面山布局向更富冲突与表现力的一河两岸布局转化，是造园史上水景去魅的一大表征，而此园以一线濠濮归于平湖，将二者熔于一炉，山水几乎并重。

图七　松江百岁坊一带清代民居

图六　松江百岁坊一带清代民居

Fig.6  The Qing Dynasty residences in Baisuifang area of Songjiang District
Fig.7  The Qing Dynasty residences in Baisuifang area of Songjiang District

## A Long Line of Chanting, the Continuous Cornice Line of the Buildings in the Garden

In order to avoid the overwhelming scale of the landscape, in the north of 'Kanlou', the main hall of the whole garden, and its corridor, a soothing waist eave was added. It stretches to the east wall of the garden, and leans against the wall to go northward, becoming a veranda running through the north and south of the entire garden. It passes through the north-east corner of the entire garden and turns to the west, directly into the independent courtyard in the north-west corner and connects with Shuzhai (Study room). The cornice line traverses mountains and rivers, flowers and brush willows, and turns ups and downs in one go, becoming the basic control line of the whole garden.

When the corridor reaches the east and middle part of the garden, a half pavilion (Qintai) protrudes to the west, facing Flat Cliff across Haopu, helping to form the focal point of the entire garden and the deep scene of 'one river with two banks'. When going to the west of the north wall, it protrudes slightly to the south as the horizontal freehand-style Chuanfang (Pavilion like a ship), facing the side of the cliff across the flat lake and helping to form a 'facing the mountain across the water' scene. The cornice line is long and sharp, interweaving and setting off, and blends the colorful buildings of the whole garden into a long chant, 'setting off the landscape with its own necessary simplicity and monotony'. It remind us the similar cornice penetration treatment of Yanguangshuige of Suzhou Yipu Garden, Deyuexuanlou of Qingpu Qushui Garden and the buildings in Suzhou Ou Garden, which have the strength of the ancient 'iron line drawing' and the simplicity of Ming style furniture, continuing the long tradition of architecture and line, and also seems to be the sound of knocking in the new era and new taste.

# 一线长吟——通贯的全园建筑檐口线条

为避免对山水景象形成尺度凌迫，全园主要厅堂『看楼』及楼廊北侧，还加施舒缓的腰檐一道，绵延至园东墙，复倚墙北行，成为贯穿全园南北的游廊，经全园东北角西折，直入西北角独立小院而与书斋相接——其檐口线条跋山涉水、分花拂柳，转折起伏、一气呵成，成为全园之基本控制线条。

此廊行至园东中部时，向西凸出半亭（琴台），隔濠濮而对平崖，助成全园焦点与一河两岸之深邃景象；行至北墙西部时，则向南稍稍凸出为横长的写意型船舫，隔平湖而仰危崖侧壁，助成隔水面山景象——该檐口线条悠长劲利，穿插映带，将全园缤纷建筑融为一气长吟，『以自身必要的质朴单调将园景映衬得明艳动人』。令人联想起苏州艺圃延光水阁、青浦曲水园得月轩楼和苏州耦园楼群的类似檐口通贯处理，如高古铁线描般的秀劲，又如明式家具般的简率，延续着悠久的建筑与线条传统，也仿佛新时代与新趣味的叩门之声。

图八 王春煦（号冶山）故居明代轿厅梁架与山雾云、由徐瑞彤绘制

Fig.8 The Ming Dynasty beam and Shanwuyun of Jiaoting of WANG Chunxu (whose pseudonym is Yeshan)'s former residence, drawn by XU Ruitong

## Miniature of 'Cutting the River and Breaking the Valley', the Rare and Precious Remains of the Late Ming Dynasty Wukang Stone Flat Cliff

There has always been a saying that the garden was designed by ZHANG Nanyuan (1587-?), the master gardener in the late Ming and early Qing Dynasties. Considering that the ZHANG family lived in the north of Junzhi Street and under the pagoda of Xilin Temple not far from the east side of this garden in the early days, the time and place were fit, so the legend is reasonable to some extent. However, ZhANG's 'realistic' gardening method advocated only intercepting nature without minimizing it, emphasizing the horizontally soothing landscape with curved bank soil slopes, and hating the drastic and grotesque changes that are over-condensed within a short distance. It was said that 'once ZHANG Nanyuan became famous, people no longer enjoyed rockery mountains and caves'. Although this mountain only intercepted a touch of Flat Cliff, solemn and boundless, but after all, it was a complete miniature. Even if it was designed by ZHANG Nanyuan, it must be a small experiment in his youth and exploration period. Two hundred years later, in the middle of the Qing Dynasty, Ge Yuliang's Huanxiu Shanzhuang's Taihu Stone Mountain also poured in from the side wall, piled up into a cliff near the pool. Although there are many peaks in the garden, it seems that it is worth saying by the scholar Hong Liangji that 'ZHANG Nanyuan and GE Dongguo, were the two best rockery mountain designers in the past three hundred years'.

The force of this mountain is leaning forward several feet to stick out Flat Cliff, and the deliberately low winding road near the water and tunnel under the cliff are actually a simple and concise classic style learned from the middle and late Ming to the early Qing. It wasn't plied up a peak, resembling the 'high-tech school' such as Shanghai Yu Garden of the late Ming Dynasty, and Yangzhou Pianshi Shanfang in the early Qing Dynasty. It wasn't divided into peaks seeking completeness, but becoming slightly trivial like Huanxiu Shanzhuang. And it especially didn't put strange stones on the top of the mountain, shaking head and smiling like Suzhou Shizilin Garden of the late Yuan Dynasty and Suzhou Wufeng Garden of the middle and late Ming Dynasty. Its artistic logic seems to be closer to gentle slope near the water in the eastern part of Suzhou Ou Garden of the early Qing Dynasty, and it is similar to the soil mountain of Jiading Guyi Garden of the middle and late Ming Dynasty and the north and south mountains of Qiuxiapu Garden.

# 缩微的『截谿断谷』——罕贵的晚明武康石平崖遗存

此园掇山向来有出自明末清初造园大师张南垣（一五八七—?）之说，考虑到张氏早期即居住于此园东侧不远处的郡治大街北、西林寺塔下，时间、地点均能合榫，则传说固有其合理性。但张氏之『现实主义』造园手法主张对自然只截取而不微缩，强调以曲岸土坡为主的横向舒缓山水，憎恶咫尺之间过度浓缩的剧烈怪诞变化，所谓『一自南垣工累石，假山雪洞更谁看』。而此山虽仅截取平崖一抹，浑成凝重，苍茫横峙，但毕竟作了截然的微缩；恐即便出自张南垣，亦属其青春探索期的小试牛刀之作。二百年后的清中叶，戈裕良之环秀山庄湖石掇山亦自侧墙涌入，临潭顿挫为峭壁，虽分峰偏多，似无愧大学者洪亮吉所云『张南垣与戈东郭，三百年来两轶群』。

此山倾千钧之力探身为平崖数丈，其崖下作刻意低伏的凌水盘道与隧函一曲，实是明中晚期至清初习见的、简约凝练的经典样式，而决不起峰似『高技派』之晚明上海豫园、清前期扬州片石山房，更不分峰求全而微微琐碎似环秀山庄，尤不于山头别立峰石、摇首诡笑如元末苏州狮子林、明中后期苏州五峰园——其艺术逻辑仿佛与清初苏州耦园东部之临池平冈更为接近，而与明中后期之嘉定古漪园之浑然土山、秋霞圃之南北两山暗通款曲。

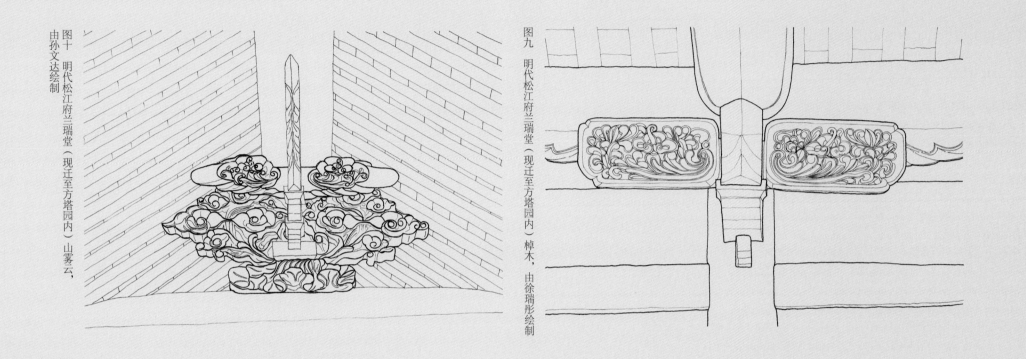

图十　明代松江府兰瑞堂（现迁至方塔园内）山雾云，由孙文达绘制

图九　明代松江府兰瑞堂（现迁至方塔园内）棹木，由徐瑞彤绘制

Fig.9　Zhaomu of the Ming Dynasty Lanruitang (now moved to Fangta Garden) in Songjiang, drawn by XU Ruitong
Fig.10　Shanwuyun of the Ming Dynasty Lanruitang (now moved to Fangta Garden) in Songjiang, drawn by SUN Wenda

In the existing Ming style Wukang Stone mountains, the north mountain of Yu Garden has a large scale but almost all details are lost. The cliffs in the east of Ou Garden are excellent, but the main peak on the west side may be added by later generations. The cliffs and valleys of Wuxi Jichang Garden, Changzhou Jin Garden, were scribbling stone gaps and even adding new stones in the process of repairing. Only in Qiuxiapu Garden of Jiading, the soil and rocks in the north scope are scattered, and the original appearance is much preserved.

Among the remote peaks, the sharp corner of the cliff is more solemn and picturesque.

现存明风武康石山中，豫园北山徒具规模而细节几乎全失；耦园东部所存崖壁极佳，但西侧主峰恐系后世添加；无锡寄畅园、常州近园之崖壁、涧谷则于修缮中滥勾石缝，甚至滥添新石，惟嘉定秋霞圃北冈土石横陈，散落成矶，保存原貌较多。

群峰邈远间，鲜明的孤崖一角，更凝重如画。

图十二 上海松江颐园武康石掇山与濠濮折桥

图十一 上海松江颐园武康石掇山与对岸琴台

图十三 上海松江颐园武康石掇山

Fig.11  Wukang Stone rockery and Qintai across the river in Songjiang Yi Garden
Fig.12  Wukang Stone rockery and Haopu Zheqiao in Songjiang Yi Garden
Fig.13  Wukang Stone rockery in Songjiang Yi Garden

## Dramatic Watery Garden and Escaped Mountain Courtyard

There are two independent courtyards in the south and north-west corners of the garden, forming different spatial levels and landscape themes.

The south courtyard is actually a continuation of the residential space, retaining a pure viewing space of the north-south axis and the sense of ritual, and has a certain visual function.

The second-storey double eave gable and hip roof opera stage 'Guanjialou' is located in the middle. With its concise 'Gongshi' beams, teapot shape ceiling, begonia shape columns, square and begonia and shoe shape pillars, cornice and roof ridge's 'Shengqi', and the gentle eaves, it maintains the overall exquisite beauty and moderate elegance. Moreover, it escapes the second-storey corridor to the south-west corner to accommodate stairs and performance preparation space, maintaining the horizontal resonance with the cornice line of the whole garden. Climbing up the building and looking southwards, you can see the egrets in the paddy fields and the smoke from the villages, to get feelings for 'village land' gardens. If the long windows of the north corridor are removed, it can directly face the building on the north side and become an opera stage.

The flat courtyard downstairs was originally a courtyard with a 'water meaning' that maintains sufficient purity of the landscape, in order to capture the beauty of 'aquatic plants staggered' that was described in the article 'Night Tour in Chengtian Temple'.

# 戏剧化的水意平庭与逸出的山庄小院

此园南部与西北角尚有两个独立院落，形成不同空间层次与景象主题。

其中南院实为宅居空间的延续，系保留着南北轴线与仪式感的纯净观演空间，而兼具一定景象功能。

二层重檐歇山顶戏台『观稼楼』居于正中，以内部简洁的贡式梁架、茶壶档轩、海棠劲柱、方形海棠纹靴形碩，以及檐、脊部俏丽的生起、平缓的出檐，保持着整体的精巧秀挺与适度的典丽；并向西南角逸出二层楼廊，以安置楼梯与表演准备空间，保持着与全园檐口线条的横逸共振。登楼南眺，可借景水田白鹭、稻浪炊烟，得『村庄地』造园之妙；而卸去北廊长窗，便可直面北侧看楼，而成盈盈戏台。

楼下平庭最初或为一保持足够景象纯度的『水意』平庭，以收『承天寺夜游』般的『藻荇交横』之妙。

图十六 上海松江颐园如意头洞门

图十四 颐园观稼楼几乎是仅存的、经碳十四精确测定的明代江南园林戏楼原构，只惜在修缮后显得较为生硬 图十五 观稼楼西北与住宅宛转相延，楼下的贝叶窗亦觉古意盎然

Fig.14 Guanjialou in Yi Garden was almost the only surviving example of a Jiangnan garden theatre building of the Ming dynasty that has been accurately determined by carbon 14, but unfortunately it has been restored properly

Fig.15 The north-west part of Guanjialou extends to the mansion, and window with the shape of shell downstairs has a sense of antiquity

Fig.16 The gate with Ruyi shape in Shanghai Songjiang Yi Garden

This building miraculously preserves the appearance of the theater space in Jiangnan area's residences during the peak period of Kunqu Opera that every family sang the popular operas in the late Ming and early Qing Dynasties, and the usual way of combining it with the landscape space of the garden. There is still a window of pattra leaves shape in the corner of the courtyard, which is like flying immortals outside the sky.

The study courtyard on the north-west side is the only escape space in the whole garden, because the south-east of the partition wall is Flat Cliff with a dark tunnel. On the east side is the flat lake, with a slight water outlet. It is guessed that it should be a mountain courtyard similar to Suzhou Yipu Garden's 'Yugou' courtyard, Yangzhou Xiaopangu Garden's 'Yunchao' courtyard, and Suzhou Liu Garden's 'Yuancuige' courtyard (now the courtyard wall was demolished). Unfortunately, the expression of its mountain meaning is insufficient. In recent years, the courtyard was decorated with Taihu Stones instead of Wukang Stones, which seemingly had weakened its links with the external Wukang Stone mountain. It also failed to embellish the spring veins and connect with the outer lake, so its current situation is relatively isolated.

此楼奇迹般地存留着晚明清初「家家『收拾起』，户户『不提防』」的昆曲巅峰时段江南豪宅戏场空间的面貌，及其与园林景象空间的惯常结合方式。院落一角尚存贝叶窗一抹，如天外飞仙，令人啧啧。

全园西北侧书房院落实为全园唯一逸出空间，因其隔墙东南为平崖，暗通隧函，东侧为平湖，微留水口，揣测原意，应为与苏州艺圃「浴鸥」、扬州小盘谷「云巢」、苏州留园「远翠阁」（今院墙被拆）类似的山庄院落。惜其山意表达不足，近年修缮时院中以湖石而非武康石点缀似削弱了其与外部武康石山林的联系。亦惜其未于院中稍缀泉脉，以暗通外湖，故其现状较为孤立。

图十七 水次仓关帝庙大殿，明天启二年（一六二二年）

图十八 建于明末的松江水次仓北的永丰桥（大仓桥），舒缓而壮美，秀映江南

图十九 西林寺塔（明代砖质塔身，腰檐平坐为近年修复，张南垣即居于西林塔下）

Fig.17 The main hall (1622) of Shuicicang Guandi Temple

Fig.18 Yongfeng Bridge (Daicang Bridge), built in the late Ming Dynasty in the north of Songjiang Shuicicang, which is entle and magnificent, famous in Jiangnan area

Fig.19 Xilin Temple Pagoda (The brick tower body of the Ming Dynasty, with a flat waist and eaves, was restored in recent years. ZHANG Nanyuan was lived under the pagoda.)

# The Corridor Connecting the Residences and the Garden: XU Zhai and ZHAO Zhai

Inferred from the current situation, circulation and surrounding neighborhoods of Yi Garden, its old house in the late Ming Dynasty should be located on its east and north sides, and its original central axis to a large extent continued to XU Wei's residence in the late Qing Dynasty.

The original XU Wei's residence had several axes, and on each axis there were seven or eight houses. Yi Garden was its courtyard on its west axis, separated from the middle axis by an alley. Today there are still a 'Bianzuo' hall and several buildings on the middle axis. The age of the main hall seems to be at least early than the mid-Qianlong period of the Qing Dynasty, and it is probably the original structure of the late Ming Dynasty. The back hall should be connected to Kanlou in Yi Garden by the corridor in the old days, and it is very likely that the original interaction between the garden and the house still remain. Qintai in the garden seems to have been closely occluded with the residential buildings. The newly built cloud wall has actually slightly destroyed the interweaving and interpenetrating relationship between the residences and the garden.

It is suspected that there is still a small Zoumalou on the east axis outside the residence of the XU family in the old days. The south building is a typical Ming style building, and there is still Baofu Caihui on the ridged purlin.

The ZHAO family's residence still exists on the east side of the XU family's, and there are still a 'Bianzuo' hall and a back building on the middle axis. The beam frame of the hall is magnificent, and it should be date back to the early Qing Dynasty.

## 宅园间的楼廊勾连：许宅与赵宅

由颐园空间现状、流线脉络及周边街区地脉推断，其明末旧宅应即在其东、北两侧，其最初的中轴面貌在很大程度上一直延续至晚清的许威宅中轴。

许威宅旧日数路并行，进深达七八进之多，颐园即其西路之庭园，与中路间仅以一避弄相隔。今其中路仍存扁作大厅及数进楼厅，其中大厅年代似不晚于清乾隆中期，很可能即晚明原构；其后部楼厅旧日应可通过楼廊与颐园看楼相连，极可能存留着最初的园宅互动关系。而园中的琴台似也曾与住宅楼厅避弄紧密咬合，今日新建之云墙实稍稍破坏了宅园交织互渗的关系。

疑似旧日许宅外东路部位尚存留小型走马楼，其南楼更系典型明式建筑，脊桁尚存留包袱彩绘。

许宅东侧尚存赵宅，今仍存其正路扁作大厅及其后楼厅各一座，大厅梁架雄壮，年代应不晚于清前期。

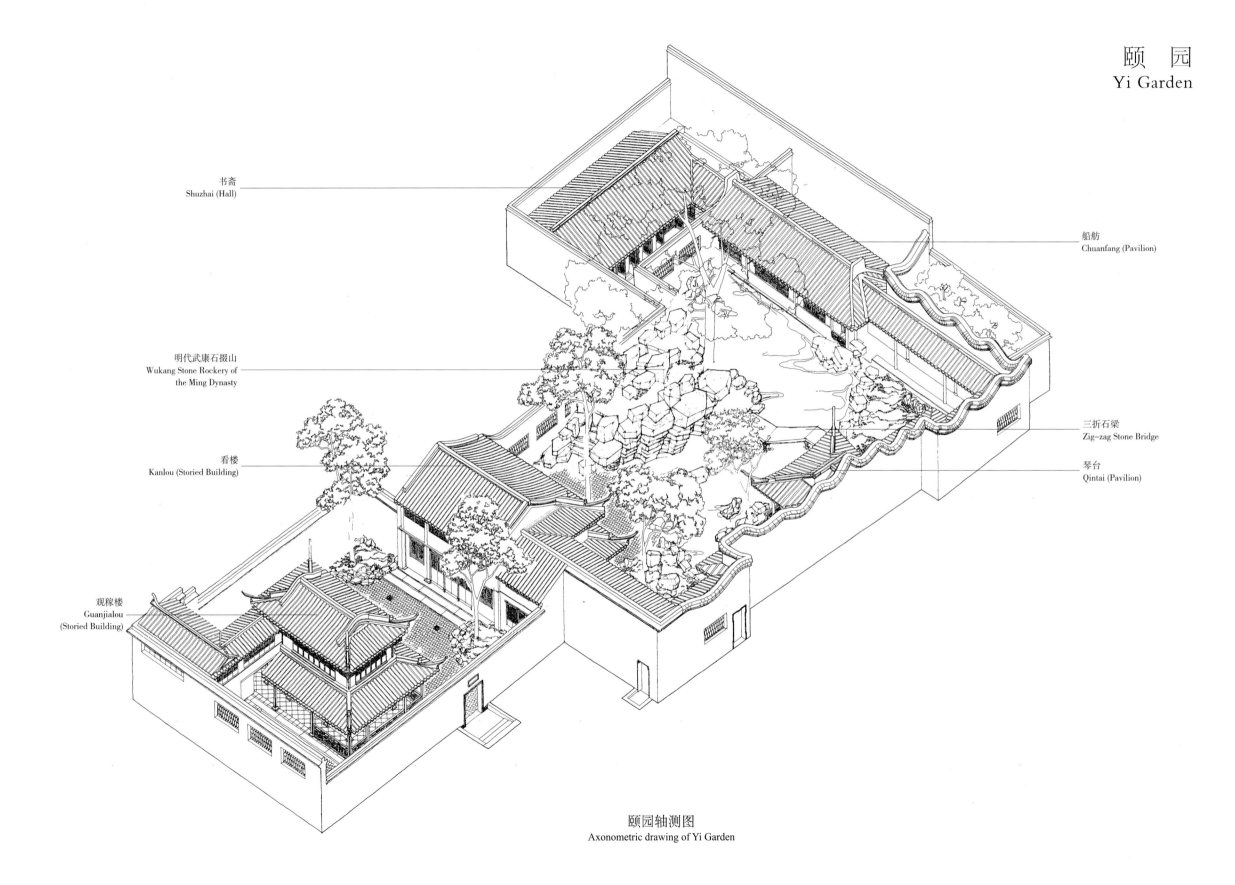

书斋
Shuzhai (Hall)

船舫
Chuanfang (Pavilion)

明代武康石掇山
Wukang Stone Rockery of
the Ming Dynasty

三折石梁
Zig-zag Stone Bridge

看楼
Kanlou (Storied Building)

琴台
Qintai (Pavilion)

观稼楼
Guanjialou
(Storied Building)

颐园轴测图
Axonometric drawing of Yi Garden

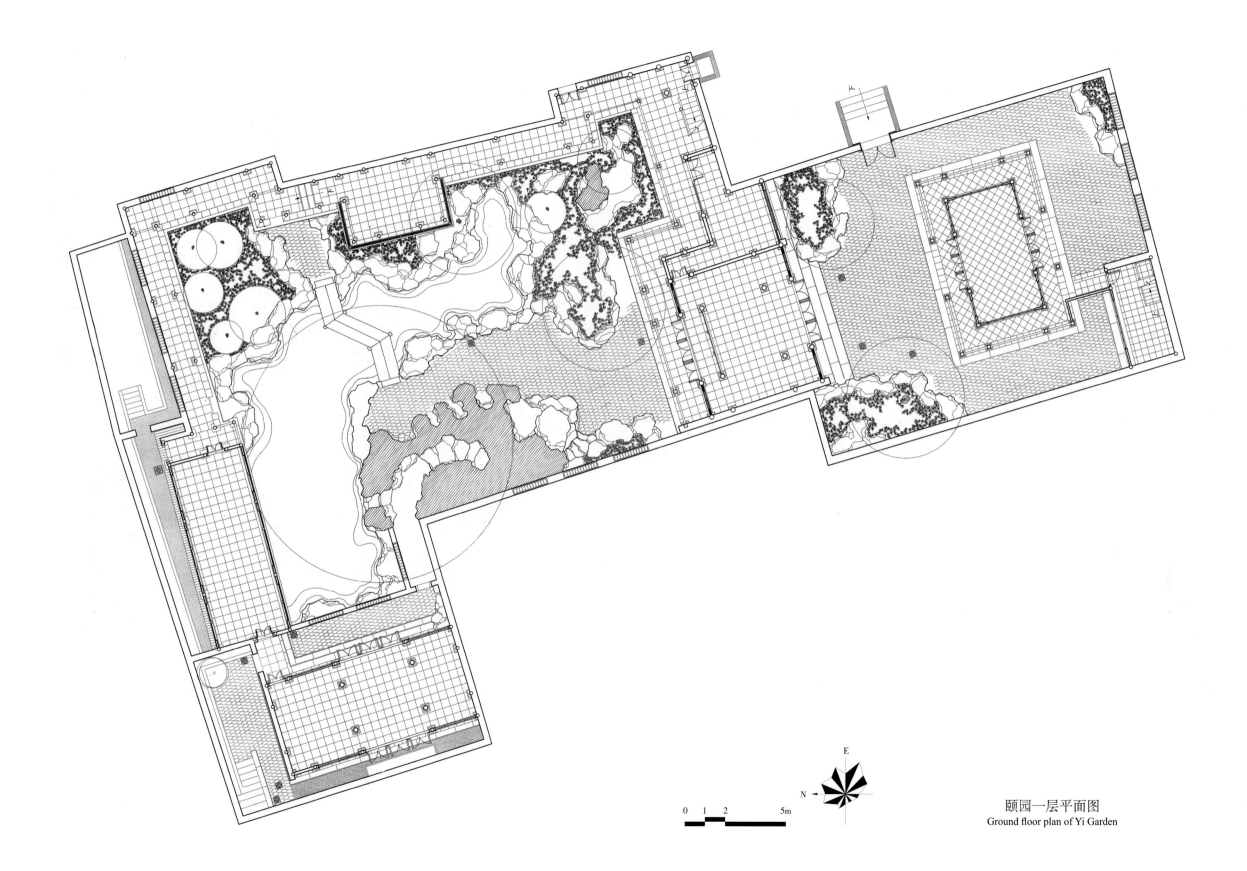

颐园一层平面图
Ground floor plan of Yi Garden

0 1 2 5m

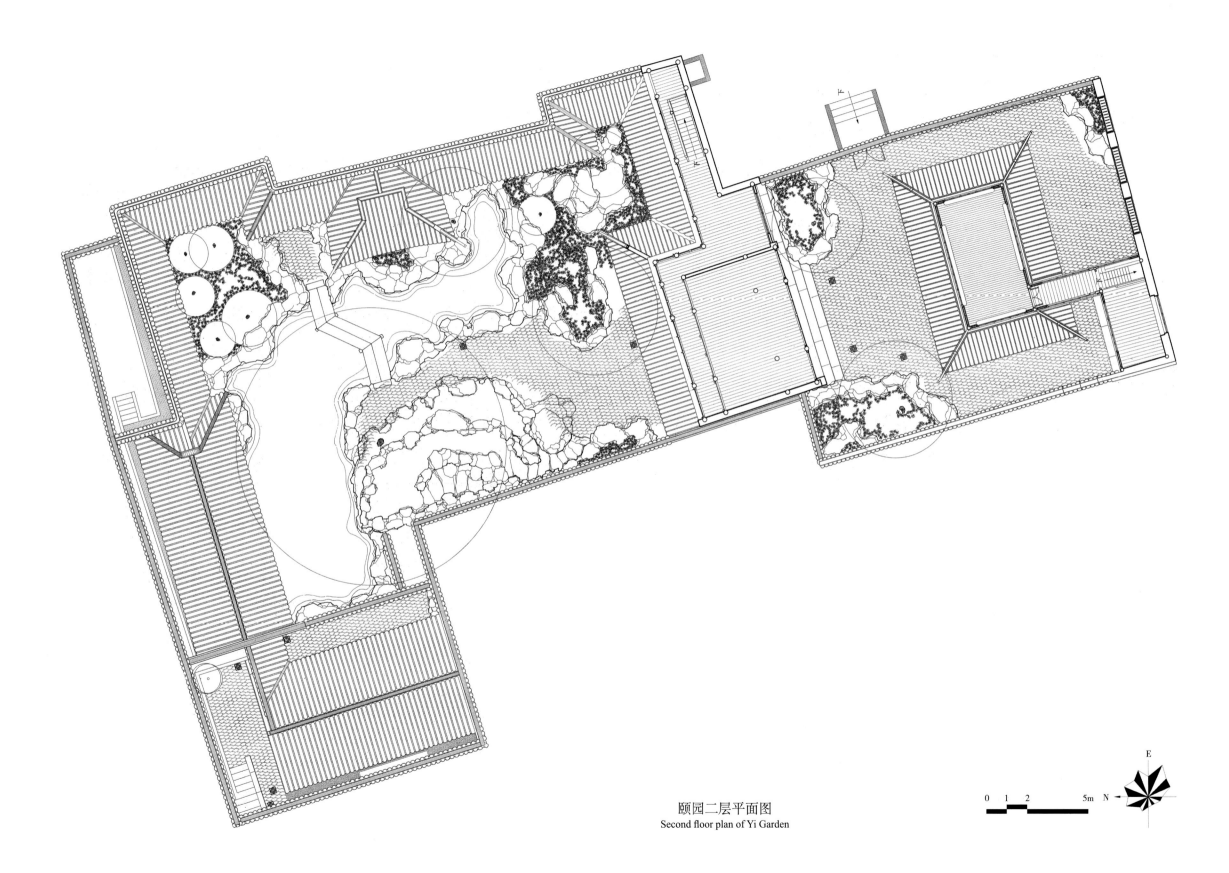

颐园二层平面图
Second floor plan of Yi Garden

0 1 2 5m N

E

N

0  1  2  5m

颐园屋顶平面图
Roof plan of Yi Garden

看楼
Kanlou (Storied Building)

观稼楼
Guanjialou (Storied Building)

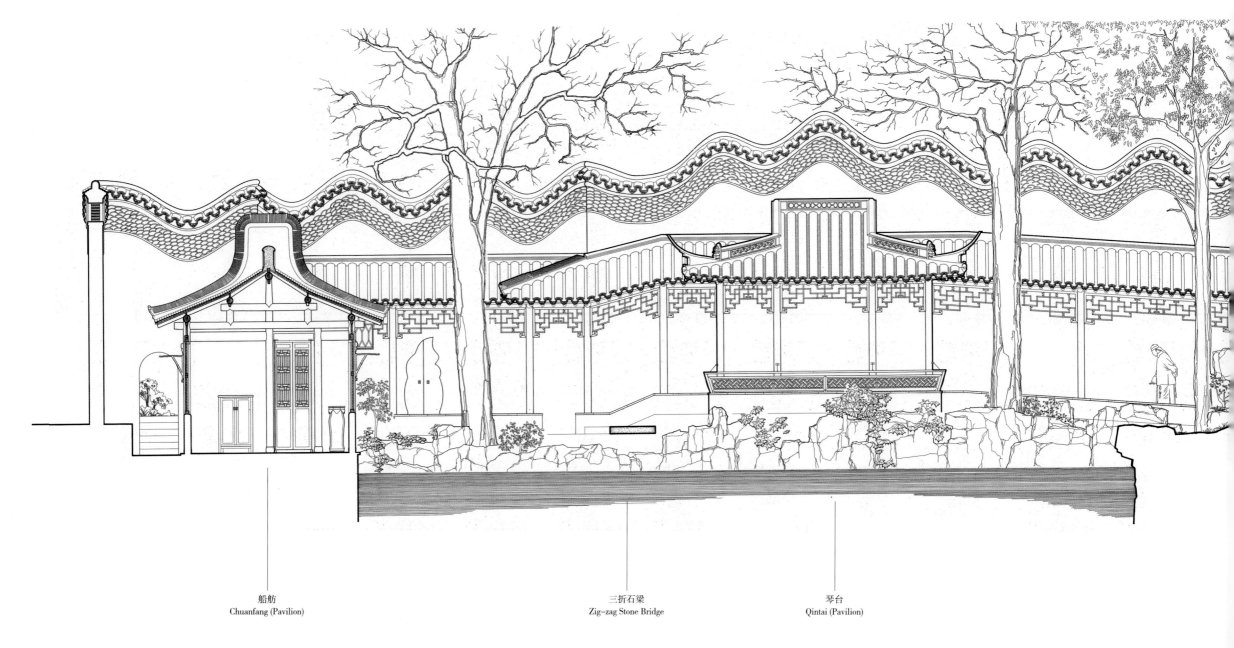

船舫
Chuanfang (Pavilion)

三折石梁
Zig-zag Stone Bridge

琴台
Qintai (Pavilion)

注：图中凡草架部分均属推测

0    0.5    1              2m

颐园南北剖面图（东岸）
South-north section of Yi Garden (the east bank)

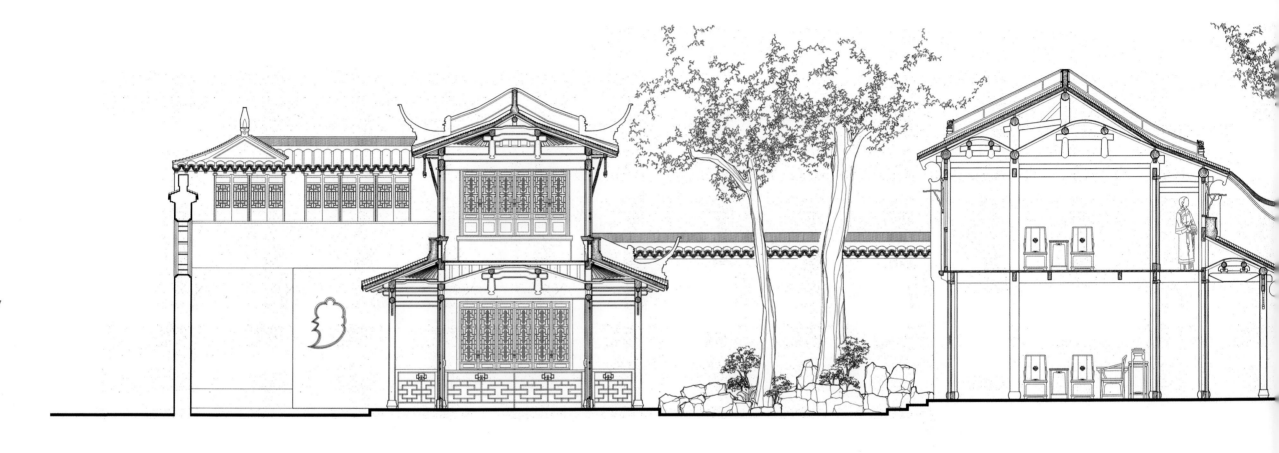

注：图中凡草架部分均属推测

0　0.5　1　　2m

颐园南北剖面图（西岸）
South-north section of Yi Garden (the west bank)

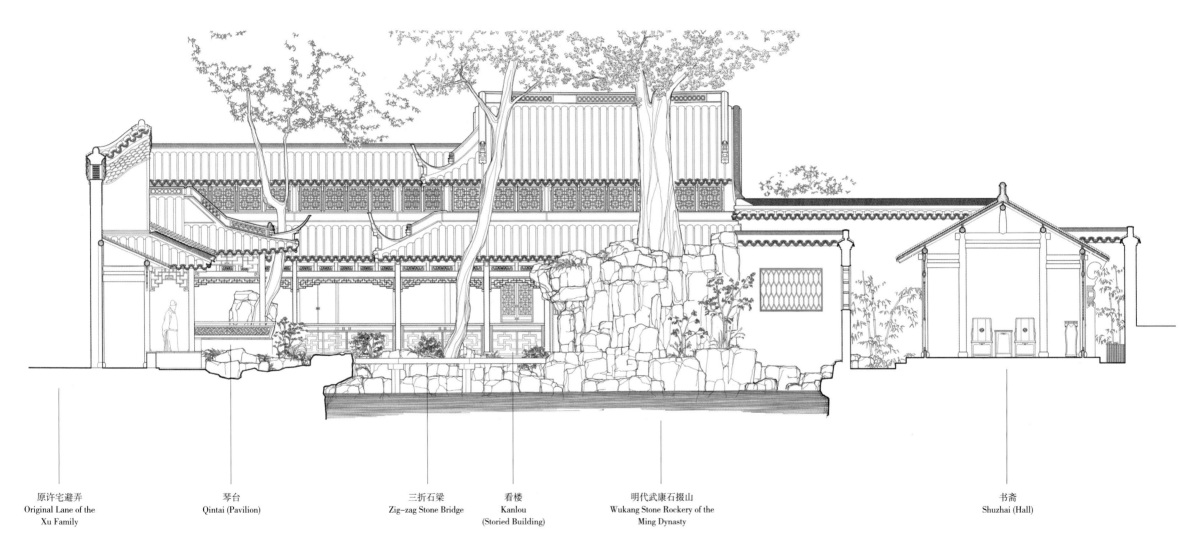

| 原许宅避弄 | 琴台 | 三折石梁 | 看楼 | 明代武康石掇山 | 书斋 |
|---|---|---|---|---|---|
| Original Lane of the Xu Family | Qintai (Pavilion) | Zig-zag Stone Bridge | Kanlou (Storied Building) | Wukang Stone Rockery of the Ming Dynasty | Shuzhai (Hall) |

注：图中凡草架部分均属推测

颐园东西剖面图（南岸）
East-west section of Yi Garden (the south bank)

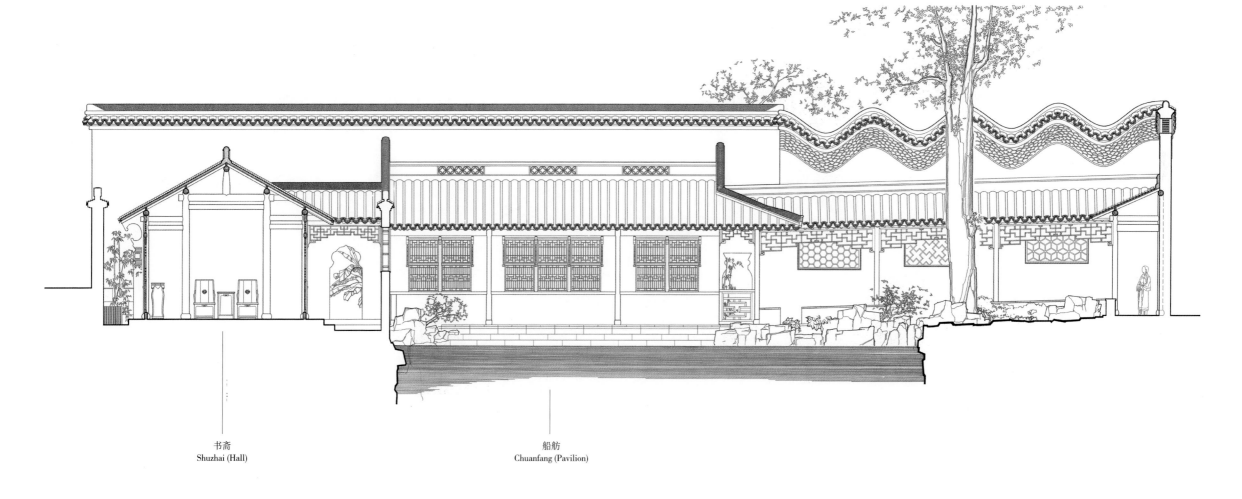

书斋
Shuzhai (Hall)

船舫
Chuanfang (Pavilion)

注：图中凡草架部分均属推测

0　0.5　1　　2m

颐园东西剖面图（北岸）
East-west section of Yi Garden (the north bank)

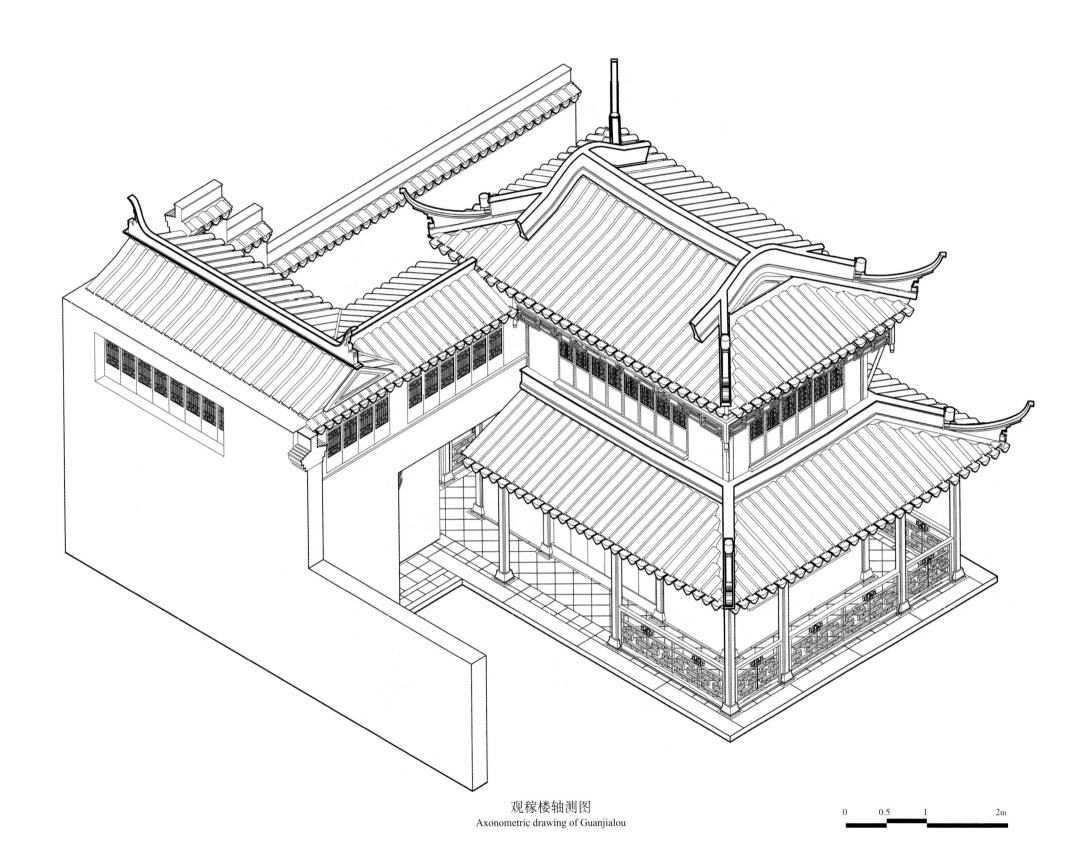

观稼楼轴测图

Axonometric drawing of Guanjialou

0    0.5    1    2m

注：观稼楼经修缮、其地坪连同院落地坪已有所抬高、院落铺
地现状为海棠芝花式

N

E

观稼楼一层平面图
Ground floor plan of Guanjialou

0　1　2　　4m

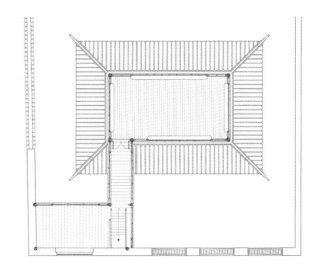

观稼楼二层平面图
Second floor plan of Guanjialou

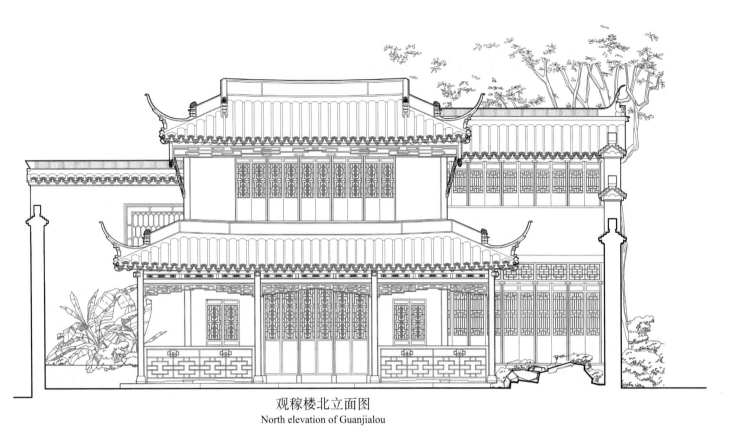

观稼楼北立面图

North elevation of Guanjialou

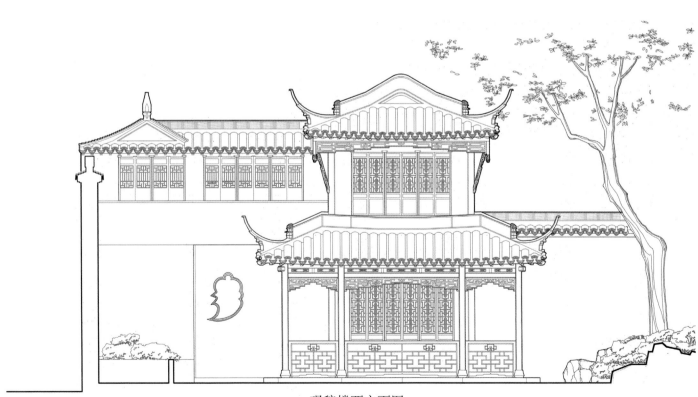

观稼楼西立面图

East-west section of Yi Garden (the south bank)

0  0.5  1  2m

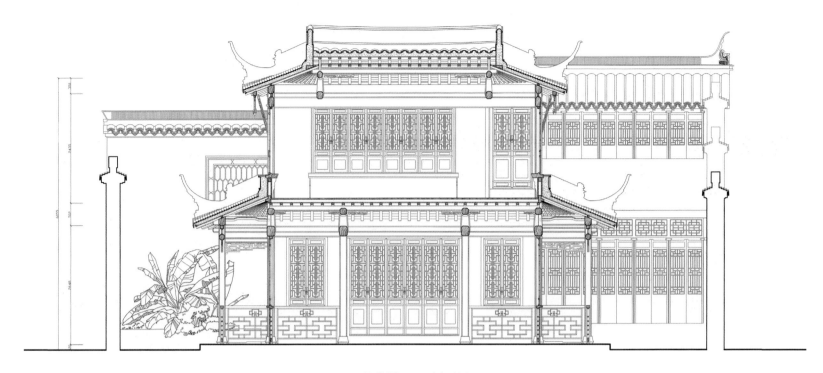

观稼楼 2−2 剖面图
Section 2-2 of Guanjialou

0  0.5  1       2m

观稼楼 1−1 剖面图
Section 1-1 of Guanjialou

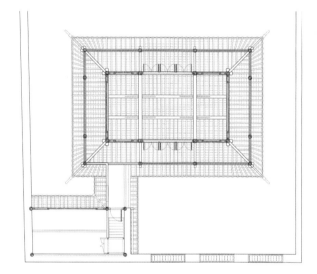

观稼楼一层屋架仰视图
Bottom view of Guanjialou ground floor's beams

0　1　2　　4m

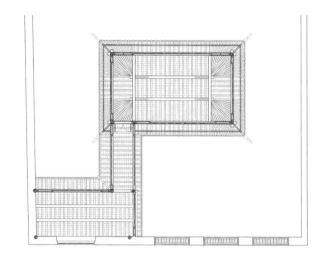

观稼楼二层屋架仰视图
Bottom view of Guanjialou second floor's beams

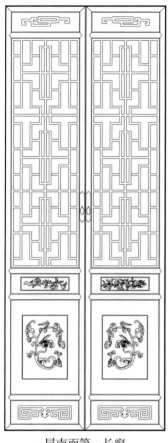

一层南面第一长窗
First Door on the south side of the first floor

二层长窗
Door on the second floor

二层西面第二半窗
Second Window on the west side of the
second floor

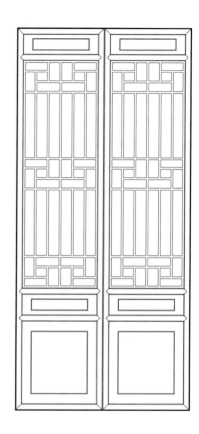

一层楼梯间长窗
Door of the staircase of the first floor

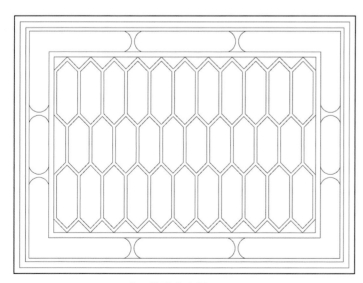

南面外墙花窗洞
Window on the south wall

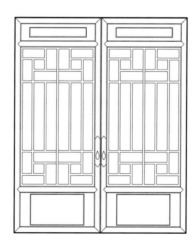

二层楼梯间半窗
Window of the staircase of the second floor

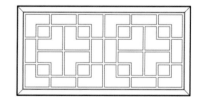

一层楼梯间北面横风窗
Window on the north side of the staircase of the first floor

观稼楼大样图
Details of Guanjialou

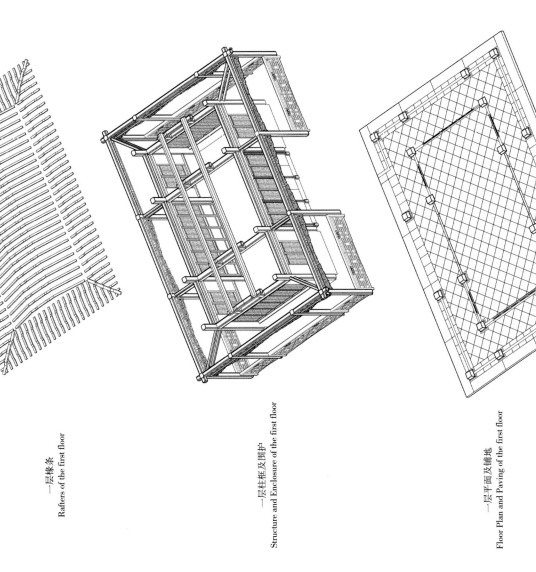

一层椽条
Rafters of the first floor

一层柱框及围护
Structure and Enclosure of the first floor

一层平面及铺地
Floor Plan and Paving of the first floor

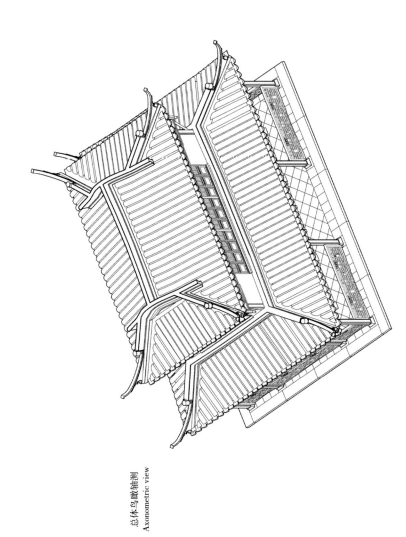

总体鸟瞰轴测
Axonometric view

观稼楼轴测解析图
Axonometric and exploded drawing of Guanjialou

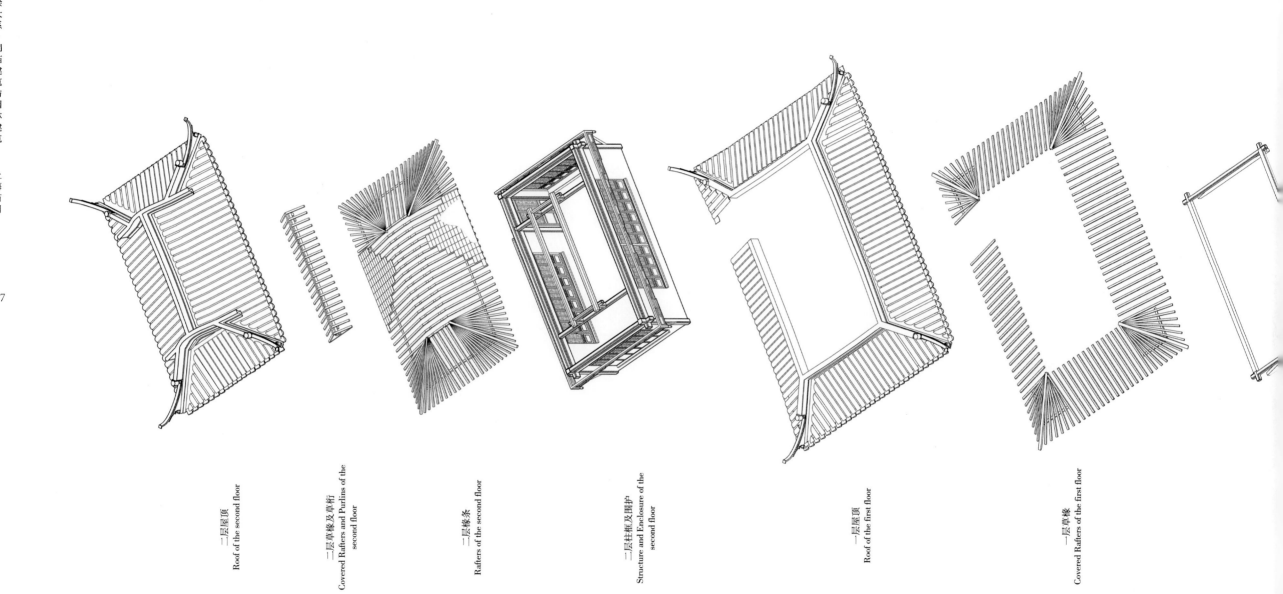

二层屋顶
Roof of the second floor

二层草椽及草望栍
Covered Rafters and Purlins of the second floor

二层椽条
Rafters of the second floor

二层柱框及围护
Structure and Enclosure of the second floor

一层屋顶
Roof of the first floor

一层草椽
Covered Rafters of the first floor

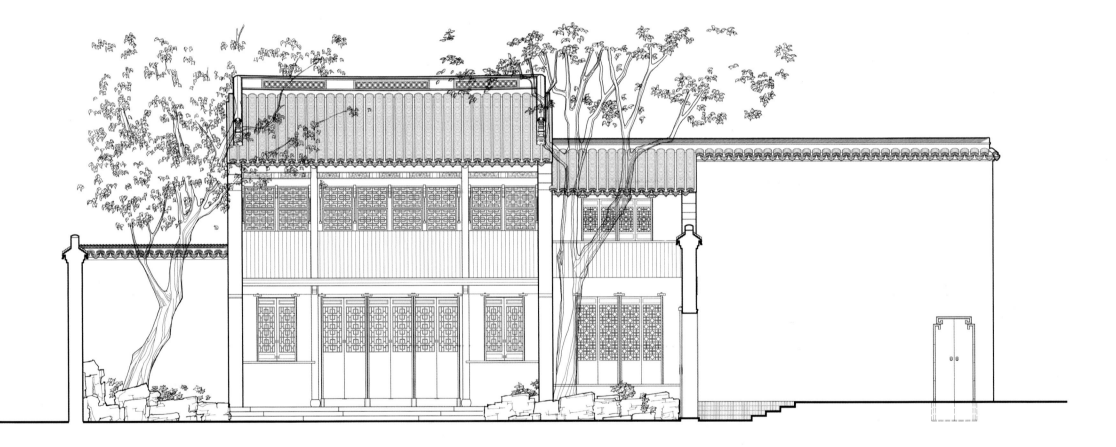

看楼南立面图
South elevation of Kanlou

0  0.5  1  2m

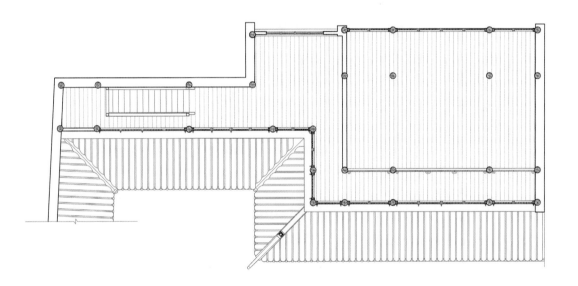

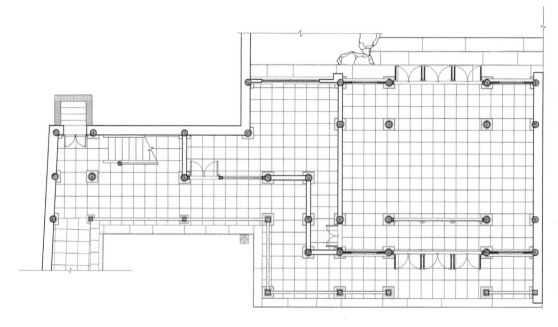

注：看楼北侧铺地现状为冰纹梅花式

看楼二层平面图
Second floor plan of Kanlou

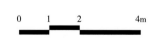

看楼一层平面图
Ground floor plan of Kanlou

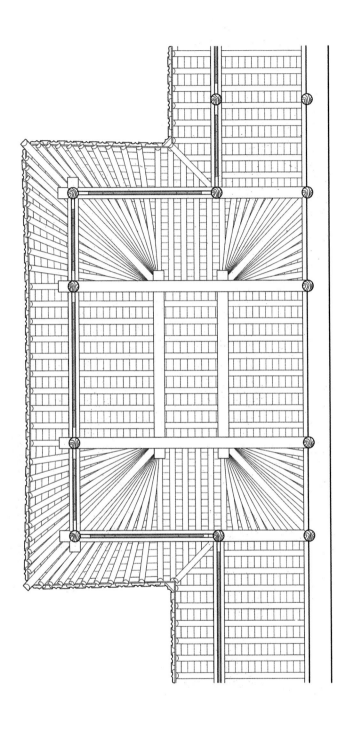

琴台屋架仰视图
Bottom view of Qintai's beams

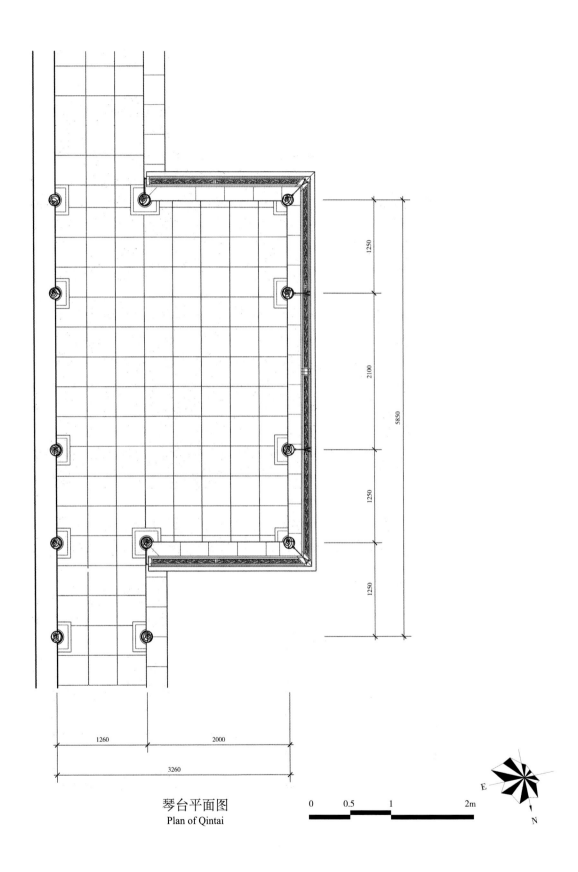

琴台平面图
Plan of Qintai

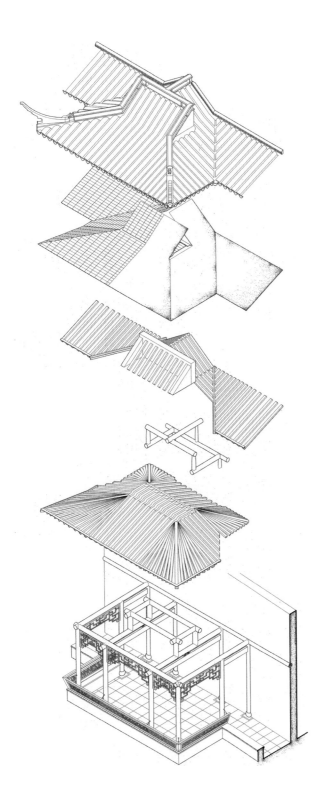

琴台分解轴测图
Axonometric and exploded drawing of Qintai

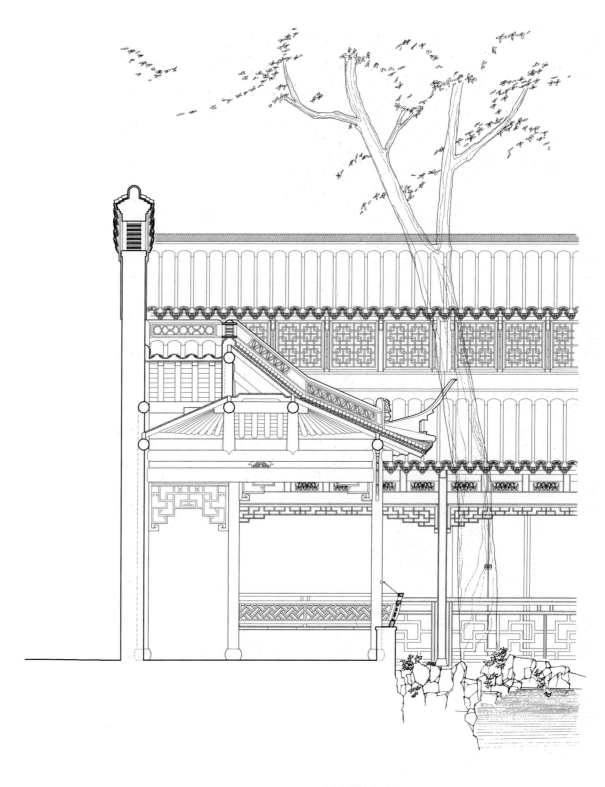

| 0 | 0.5 | 1 | | 2m |

琴台横剖面图
Cross-section of Qintai

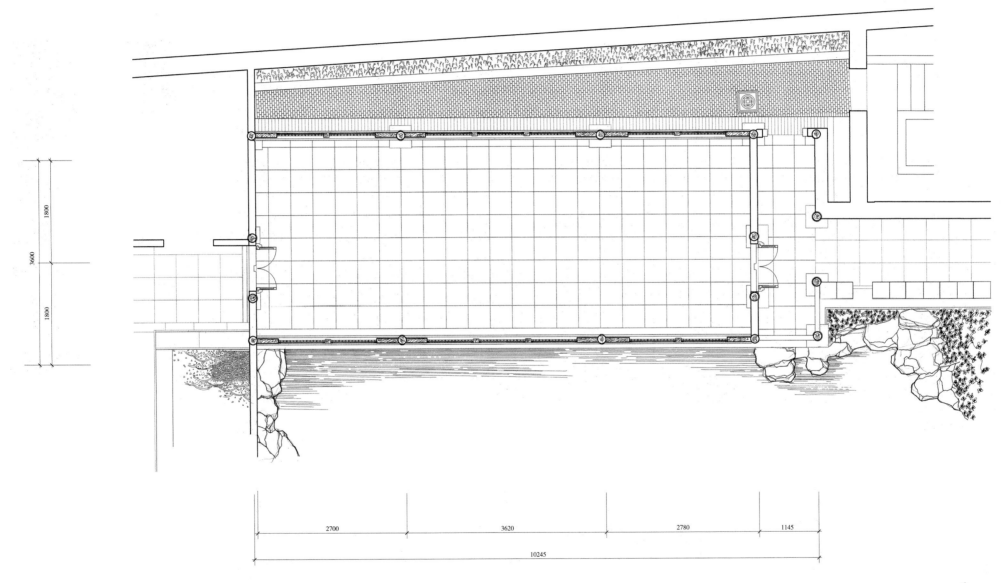

1800

3600

1800

2700　　　3620　　　2780　　1145

10245

船舫平面图
Plan of Chuanfang

0　0.5　1　　2m

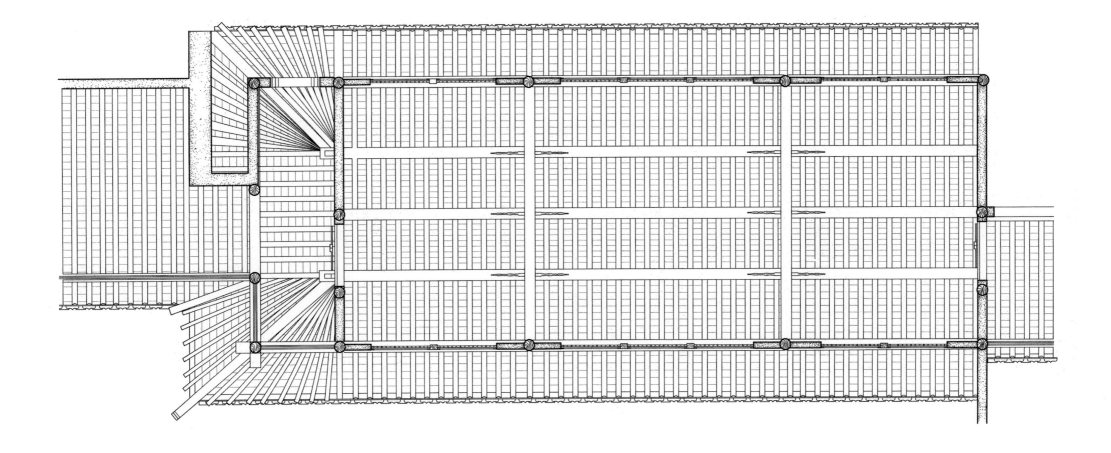

0　0.5　1　2m

船舫屋架仰视图
Bottom view of Chuanfang's beams

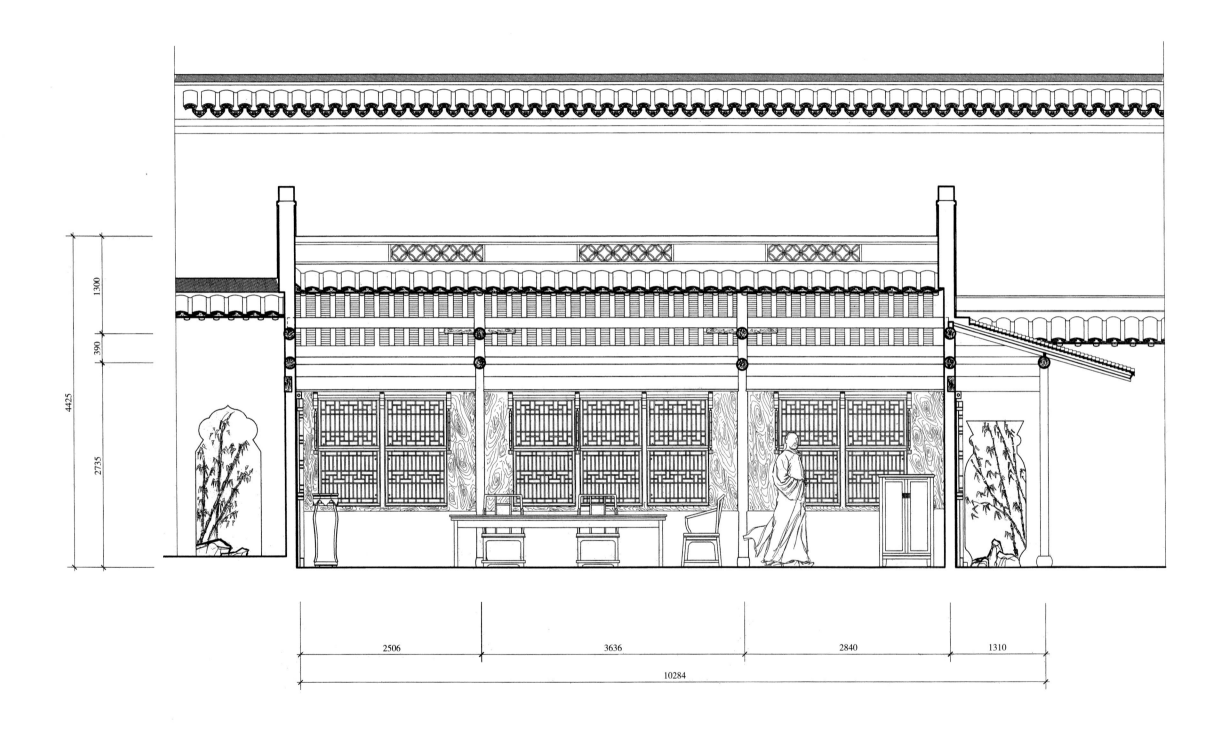

船舫横剖面图
Cross-section of Chuanfang

0 0.5 1 2m

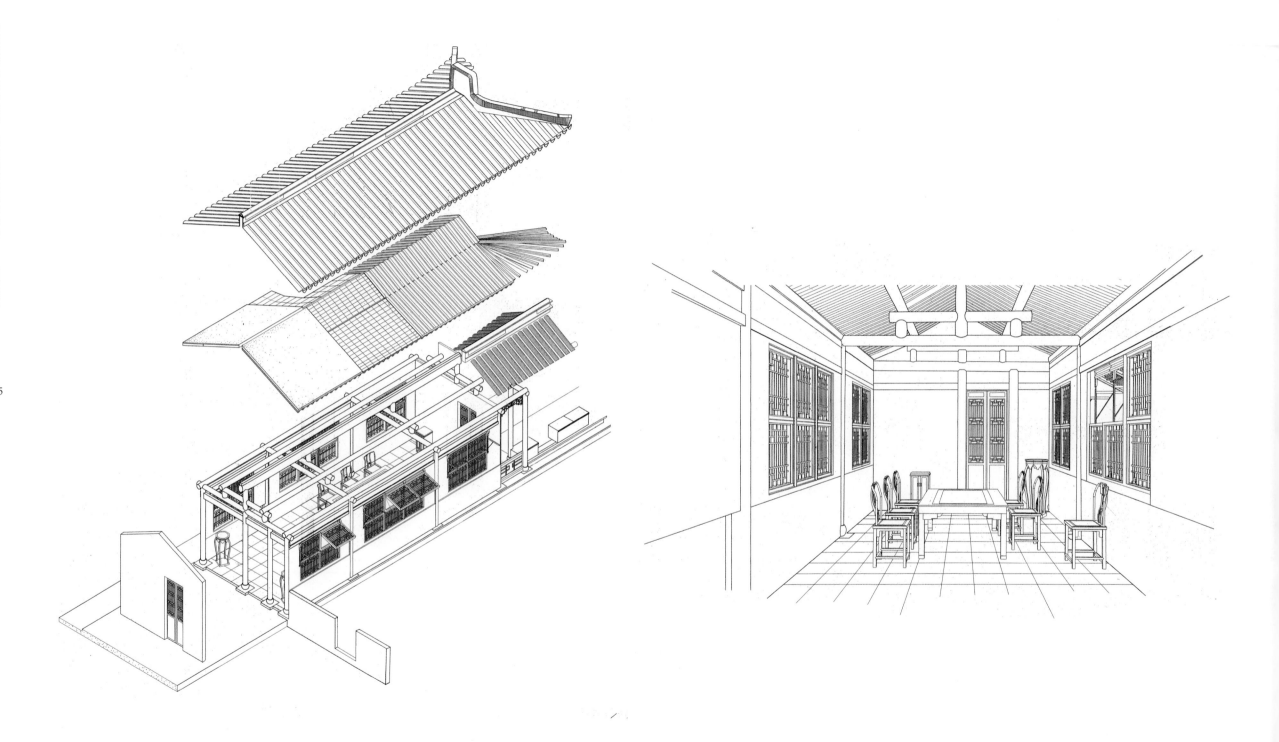

船舫分解轴测图
Axonometric and exploded drawing of Chuanfang

船舫室内透视图
Perspective view of Chuanfang's inner space

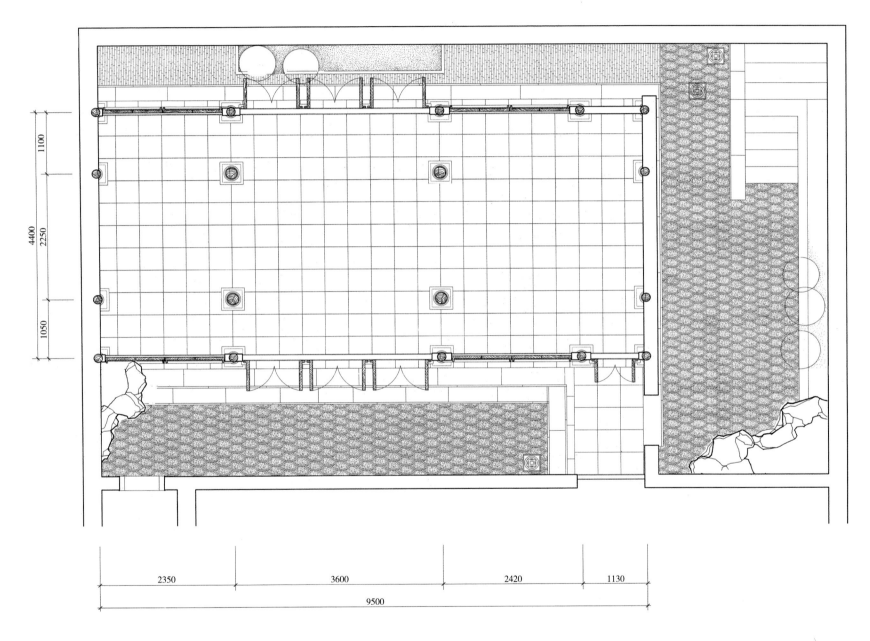

1100
2250
4400
1050

2350　3600　2420　1130

9500

书斋平面图
Plan of Shuzhai

0　0.5　1　2m

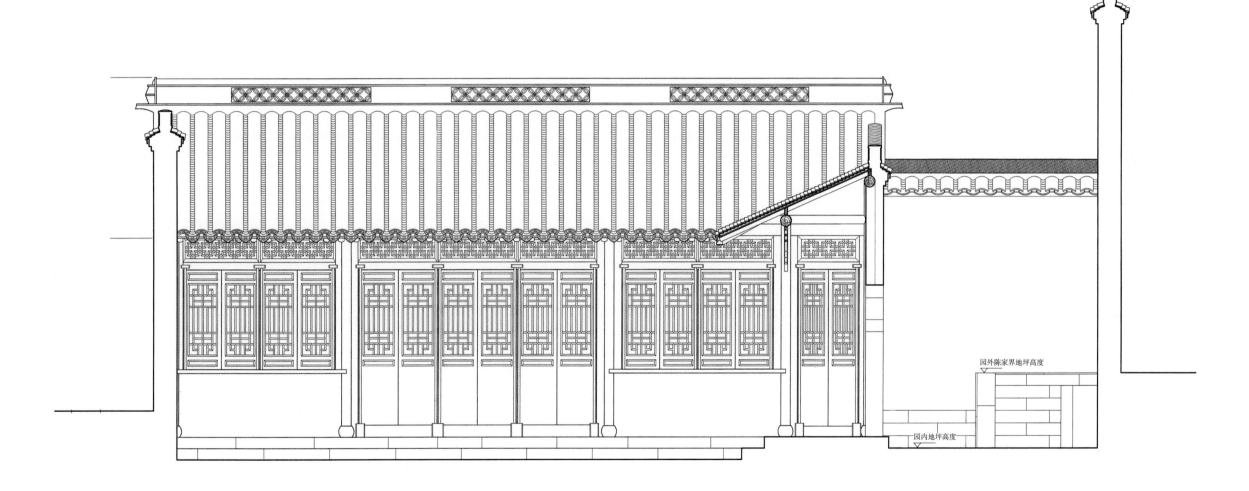

园外陈家界地坪高度

园内地坪高度

0　0.5　1　　2m

书斋东立面图
East elevation of Shuzhai

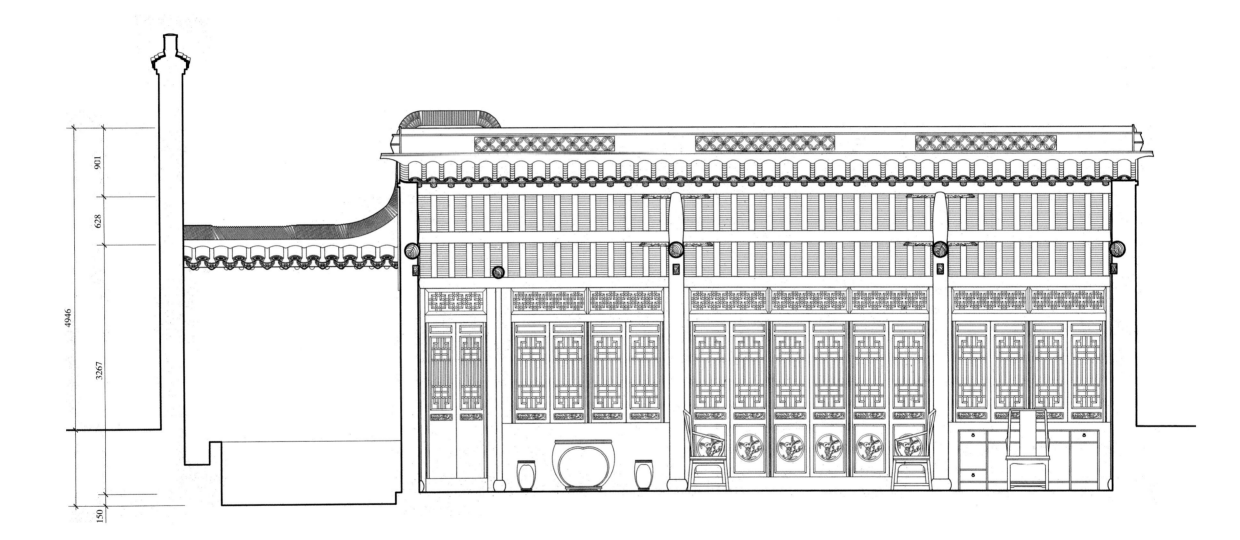

书斋纵剖面图
Longitudinal section of Shuzhai

0    0.5    1    2m

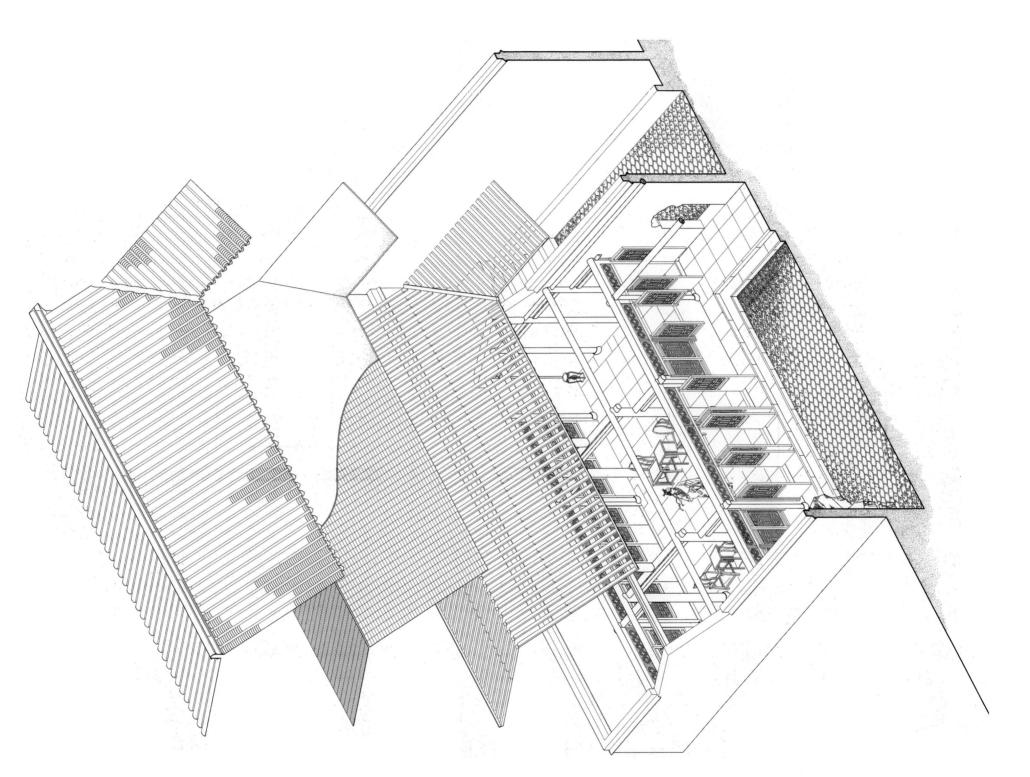

书斋分解轴测图

Axonometric and exploded drawing of Shuzhai

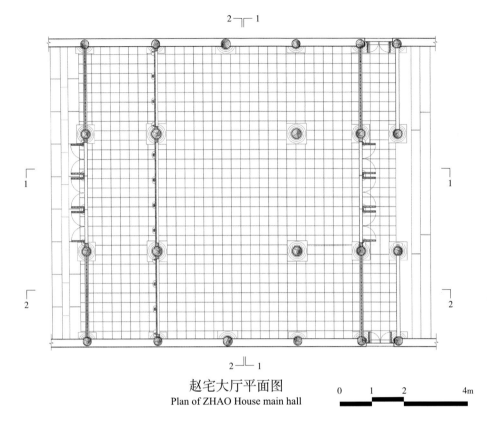

赵宅大厅平面图
Plan of ZHAO House main hall

0 1 2 4m

赵宅大厅梁架仰视图
Bottom view of ZHAO House main hall's beams

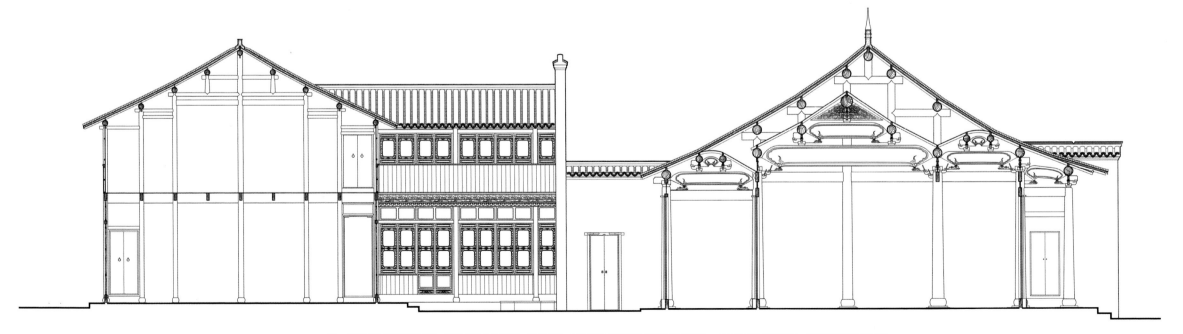

注：赵宅经过维护与整修后，现已作为仓库所用，所有门柏、窗柏以及挂落均为推测后的复原状，大厅前后阶条石仍存，其下台阶系推测复原

赵宅大厅及后楼中轴横剖面图
Cross-section of ZHAO House main hall and back building

0 0.6 1.2 3m

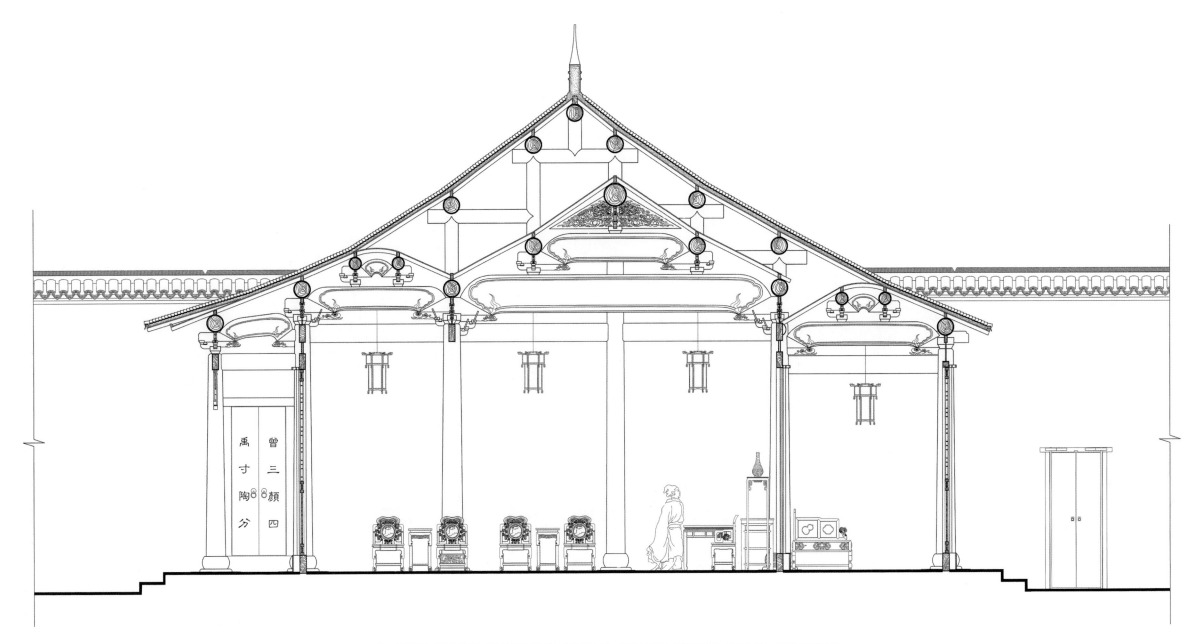

禹
寸
陶
分

曾
三
颜
四

注：草架部分系据山墙定点高度测量推测所得，因年代久远，整栋建筑发生较严重沉降，建筑前后台阶均为推测复原

0　0.5　1　　2m

赵宅大厅 1-1 横剖面图
Cross-section 1-1 of ZHAO House main hall

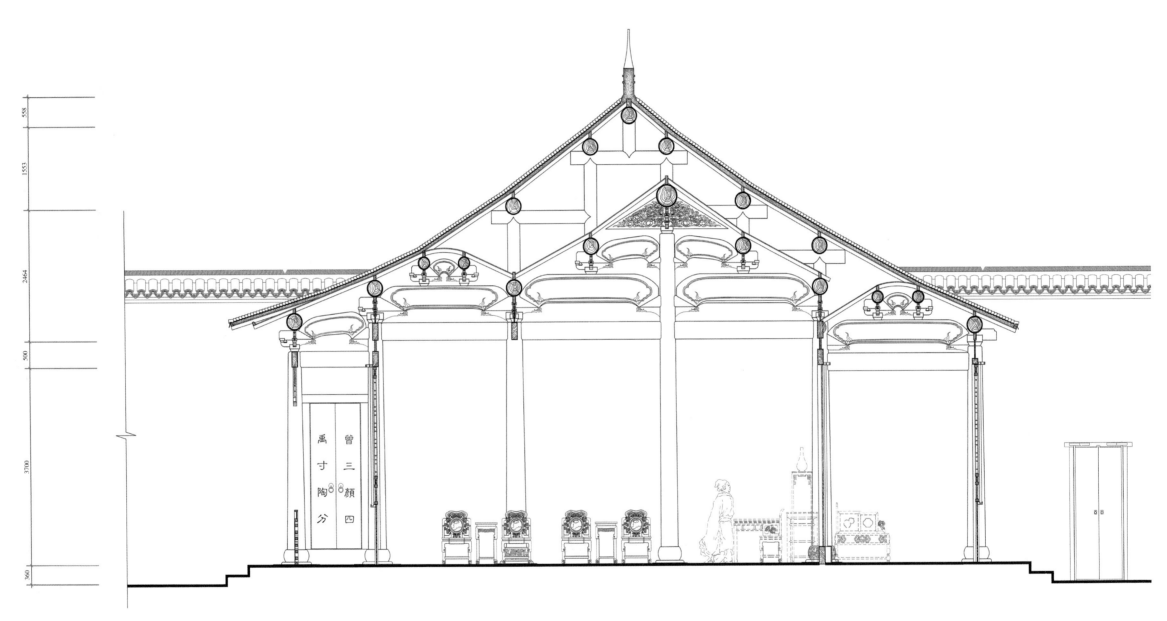

曾三颜四

禹寸陶分

注：草架部分系据山墙定点高度测量推测所得，因年代久远，整栋建筑发生较严重沉降，建筑前后台阶均为推测复原

**赵宅大厅 2-2 横剖面图**
Cross-section 2-2 of ZHAO House main hall

0    0.5    1    2m

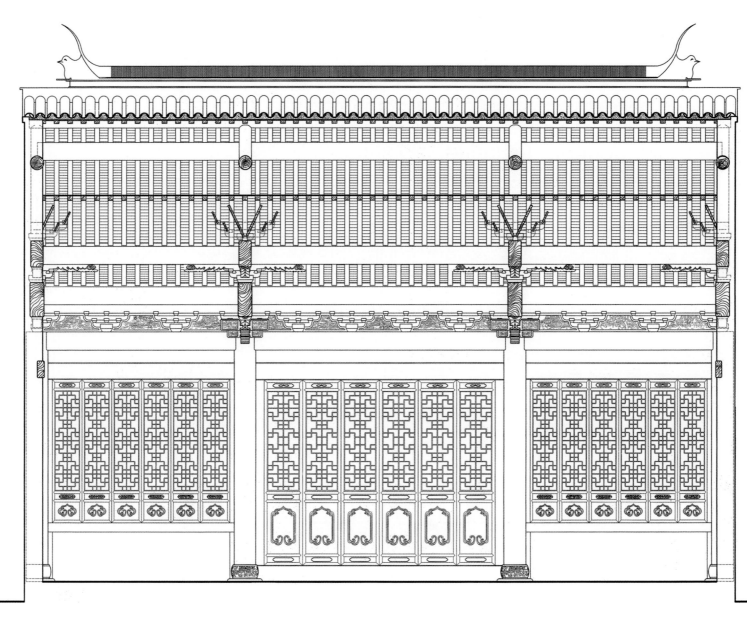

注：草架部分系据山墙定点高度测量推测所得，因年代久远，整栋建筑发生较严重沉降，建筑前后台阶均为推测复原

0    0.5    1        2m

赵宅大厅 1-1 纵剖面图
Longitudinal section 1-1 of ZHAO House main hall

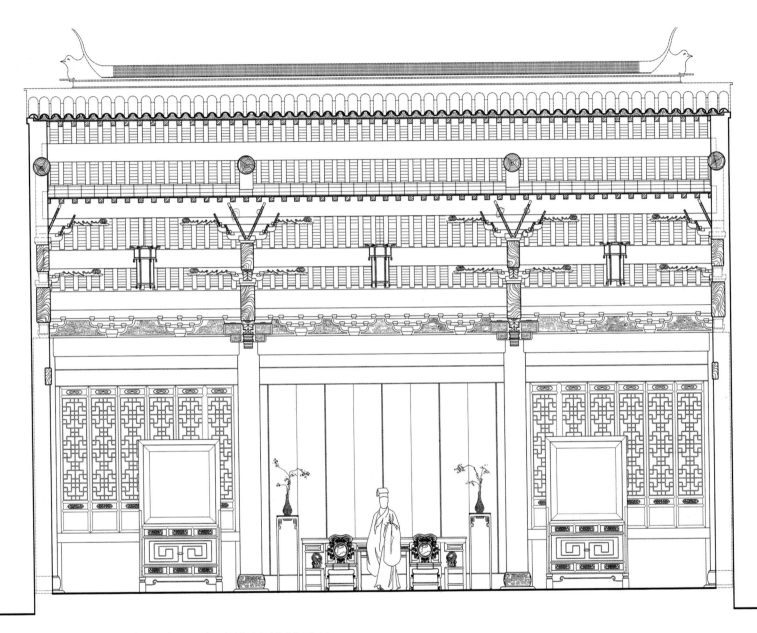

注：草架部分系据山墙定点高度测量推测所得，因年代久远，整栋建筑发生较严重沉降，建筑前后台阶均为推测复原

赵宅大厅 2-2 纵剖面图
Longitudinal section 2-2 of ZHAO House main hall

0    0.5    1         2m

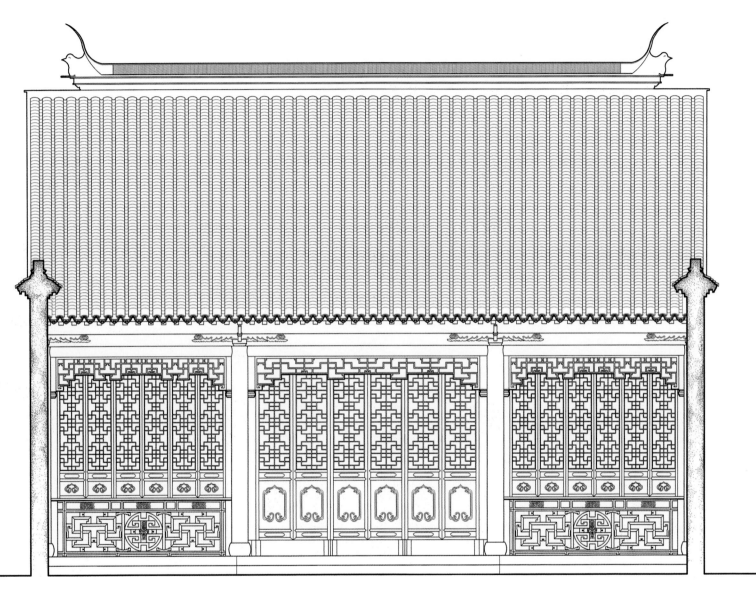

注：此厅栏杆尚存，半窗系据残件推测复原；长窗据半窗复原；挂落系推测所得

0    0.5    1         2m

赵宅大厅南立面图
South elevation of ZHAO House main hall

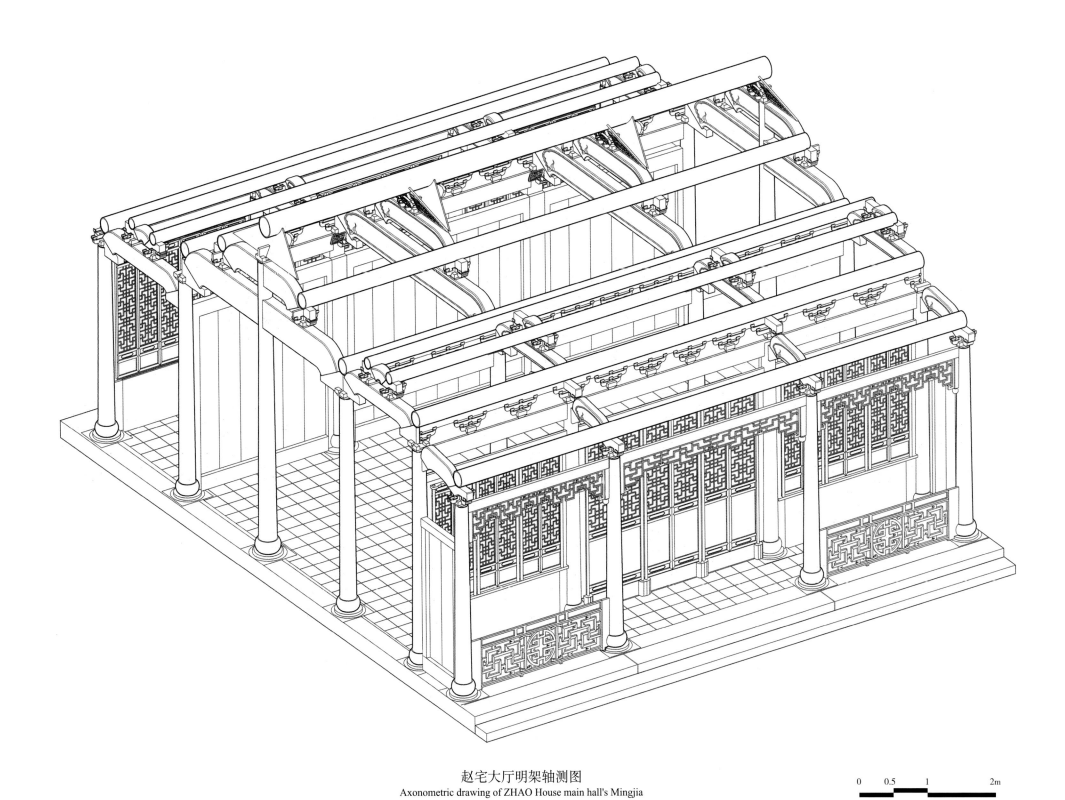

赵宅大厅明架轴测图
Axonometric drawing of ZHAO House main hall's Mingjia

0  0.5  1  2m

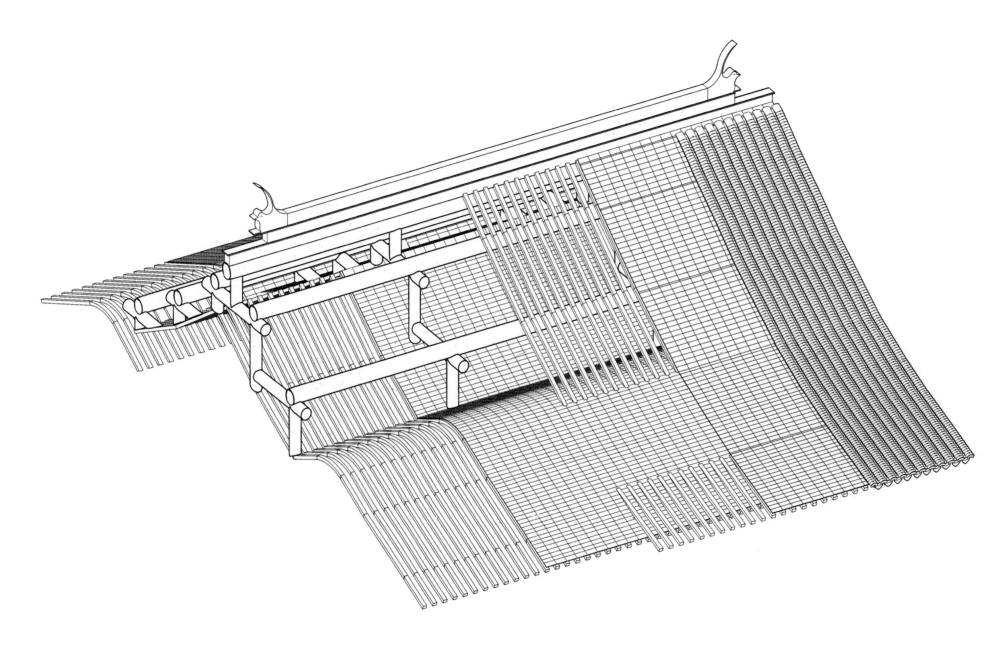

注：图中隐匿部分系推测所得

赵宅大厅草架轴测图
Axonometric drawing of ZHAO House main hall's Caojia

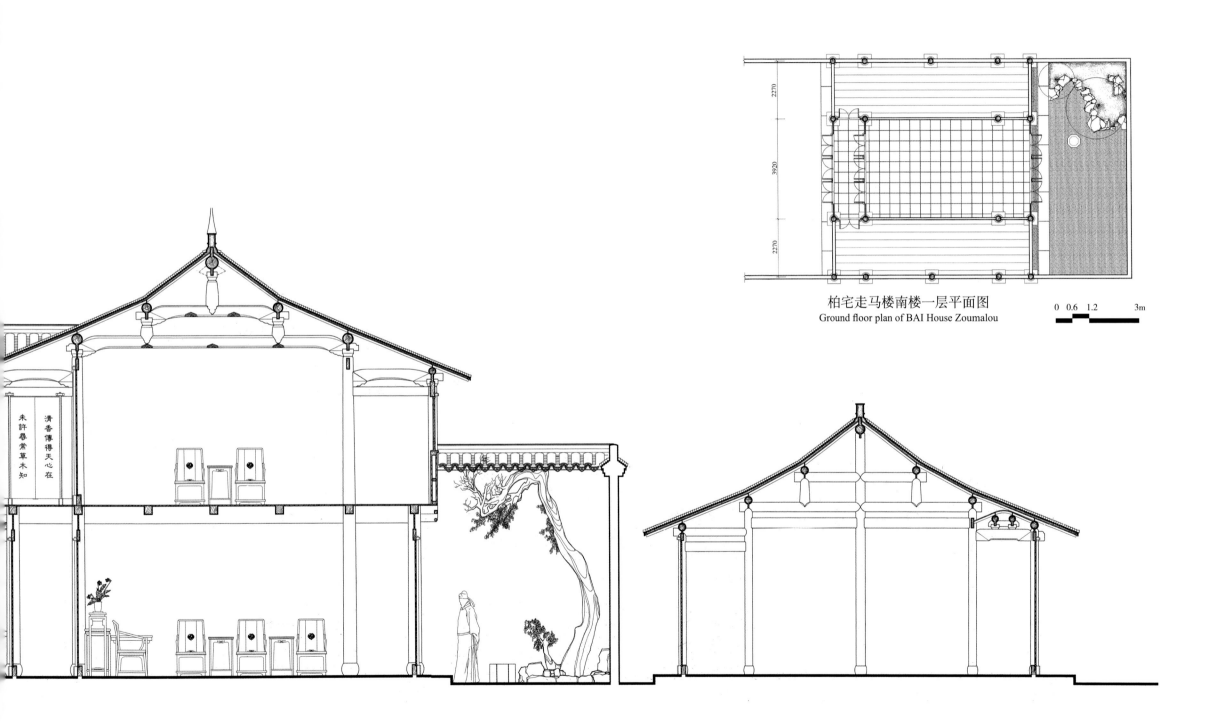

柏宅走马楼南楼一层平面图
Ground floor plan of BAI House Zoumalou

0  0.6  1.2       3m

0 0.05 0.1 0.25m

南楼脊桁架底部正心彩绘大样图
Zhengxin Caihui of the south building's Jiheng

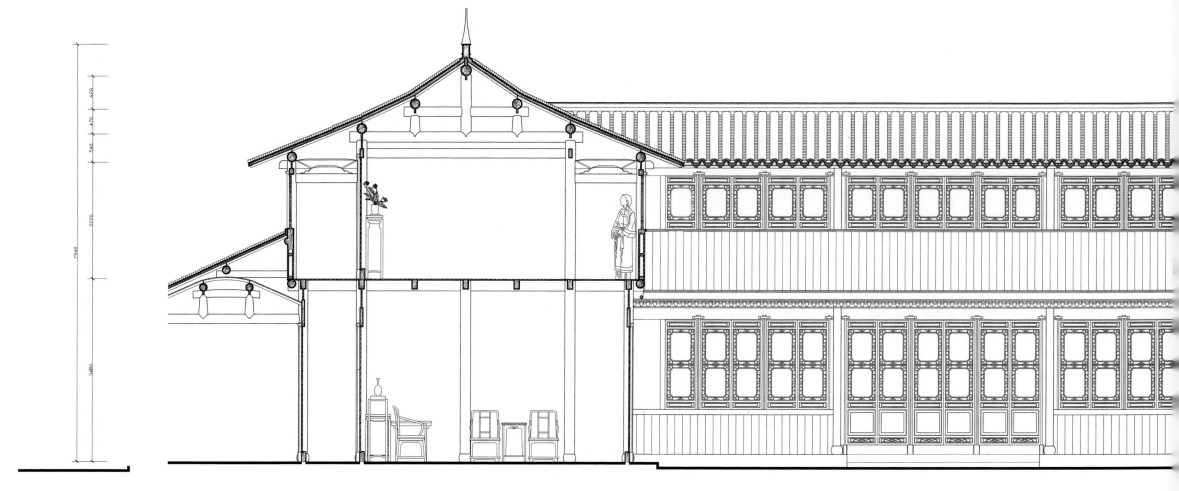

0 0.5 1 2m

柏宅走马楼中轴横剖面图
Cross-section of BAI House Zoumalou

# 主要参考文献

一 （明）陈威，顾清纂修·正德松江府志[M]·明正德七年（1512）刻本·

二 （明）郑洛书，高企纂修·清徐渭仁跋·嘉靖上海县志[M]·明嘉靖三年（1524）刻本·

三 （明）韩浚，张应武纂修·万历嘉定县志[M]·明万历三十三年（1605）刻本·

四 （明）方岳贡，陈继儒纂修·崇祯松江府志[M]·明崇祯三年（1630）刻本·

五 （明）焦竑著·国朝献征录[M]·台湾：台湾学生书局，1984·

六 （清）郭廷弼修·康熙松江府志[M]·清康熙二年（1663）刻本·

七 （清）赵昕，苏渊纂修·康熙嘉定县志[M]·康熙十二年（1673）刻本·

八 （清）史彩修；（清）叶映榴纂·康熙上海县志[M]·清康熙二十二年（1683）刻本·

九 （清）程国栋纂修·乾隆嘉定县志[M]·清乾隆七年（1742）刻本·

十 （清）孙凤鸣修；（清）王昶纂·乾隆青浦县志[M]·清乾隆五十三年（1788）刻本·

十一 （清）王大同修；（清）李林松纂·嘉庆上海县志[M]·清嘉庆十九年（1814）刻本·

十二 （清）宋如林修；（清）孙星衍，莫晋纂·嘉庆松江府志[M]·清嘉庆二十二年（1817）刻本·

十三 （清）应宝时修；（清）俞樾，方宗诚纂·同治上海县志[M]·清同治十年（1871）刻本·

十四 （清）汪祖绶纂；熊其英纂·光绪青浦县志[M]·清光绪五年（1879）·

十五 （清）程其珏修；杨震福纂·光绪嘉定县志[M]·清光绪七年（1881）刻本·

十六 （清）顾大申撰·顾大申自订年谱手稿[M]·北京：国家图书馆出版社，2016·

十七 （民国）于定修·刘凤桥点校·民国青浦县续志[M]·民国23年（1934）刻本·

十八 陈从周·金咏榴纂·上海的豫园与内园[J]·文物，1957，6·

十九 陈从周·嘉定秋霞圃和海宁安澜园[J]·文物，1963·

二十 郭俊纶·金利源码头并非外商所建[J]·上海滩，1988，12·

二十一 朱宇晖·上海传统园林研究[D]·上海：同济大学，2003·

二十二 朱宇晖·歧路梦寻 苏州留园的历史与空间脉络新析[J]·城市与设计学报，2008，20·

二十三 刘锦·上海本邑绅商沙船主朱氏家族研究[J]·中国社会经济史研究，2012，3·

二十四 张建华，陶继明主编·嘉定区地方志办公室，嘉定博物馆编·嘉定碑刻集[M]，2012·

二十五 朱宇晖，路秉杰·上海传统园林研究[M]·上海：同济大学出版社，2019·

二十六 朱宇晖，徐瑞彤，孙文达·湖心叠漪——上海豫园街区历史层理保护与再生畅想[J]·H+A华建筑，2020，4·

# References

1. (Ming) CHEN Wei, GU Qing. *Zhengde Songjiangfuzhi*. 1512.

2. (Ming) ZHENG Luoshu, GAO Qi; XU Weiren. *Jiajing Shanghaixianzhi*. 1524.

3. (Ming) HAN Jun, ZHANG Yingwu. *Wanli Jiadingxianzhi*. 1605.

4. (Ming) FANG Yuegong, CHEN Jiru. *Chongzhen Songjiangfuzhi*. 1630.

5. (Ming) JIAO Hong. *Guochao Xianzheng Lu*. Taiwan: Taiwan Xuesheng Shuju, 1984.

6. (Qing) GUO Tingbi. *Kangxi Songjiangfuzhi*. 1663.

7. (Qing) ZHAO Xin, SU Yuan. *Kangxi Jiadingxianzhi*. 1673.

8. (Qing) SHI Cai; (Qing) YE Yingliu. *Kangxi Shanghaixianzhi*. 1683.

9. (Qing) CHENG Guodong. *Qianlong Jiadingxianzhi*. 1742.

10. (Qing) SUN Fengming; WANG Chang. *Qianglong Qingpuxianzhi*. 1788.

11. (Qing) WANG Datong; (Qing) LI Linsong. *Jiaqing Shanghaixianzhi*. 1814.

12. (Qing) SONG Rulin; (Qing) SUN Xingyan, MO Jin. *Jiaqing Songjiangfuzhi*. 1817.

13. (Qing) YING Baoshi; (Qing) YU Yue, FANG Zongcheng. *Tongzhi Shanghaixianzhi*. 1871.

14. (Qing) WANG Zushou; XIONG Qiying. *Guangxu Qingpuxianzhi*. 1879.

15. (Qing) CHENG Qijue; YANG Zhenfu. *Guangxu Jiadingxianzhi*. 1881.

16. (Qing) GU Dashen; LIU Fengqiao. *Gu Dashen Ziding Nianpu Shougao*. Beijing: National Library Press, 2016.

17. (RC) YU Dingxiu; JIN Yongliu. *Minguo Qingpuxianxuzhi*. 1934.

18. CHENG Congzhou. *Shanghai Yu Garden and Nei Garden*. Cultural Relics,1957-6.

19. CHENG Congzhou. *Jiading Qiuxiapu Garden and Haining Anlan Garden*. Cultural Relics, 1963-2.

20. GUO Junlun. *Jinliyuan Wharf was not built by foreign investors*. Shanghaitan, 1988-12.

21. ZHU Yuhui. *The Research of Shanghai Traditional Gardens*. Shanghai:Tongji University,2003.

22. ZHU Yuhui. *Seeking the Dream in twists and turns-Exploring the Historical and Spacial Expression of Lingering Garden in Suzhou*. Journal of Urban and Design,2008-20.

23. LIU Jin. *A Study on the Zhu Family of Shanghai Gentry Merchants Shipowners*. Research on Chinese Social Economic History, 2012-3.

24. ZHANG Jianhua, TAO Jiming; Jiading District Local History Office, Jiading Museum. *Jiading Inscriptions Collection*, 2012.

25. ZHU Yuhui, LU Bingjie. The Research of Shanghai Traditional Gardens. Shanghai:Tongji University Press, 2019.

26. ZHU Yuhui, XU Ruitong, SUN Wenda. *The Multiplex Ripple of Huxin Pavilion Imagination on the Protection and Regeneration of Historical Stratification in Yuyuan District*, Shanghai. H+A Hua Architecture, 2020-04.

参与上海地区庙园建筑测绘及图稿整理的人员名单

## 1. 上海县城隍庙及庙园（豫园）：2004年，2005，2013年年测绘

2004年指导教师：常青 李渡 朱宇晖

2004年研究生助教：李彦伯 蔡燕歆 赵杰 黄金玉 阮萍

2004年协助测绘：须鼎兴 张春福 马贤成（同济大学测量系）

2004年测绘学生：张竟一 王品 余洋 李鹏 常兴 陈嘉卿 沈子卿 高磊 夏志刚
李欣 邹旦妮 王瑾瑾 李文杰 葛小康 邢杰 吴芬芳 王佳 施柳柳
潘佳力 严律己 江家畅 徐文博 祝婺韵 蔡慧明 顾玉婷 任思颖 边克举
万筱璐 仲曼晔 方一帆 付忠强 闵立 姚迅 葛斌 张文良
周乐 孙琦 肖俊瑰 闫玮 孙慧芳 袁莹 黄培肄 李欢欢 黄一帆
高峰 蔡筱璐 周烨恒 宋冰清 蒋燕 尚烨 王廊 丁旭芬 王欣

2005年指导教师：常青 邵陆 梅青 朱宇晖

2005年研究生助教：胡涛 李颖春 宋庆 翟海林 郑君彧

2005年测绘学生：帅佳妮 戴伟 池文俊 黄正骊 陈磊 刘伟 齐莹 张琳 马亮
陈意鸣 倪丹凤 刘俊洁 李培力 陈彦 邓洪波 刘丛 黄俊 汪颖叶芸
冯晶 陈轶洁 韩宜欣 狄翔杰 王龙海 李莹 卢捷 常成 李文杰
张拓 曾鹏 吕茶 曾哲 程剑 石楠 钟丽佳 吕东旭 卢宝生
颜晋琳 孙诣钦 赵愉 王芳 宋绍佳 钱维聪 缪海琳 贡博云 朱峰
王舒轶 邹立扬 顾云端 王懿宪 詹欢 董之平

2013年指导教师：董屹 朱宇晖

2013年测绘学生：陈伯良 黄闽君 孙伟 陆叶 罗瑞华 魏子凝 胡裕庆 解天缘 柯心然
周鉴云 刘嘉纬 于圣飞 王懿珏 魏逸飞 郭纯一 徐政 曾思源 褚莹斐
桂铭泽 卢欣杰 袁野 张凤嘉 谢嶷

## 2. 嘉定县城隍庙及庙园（秋霞圃与沈氏园）：1998年测绘

指导教师：常青

研究生助教：崔勇 张鹏 朱宇晖

测绘学生：孙田 龚辰 张韵 过俊 康雷 朱为为 姜江 陈勇 罗震华
胡晓翔 红华 曾浩 陈春燕 何俊 施丁平 张劲松 官霄龑 雍俊
项炯 杨雷贤 孙艳丽 沈晏 王承 梁雅娜 王爱峰 胡斌 吴晓东
朱亮 冯昕 舒文 叶兵 邵长海 刘雪蕾 汪启颖 张栩 马长宁
王建峰 郑华 陈宇华 蔡振宇

# List of Participants Involved in Surveying and Related Works

**Survey in 2004, Shanghai County Chenghuang Temple and Temple Garden (Yu Garden)**

Advisors: CHANG Qing, LI Zhen, ZHU Yuhui. XU Dingzing, ZHANG Chunfu, MA Xiancheng. LI Yanbo, CAI Yanxin, ZHAO Jie, HUANG Jinyu, RUAN Ping.

Undergraduate students: ZHANG Jingyi, WANG Pin, YU Yang, CHANG Xing, CHEN Jiaqing, SHEN Ziqing, GAO Lei, XIA Zhigang, LI Xin, ZOU Danni, WANG Jinjin, LI Wenjie, GE Xiaokang, XING Jie, WU Fengang, WANG Jia, SHI Liuliu, PAN Jiali, YAN Lvji, JIANG Jiachang, XU Wenbo, ZHU Wuyun, CAI Huiming, GU Yuting, REN Siying, BIAN Keju, WAN Xiaolu, ZHONG Manhua, FANG Yifan, FU Zhongqiang, MIN Li, YAO Xun, GE Bin, ZHANG Wenliang, ZHOU Le, SUN Qi, XIAO Jungui, YAN Wei, SUN Huifang, YUAN Ying, HUANG Peiyi, LI Huanhuan, HUANG Yifan, GAO Feng, CAI Xiaolu, ZHOU Yeheng, SONG Bingqing, JIANG Yan, SHANG Ye, WANG Kuo, DING Xufen, WANG Xin.

**Survey in 2005, Shanghai County Chenghuang Temple and Temple Garden (Yu Garden)**

Advisors: CHANG Qing, SHAO Lu, MEI Qing, ZHU Yuhui. HU Tao, LI Yingchun, SONG Qing, LIU Wei, QI Ying, ZHAI Hailin, ZHENG Junyu.

Undergraduate students: SHUAI Jiani, DAI Wei, CHI Wenjun, HUANG Zhengli, CHEN Lei, HU Yanzhe, LIU Cong, ZHANG Lin, MA Liang, CHEN Yiming, NI Danfeng, LIU Junjie, LI Peili, CHEN Yan, DENG Hongbo, HUANG Jun, WANG Ying, YE Yun, FENG Jing, CHEN Yijie, HAN Yixin, DI Xiangjie, WANG Longhai, LI Ying, LU Jie, CHANG Cheng, LI Wenjie, ZHANG Tuo, ZENG Peng, LV Cha, ZENG Zhe, CHENG Jia, SHI Nan, ZHONG Lijia, LV Dongxu, LU Baosheng, YAN Jinlin, SUN Yiqin, ZHAO Yu, WANG Fang, SONG Shaojia, QIAN Weicong, MIU Hailin, GONG Boyun, ZHU Feng, WANG Shuyi, ZOU Liyang, GU Yunduan, WANG Yixian, ZHAN Huan, DONG Zhiping.

**Survey in 2013, Shanghai County Chenghuang Temple, Chengxiangge and Shichuntang**

Advisors: DONG Yi, ZHU Yuhui.

Undergraduate students: CHEN Boliang, HUANG Minjun, SUN Wei, LU Ye, LUO Ruihua, WEI Zining, HU Yuqing, XIE Tianyuan, KE Xinran, ZHOU Jianyun, LIU Jiawei, YU Shengfei, WANG Yijue, WEI Yifei, GUO Chunyi, XU Zheng, ZENG Siyuan, CHU Yingfei, GUI Mingze, LU Xinjie, YUAN Ye, ZHANG Fengjia, XIE Ni.

**Survey in 1998, Jiading County Chenghuang Temple and Temple Garden (Qiuxiapu Garden and Shenshi Garden)**

Advisors: CHANG Qing. CUI Yong, ZHANG Peng, ZHU Yuhui.

Undergraduate students: SUN Tian, GONG Chen, ZHANG Yun, GUO Jun, KANG Lei, ZHU Weiwei, JIANG Jiang, CHEN Yong, LUO Zhenghua, HU Xiaoxiang, HONG Hua, ZENG Hao, CHEN Chunyan, HE Jun, SHI Dingping, ZHANG Jingsong, GUAN Xiaoyan, YONG Jun, XIANG Jiong, YANG Leixian, SUN Yanli, SHEN Yan, WANG Cheng, LIANG Yana, WANG Aifeng, HU Bin, WU Xiaodong, ZHU Liang, FENG Xin, SHU Wen, YE Bing, SHAO Changhai, LIU Xuelei, WANG Qiying, ZHANG Xu, MA Changning, WANG Jianfeng, ZHENG Hua, CHENG Yuhua, CAI Zhenyu.

**Survey in 1999, Qingpu County Chenghuang Temple and Temple Garden (Qushui Garden)**

Advisors: CHANG Qing. SHAO Lu, XU Feng.

Undergraduate students: DONG Yiping, BAI Yun, CHEN Zhuo, CHEN Jiong, CHEN Wei, CHEN Linglei, CHENG Xuesong, HU Junfeng, HUANG Bei, DONG Yi, GU Beibei, HOU Binchao, GUO Ge, DAI Xiaolin, HAN Zhigang, CHEN Zhihao, CHEN Ye, FU Chen, JIANG Haibin, HUA Jia, ZANG Min.

**Survey in 1997, Jiading County Xuegong and Xuegong Garden (Yingkui Mountain and Huilong Pond)**

Advisors: CHANG Qing. ZHANG Yaxiang, LI Yunbing, ZHANG Peng, LIU Jie, GU Lihua, ZHU Yuhui.

Undergraduate students: CHEN Ye, AN Le, CHEN Dongxing, CHEN Yi, CHEN Shuang, GAN Jing, HUANG Tan, GAO Baohua, HAN Bing, JIN Kewu, LE Yin, LI Donglei, LIN Liuli, GAO Bei, LI Xuyang, LIN Yun, GAO Song, LI Fangmin, LI Yunchao, DING Ming, Dai Yucheng, Luo Xiaohua, Jin Jingyu, JI Yonglei, REN Yigang, GU Jialing, LI Qing, YAN Fan, XU Yuan.

**Survey in 1999, Songjiang County Shantang and Gardens (Zuibaichi Garden)**

Advisors: CHANG Qing. XU Yihong, HUANG Song, ZHU Yuhui.

Undergraduate students: MENG Xin, LIU Wei, LI He, LIN Chuanhua, LIU Chenjie, LU Xinhua, MO Jialing, LIANG Feng, WANG Lei, WU Xin, WU Xiaokang, WU Jiming, XIAO Jun, XIAO Shenjun, WANG Junliang, XIE Yonghao, XIN Donghua, TAO Xianhua, WANG Wenting, WANG Yan, WANG Yuluo, SHEN Li, SHEN Jun, SONG Hao, TANG Wei, SUN Wei, SUN Lei, XIE Jun, WANG Wenting, WANG Hui, QIN Zhen, QIN Lei, QU Feng, WANG Guanzhong, XU Fang.

**Survey in 2005, Shanghai County Guozhai East Garden (Shuyin Building)**

Advisors: CHANG Qing, SHAO Lu, MEI Qing, ZHU Yuhui. CHEN Jie, ZOU Xun, ZHAI Hailin.

Undergraduate students: HUA Ke, WANG Wenjun, DU Daini, GUO Wenqing, WANG Hua, LIN Jing, ZHANG Beiyan, HE Xiaoshu, JI Jiexin, LI Qian, YUAN Jing, CHEN Huilin, BAI Xiaofeng, XIONG Zhidan, ZHAN Xiang, LI Ran, LI Dukui, HE Yinfeng, MAO Mingqian, ZHOU Ji, LIU Jian, LIN Liping, YIN Liangshun, ZHANG Sheng.

**Survey in 2006, Songjiang County Yi Garden and Nearby Houses**

Advisors: CHANG Qing, ZHU Yuhui. WANG Yan, YAO Yun, CHANG Rongjie.

Undergraduate students: CHEN Xuan, CHEN Xinsi, DING Chun, LI Qing, MA Liang, WANG Yu, XU Ye, YU Guopu, YU Xiao, YU Zhongqi, ZHOU Chen, ZHOU Wengang, ZHOU Ye, ZENG Yi, ZHOU Zhihua.

**Name list of participants in reorganization of survey drawings**

Text organizers: ZHU Yuhui

Survey drawing organizors: SHI Ruilin, XU Ruitong, HUANG Yiqun, MAO Yan, GU Jinyi, SUN Wenda.

English translator: XU Ruitong.

图书在版编目（CIP）数据

上海庙园 = THE CITY TEMPLES AND ITS GARDENS IN
SHANGHAI：汉英对照 / 常青，朱宇晖主编；同济大学
建筑与城规学院编写. —北京：中国建筑工业出版社，
2019.12
（中国古建筑测绘大系·祠庙建筑与园林建筑）
ISBN 978-7-112-24544-4

I.①上…　II.①常…②朱…③同…　III.①古典园
林—建筑艺术—上海—图集　IV.①TU-092.2

中国版本图书馆CIP数据核字（2019）第284365号

丛书策划 / 王莉慧

责任编辑 / 李　鸽　刘　川

英文审稿 / 刘仁皓

书籍设计 / 付金红

责任校对 / 王　烨

中国古建筑测绘大系·祠庙建筑与园林建筑

**上海庙园**

同济大学建筑与城规学院　编写

常　青　朱宇晖　主编

Traditional Chinese Architecture Surveying and Mapping Series:
Shrines and Temples Architecture & Garden Architecture
**THE CITY TEMPLES AND ITS GARDENS IN SHANGHAI**
Compiled by College of Architecture and Urban Planning, Tongji University
Edited by CHANG Qing, ZHU Yuhui

\*

中国建筑工业出版社出版、发行（北京海淀三里河路9号）

各地新华书店、建筑书店经销

北京雅盈中佳图文设计公司制版

北京雅昌艺术印刷有限公司印刷

\*

开本：787毫米×1092毫米　横1/8　印张：60$\frac{1}{2}$　字数：1563千字
2022年4月第一版　2022年4月第一次印刷
定价：**458.00**元
ISBN 978-7-112-24544-4
（35232）

**版权所有　翻印必究**

如有印装质量问题，可寄本社图书出版中心退换

（邮政编码100037）